SHOTS IN THE DARK

Shots in the Dark

THE WAYWARD SEARCH FOR AN AIDS VACCINE

Jon Cohen

W. W. NORTON & COMPANY
NEW YORK LONDON

Printed in the United States of America
First Edition

For information about permission to reproduce selections from this book, write to Permissions, W. W. Norton & Company, Inc., 500 Fifth Avenue, New York, NY 10110

The text of this book is composed in Sabon with the display set in Sabon and Gill Sans
Composition by Thomas Ernst
Manufacturing by the Haddon Craftsmen, Inc.
Book design by BTDnyc

LIBRARY OF CONGRESS CATALOGING-IN-PUBLICATION DATA

Cohen, Jon, 1958–
Shots in the dark : the wayward search for an AIDS vaccine / Jon Cohen.
p. cm.
Includes bibliographical references and index.
ISBN 0-393-05027-0
1. AIDS vaccines—Popular works. I. Title.

QR189.5.A33 C64 2001
616.97'9206—dc21 00-061635

W. W. Norton & Company, Inc., 500 Fifth Avenue, New York, N.Y. 10110
www.wwnorton.com

W. W. Norton & Company Ltd., 10 Coptic Street, London WC1A 1PU
1 2 3 4 5 6 7 8 9 0

To Shannon and Erin

Contents

ACKNOWLEDGMENTS ix
PROLOGUE xv
CHAPTER 1 Fast Predictions 3
CHAPTER 2 The Next Breakthrough 16
CHAPTER 3 Empiricists Versus Reductionists 43
CHAPTER 4 Moving into Humans 64
CHAPTER 5 Animal Illogic 78
CHAPTER 6 Market Forces 102
CHAPTER 7 Unwanted: Dead or Alive 119
CHAPTER 8 All MicroGeneSys's Men 139
CHAPTER 9 A Manhattan Project for AIDS 175
CHAPTER 10 The Dairymaids of AIDS 200
CHAPTER 11 Perpetual Uncertainty 227
CHAPTER 12 New World Order 258
CHAPTER 13 Running in Place 295
CHAPTER 14 Better Ways 317
CHAPTER 15 Disparate Measures 333
EPILOGUE 358
POSTSCRIPT: BREAKING THE SILENCE 361
NOTES 370
GLOSSARY 419
ACRONYMS 428
INDEX 430

Acknowledgments

Friends stopped asking me about this book years ago. "I'm almost done" became the joke.

This started off as a project to write about one year in the search for an AIDS vaccine. The year I chose was 1990. During an interview with Bob Gallo that May, I asked him whether he thought the vaccine idea that most interested him would move into animal tests or humans anytime soon. "Not soon, not like six months, not like the year that you're worried," said Gallo in his classic mangled prose. "Not like the year that you're worried." It stuck in my head for a long time: one year doesn't mean anything to AIDS vaccine researchers.

I decided to shift focus, and instead write a book about the first vaccine that came to market, charting the ups and downs of the field over time. It was an ambitious plan and, as I suddenly realized in 1997, too ambitious.

My wake-up call came from Bill Clinton, who in his famous 1997 speech set a goal of finding an AIDS vaccine in 10 years. Clinton's declaration crystallized for me something I had thought from the time I first started researching the book: the field had no real leadership, no real funding, no real sense of urgency. So my first thank-you goes to Bill Clinton, because his pronouncement, naïve as I think it was, shifted my focus again, and now, 10 years since beginning this project, I am completing it.

It would be a fool's errand for me to name the scientists who have given generously of their time during the past decade, because the list would begin to resemble a phonebook. Put it this way: few researchers have failed to share with me however much time I have requested. This generosity becomes even more remarkable after factoring in that many of these scientists run institutions, companies, large laboratories, and the like. Many have international reputations, with all the attendant press and public demands conferred by that status. Many even think me a pest or worse. Still, they took my calls, answered my letters and e-mails, and, in some cases, let me stay in their labs for days on end.

A few scientists do deserve extra thanks from me, though. I asked only one researcher to critique this manuscript, Steven Wolinsky. I chose Wolinsky because he has exquisite scientific taste, he is precise, he has a deep knowledge of the subject, and I figured he would disagree with many of my main arguments. He did not let me down. John Moore, a journalist-oriented scientist (if there are science-oriented journalists, why not the reverse?), has constantly filled my head with interesting information about the field, and some of it even turned out to have significance. Richard Trauger gave me an inside look at the life of a bench scientist at a biotechnology company (Immune Response Corporation) working on an AIDS vaccine. He always was available to argue the relative merits of one theory or another—and to keep me in touch with my ignorance. Paul Shapiro frequently has fetched papers for me, listened to my confused descriptions of mechanisms of this or that, and made me see the field from the perspective of someone who is not part of it (he does Alzheimer's research).

AIDS activists, people infected with HIV, and people at high risk of becoming infected, in many different countries and from varied economic backgrounds, have provided another critical component of my education about this disease. Many have brought me into their homes and freely shared the most intimate details of their lives, including recounting the painful stories of those they have loved and lost, and describing their own frustrations and fears. How do you appropriately thank people for that? A few activists who have pushed the vaccine research agenda, however, do stand out as particularly helpful to this project: Bill Snow, David Gold, Sam Avrett, David Barr, Garance Frank-Ruta, Mark Harrington, Gregg Gonsalves, Martin Delaney, Jesse Dobson (deceased), Derek Hodel, and the father of the activist movement, Larry Kramer.

On the editing front, I owe deep gratitude to three of the best editors a writer could ever hope to work with: Jack Shafer, John Benditt, and Colin Norman. Jack Shafer was my editor at the *Washington City Paper* when I first became interested in AIDS vaccine research, and, without his initial encouragement, I never would have written the first article that led me to conceive of the book. Without Jack's continued friendship and relentless prodding ("Are you done yet?"), I might not have stuck with it, either. John Benditt, my former editor at *Science*, gave me my first story assignment after I phoned him out of the blue. John, who also became a close friend, introduced me to the world of science, which, even though I have a degree in science writing, I had covered only peripherally until then. Finally, Colin Norman, another editor at *Science* who has grown into a close friend, has, week in and week out, made me come off as smarter, wittier, and more clearheaded than I am in real life.

Another of my closest friends, photojournalist Darrow Montgomery, graciously sifted through his files and allowed me to reprint a few of his distinctive images here. Darrow, who lived in Thailand as a child, defines discretion and was especially helpful in 1995 when we did sensitive interviews with HIV-infected people and caregivers in and around his old hometown of Chiang Mai.

Doron Weber and the Alfred P. Sloan Foundation, which he works for, made this latest incarnation of the book possible. I cold-called Doron on a Friday afternoon, he said, "get me a proposal by Monday," and six months later I had a Sloan Foundation grant that allowed me to take a year off from *Science* and concentrate on how best to communicate what I thought and why. Doron also had the good sense to hire the strongest critic of my grant proposal, Robert Pool, to help me shape the book from outline through manuscript. Robert's advice led me to lose tons of "great" material, explore concepts that I never should have overlooked, and to never, never forget the points I was attempting to make.

Gail Ross, my agent, has encouraged me throughout, with a faith in the project that other agents, I suspect, would have lost years ago. Angela von der Lippe, my editor at Norton, also committed herself to this book when many other editors came up with long lists of reasons why they could not publish it. Her assistant, Stefanie Diaz, also has been a great help, as has the astonishingly meticulous copy editor, Susan Middleton.

Any mistakes in this book are of course my own, and no one who has read drafts has offered anything other than advice, which I was free to take or ignore.

For earlier incarnations of the book, thanks to Hampton Sides, Mark Nollinger, Richard Carter, and Mark Feldstein for their unsparing critiques. Donald Burke also kindly read an early version of a chapter on vaccine history. Jody Rein, Emily Reichert, and Tom Spain at Delacorte Press all gave me sage directions at one point or another.

I also owe my gratitude to the many editors at various publications who have helped make the AIDS stories I've written for them (and reconfigured here) clearer, sharper, and better looking: at *Science*, Tim Appenzeller, Betsy Carpenter, Eliot Marshall, Jean Marx, and the big cheese, Ellis Rubinstein;

Henry Finder at *The New Yorker*; Carl Zimmer at *Discover*; and Anne Goodwin-Sides at National Public Radio.

While journalism trains reporters in the art of dodging press flacks, a few have made my life much, much better. At the National Institutes of Health, Wendy Wertheimer from the Office of AIDS Research has steered me through the maze of AIDS policy-making officials, committees, and lawmakers, and has always worked overtime to fulfill my requests, no matter how urgent or ridiculous. Laurie Doepel and Greg Folkers at the National Institute of Allergy and Infectious Diseases have endured my most demanding questions with good humor; their understanding of HIV and AIDS exceeds that of some scientists in the field. David Corkery and Karen O'Malley at Fenton Communications greatly reduced the difficulty of covering the enormous AIDS conferences, where they handled the press, with invaluable daily packages of press clippings about the meetings, as well as a sense of calm and order that otherwise would not have existed. Lapsed reporter Victor Zonana, now with the International AIDS Vaccine Initiative and formerly with the Department of Health and Human Services, has a keen sense, too, of what it takes to do this job, and has assisted me too many times to count. Beth Waters of Cooney Waters has the odd distinction of representing companies that both make polio vaccines and do AIDS vaccine R&D. She not only knows these topics; she cares about them and has never steered me wrong.

A few AIDS reporters deserve extra thanks for swapping ideas—and challenging them. Laurie Garrett at *Newsday*, Mark Schoofs at the *Village Voice*, and the *New Scientist*'s Phyllida Brown have helped refine my thinking over the years.

Several individuals freely shared documents with me from their personal archives and deserve my thanks too. Bruce Weniger gave me hundreds of pages of official meeting notes from the Presidential Advisory Council on HIV/AIDS, which he sat on. Don Francis supplied me with old memos from his time at the Centers for Disease Control and Prevention. Irvin Chen, Jack Nunberg, and many others shared "confidential" copies of the "pink sheets" from study sections that critiqued their AIDS vaccine grant applications. Peter Lurie of Public Citizen sent me the complete files about the U.S. military's investigation of Robert Redfield, which Lurie had received after bringing suit against the Department of Defense for wrongly deleting much of the information in earlier requests for the files made by me, Lurie, and others.

Finally, I thank my family, profoundly. My father, Avshalom, is a songwriter who taught me the joy of creating. My mother, Esther, pushed me to become a doctor (sorry), but cheerleads everything and anything I do, and simultaneously offers loads of entertaining, unsolicited advice. My older brother, Ronnie, a playwright, poet, and drummer, helped give me the nerve to chart my own course.

I thank my daughter, Erin, for helping me cultivate sources. I started this project when she was born, and for the first few years, she shared my home office with me, charming whomever I was attempting to interview with her babbles and screams. Erin, who fortunately still sees me as a playmate, also has learned to share time with this book, standing outside my office on many, many afternoons and weekend days, skateboard or soccer ball in hand, perfectly mocking my "I'm almost done."

And then there's my wife, Shannon Bradley. Shannon has steadfastly believed in me and this project, even during the few lean years when my financial contributions to the family coffers didn't support my own room and board. On top of that, she is a journalist with an excellent eye for detail and has combed through more drafts than anyone could expect a spouse to read and has caught many of my errors, oversights, and boneheaded attempts to explain the intricacies of science to a lay audience.

Ultimately it is because of Shannon that now, finally, I can say I'm done.

Prologue

I cannot truly comprehend the magnitude of the statistic that 16,000 people become infected with HIV each day. I, of course, know that is a lot, and, having reported on AIDS for more than 10 years and witnessed the disease in the United States, Europe, Africa, Thailand, Japan, Mexico, and Canada, I am well aware of the suffering caused by even one infection. Yet, like many people who do *not* follow the disease closely, I become numb to numbers.

But the numbers emerging from studies in Africa as we enter a new millennium startle me each time I look at them.

More than one in every three adults who live in Botswana or Zimbabwe is infected with HIV. Let that number sink in: 35.8% of the schoolteachers, doctors, cement mixers, mothers, engineers, truck drivers, seamstresses, and politicians are infected with a lethal virus that, in most cases, will cut their lives short by decades. Lesotho, Namibia, South Africa, Swaziland, Zambia, and Zimbabwe are only slightly better off, where the number is one in five adults. In five other countries in sub-Saharan Africa, it is more than one in ten. All told, sub-Saharan Africa now accounts for two-thirds of the estimated 40 million HIV-infected people in the world, and AIDS on that continent is the leading cause of death for adults. Hundreds of thousands of children are infected, and even more have been orphaned. Life expectancy, which steadily had been climbing during the past several decades, is rapidly *de*clining: In Botswana, for example, the World Health Organization (WHO) says the average child born today will live to be about 40; the same child could have expected to live to 70 in the absence of HIV.

While HIV crumbles sub-Saharan Africa, shaking it like an earthquake that will not stop, Americans largely have lost interest in AIDS. With the advent of powerful cocktails of anti-HIV drugs in the late 1990s, journalists began trumpeting the idea that AIDS was history, which the American public readily accepted. Death rates from AIDS plummeted for a time, furthering the notion that the epidemic has abated and will soon disappear altogether. AIDS activists, politically savvy and effective rabble-rousers who for a decade drew attention to the disease and sped development of treatments, have become as anachronistic as hippies.

Most of the world cannot afford the wonder drugs that have staved off disease and death for infected people in wealthier countries. And even in the United States, the advances celebrated a few short years ago are slipping, in no small part because the virus eventually develops resistance to every drug thrown at it. Look at AIDS deaths alone. In 1997, the year after the advent of the powerful drug cocktails, the U.S. death rate from AIDS plunged by 42%. The next year, the drop was only 20%, and the number of new AIDS cases had also nearly doubled from 1997. This is, in short, no time for any country to let down its guard against HIV, especially the United States, which spends more money on AIDS research than all other countries combined.

There are many ways to slow the spread of HIV. Using condoms and clean needles, abstaining from sex, treating other sexually transmitted diseases, male circumcision, and screening blood products all have proved effective—to a point. But the best hope the world has to thwart this virus is the same weapon effectively used against smallpox, polio, hepatitis B, rabies, and other devastating viruses: a vaccine.

Much effort indeed has gone into finding an AIDS vaccine, and the scientific problems are staggering. Yet there are problems that supercede science. This book describes in detail how scientists came to the impasse they are facing, explaining plainly the complicated mixture of scientific, business, political, and ethical forces that haphazardly steer the enterprise. I also offer a solution to the nonscientific problems that, I believe, ultimately would speed the search for a vaccine, possibly shaving years off the time it takes to slow HIV. My proposal is neither revolutionary nor unique: I argue that scientists should approach the problem in a logical, stepwise fashion, testing vaccines in meaningful animal experiments and following up on all promising leads. This book documents how, amazingly, this has not happened to date, and shows the strong forces in the U.S. scientific culture that resist such an approach. And I am convinced that similar problems dog the search for vaccines against malaria, tuberculosis, and hepatitis C—and will hamper vaccine research for new diseases that in the future will suddenly assault populations the way AIDS did in the 1980s.

I doubt that my book, which I began researching in 1989 and completed with help from a Sloan Foundation grant, will single-handedly change the U.S. scientific culture. But maybe, when the public and scientists alike take a careful look at the shortcomings of the AIDS vaccine research effort, a momentum will build to attack the problem with all of the might the world can muster. If one in four adults in the United States were infected by HIV, I am confident that such a critical rethinking of the scientific community's battle plan already would have happened.

SHOTS IN THE DARK

CHAPTER 1

Fast Predictions

THE WHITE HOUSE
Office of the Press Secretary

PRESS BRIEFING BY
LARRY SPEAKES
October 15, 1982
The Briefing Room
12:45 P.M. EDT

ΛΛΛΛΛΛΛΛΛΛΛΛΛΛΛΛΛΛΛΛΛΛΛΛΛΛΛΛΛΛΛΛΛΛΛΛΛΛΛΛΛΛΛΛ

Q: Larry, does the President have any reaction to the announcement—the Centers for Disease Control in Atlanta, that AIDS is now an epidemic and have over 600 cases?

MR. SPEAKES: What's AIDS?

Q: Over a third of them have died. It's known as "gay plague." (Laughter.) No, it is. I mean it's a pretty serious thing that one in every three people that get this have died. And I wondered if the President is aware of it?

MR. SPEAKES: I don't have it. Do you? (Laughter.)

Q: No, I don't.

MR. SPEAKES: You didn't answer my question.

Q: Well, I just wondered, does the President—

MR. SPEAKES: How do you know? (Laughter.)

Q: In other words, the White House looks on this as a great joke?

MR. SPEAKES: No, I don't know anything about it, Lester.[1]

Q: Does the President, does anybody in the White House know about this epidemic, Larry?

MR. SPEAKES: I don't think so. I don't think there's been any—

Q: Nobody knows?

MR. SPEAKES: There has been no personal experience here, Lester.

Q: No, I mean, I thought you were keeping—

MR. SPEAKES: I checked thoroughly with Dr. Ruge this morning and he's had no—(laughter)—no patients suffering from AIDS or whatever it is.

Q: The President doesn't have gay plague, is that what you're saying or what?

MR. SPEAKES: No, I didn't say that.

Q: Didn't say that?

MR. SPEAKES: I thought I heard you on the State Department over there. Why didn't you stay there? (Laughter.)

Q: Because I love you, Larry, that's why. (Laughter.)

MR. SPEAKES: Oh, I see. Just don't put it in those terms, Lester. (Laughter.)

Q: Oh, I retract that.

MR. SPEAKES: I hope so.

Q: It's too late.

On October 15, 1982, Larry Speakes, then the spokesperson for President Ronald Reagan, like most people in the world, had no idea that a new epidemic disease, acquired immunodeficiency syndrome, was racing around the United States, Europe, and Africa.

The disease that eventually became known as AIDS likely has existed in Africa for many decades, if not many centuries. The most probable scenario is that humans caught it from chimpanzees when trapping or butchering the animals. Because people with AIDS die from complications like unchecked tuberculosis, wasting, or pneumonia, no alarm bells would have sounded, as nothing would have looked unusual. This is especially true given the sorry state of health care for most Africans, where many children die young and, in some countries, the majority of adults never reach their 50th birthday.

It is easy to see how, as Africans began migrating to burgeoning cities and syringes became more common in public health campaigns, AIDS could have flourished but still remain undetected. Air travel then brought the disease to Europe, Haiti, and the United States in the late 1960s and early 1970s. The sexual promiscuity of gay men in these liberated years, where it was not uncommon for people to have 50 partners in a year, offered the disease a chance to firmly establish itself on these more developed continents. And these were places where doctors had been trained to take note of odd cases, such as young men dying from pneumocystis carinii pneumonia, a normally wimpy disease.

Five cases of this rarely fatal disease in gay men indeed caught the attention of a Los Angeles physician in 1981, and he reported his findings on June 5, 1981, in the *Morbidity and Mortality Weekly Report*, a no-frills, but widely read publication issued by the U.S. Centers for Disease Control (CDC) to alert the medical community to public health issues. By October 1982, when Speakes joked about the emerging epidemic, *MMWR* had published 11 reports on AIDS. Articles had appeared in the major medical journals. True, the popular media had been shy about covering the story, but even by that point it had appeared on the network news and on front pages of major daily newspapers. Larry Speakes, in short, should at least have known that AIDS existed. And his ignorance highlighted the Reagan Administration's indifference—and even outright hostility—toward homosexuals. It also foreshadowed how unprepared the people at the top were to lead the battle against this burgeoning epidemic, especially when it came to organizing the search for a vaccine.

Presidents *can* make enormous differences when it comes to conquering disease. Fifty years earlier, when thousands of children in the United States were being crippled each year by polio, President Franklin Delano Roosevelt organized charity balls on his birthday "to dance so others may walk." These balls evolved into the National Foundation for Infantile Paralysis, colloquially known as the March of Dimes, which would successfully lead the campaign to research and develop the first polio vaccines. But then, FDR had polio himself. He also appointed his former law partner, Basil O'Connor, to run the March of Dimes. Ronald Reagan for many years refused even to say the word "AIDS" in public. And Reagan's top aides had as difficult a time with AIDS as he did. Consider this exchange between Speakes and journalists at a White House Press briefing on June 13, 1983:

> Q: Larry, does the President think that it might help if he suggested that the gays cut down on their "cruising"? (Laughter.) What? I didn't hear your answer, Larry.
>
> MR. SPEAKES: I just was acknowledging your interest—
>
> Q: You were acknowledging but—
>
> MR. SPEAKES: —interest in this subject.
>
> Q: —you don't think that it would help if the gays cut down on their cruising—it would help AIDS?

MR. SPEAKES: We are researching it. If we come up with any research that sheds some light on whether gays should cruise or not cruise, we'll make it available to you. (Laughter.)

Q: Back to fairy tales.

Could the person who handled the press for FDR publicly have made jokes about polio and lived to see the inside of the White House again?

• • •

AT THE TIME Larry Speakes was making fun of gay cruising, Margaret Heckler had been the Administration's top health official for three months and was preparing for a speech the next day to the U.S. Conference of Mayors, at which she would declare that AIDS was the "number one health priority."

A lawyer by training, Heckler had represented Massachusetts in the Congress for 16 years before Reagan appointed her to head the Department of Health and Human Services (HHS), a mammoth agency tasked with overseeing both the National Institutes of Health (NIH) and the CDC. Heckler's chief of staff, C. McClain "Mac" Haddow, recalled that his boss knew little about AIDS when she took the HHS post. So Haddow, a former Utah state representative, and other staffers scheduled a briefing for her.

To educate Heckler about the disease, they brought in James Mason, head of the CDC in Atlanta, Georgia. Edward Brandt Jr., the assistant secretary for Health, also came to the gathering in her office, as did a gay man who had been a close aide to Heckler since her time on Capitol Hill. Mason walked Heckler through the basics of how AIDS appeared to be transmitted, explaining that it focused on homosexuals and why that might be. "He was talking about anal intercourse and Heckler got this shocked look on her face and said, 'What?' " recalled Haddow. "Jim Mason, who's a very straight-laced Mormon and was having a little difficulty with this, said the rectal wall was more fragile than the vaginal wall."

"Anal intercourse?" said Heckler, looking at her trusted, gay staffer. "You do *that*?"

Ed Brandt looked at Haddow. "I think we better come back and discuss this a little later," said Brandt.

Haddow said that everybody in the room was stunned. "Mrs. Heckler didn't understand that homosexuals engaged in anal intercourse."

• • •

ON APRIL 23, 1984, MARGARET HECKLER summoned Robert Gallo into her office at the Hubert H. Humphrey Building in Washington, D.C. A prominent and controversial researcher at the National Cancer Institute (NCI), Gallo had just raced home from a scientific conference in Italy to attend what would turn out to be one of the most awkward, confused—and talked about—scientific press conferences in history: in a few hours' time, Heckler would announce that Gallo's lab had found the cause of AIDS, firing the starting gun for the vaccine search. But before she made this grand declaration, she wanted Gallo's input about the amount of time it would take to translate his laboratory's discovery into an AIDS vaccine.

Scientists who make discoveries in basic research long have shied away from offering predictions about practical outcomes, and reporters long have hounded scientists to offer just such predictions. Mac Haddow wanted Heckler to be able to address this question head on, and for days he had been pressuring the NIH to give him a time line. "The NIH public affairs office was asked, and they came up with a whole list of possible answers," said Haddow. "The bottom line was it was going to have to be Gallo's call. And he would have to be comfortable with it."

When Haddow first began seeking an answer to this question, he thought he would have several weeks to find an answer. Plans called for announcing the explosive news of the Gallo lab's discovery in concert with the May 4 issue of *Science*, which would contain four back-to-back papers describing how the lab had fingered a new virus as the culprit and grown large quantities of it. But Haddow's PR plans became fouled up when news stories hinting at the discovery began to dribble out during the second week of April. Creating further grief for HHS, the White House had high hopes of using the news during this election year to confront criticisms about the Reagan Administration's AIDS record. At Ed Brandt's urging, Gallo and NIH director James Wyngaarden, who had traveled together to the conference in Italy, both hastily returned home to attend this Monday afternoon press conference.

When Gallo, Wyngaarden, and NCI director Vincent DeVita Jr. arrived at Heckler's office that morning, Haddow quickly dressed them down for a story that had appeared the day before on page 1 of the *New York Times*. Headlined "Federal Official Says He Believes Cause of AIDS Has Been Found," the article was not about the findings in Gallo's lab. Rather, it quoted James Mason, head of the CDC, describing the work of researchers at the Pasteur Institute in France, and Mason said he had not yet seen the papers from Gallo's lab. Haddow was livid about the perception that the CDC and the NIH—both of which are part of HHS—were at odds with each other, and contended that the NIH was to blame for some of the tension between the institutes. After the bickering about who was at fault died down, Haddow asked how long it would take to develop a vaccine. The question caught Gallo unawares. "I hadn't thought about it before we went in there," said Gallo. "I didn't even anticipate the question."

DeVita and Wyngaarden attempted to convince Haddow that the idea of offering a time line was a mistake. "People have tried to make vaccines against all sorts of things, and the record is generally one of failure," Wyngaarden said. Wyngaarden had little respect for Haddow, who he thought was "authoritarian" and "gave the scientific point of view no shrift." The NIH director, who personally oversaw a $4.5 billion budget (1% of which went to AIDS) and had extensive clout on Capitol Hill, also had little respect for Heckler. "The woman was one of the worst listeners I've ever encountered in any capacity," he said. "She'd listen to one sentence or two, and then her mouth wouldn't close again."

By all accounts, Gallo came up with the official line: within two years, an AIDS vaccine would be ready for human testing. "I assume as much credit or blame for that as anyone," Gallo later acknowledged. Haddow recalled having "a lengthy discussion about why it was defensible." Said Haddow: "I wanted to be sure before Heckler said it and was brought into it that there was a reasonable expectation that we could accomplish that goal."

Brandt, who was also part of these discussions, put great faith in Gallo's enthusiasm. "He was convinced that he had all the makings of a vaccine, that he was fully prepared to move on that very quickly, as a matter of fact," said Brandt. "I really had great confidence in Gallo's ability to pull that off."

Heckler, who had a severe case of laryngitis that day, invited Gallo into her office for a one-on-one chat.

"Will we be able to get a vaccine?" she asked in her scratchy voice.

Gallo said he believed so. "I just said, 'I think, you know, due to the fact we have an unlimited amount of virus now and we're sure it causes the disease, I can't think that this should take that long.' I didn't perceive the difficulties."

• • •

HUNDREDS OF JOURNALISTS jammed into the first-floor auditorium of the Hubert H. Humphrey Building for the 1:15 P.M. press conference. Heckler stepped up to the lectern first. She had a dyed blonde head of helmet hair, and its cartoonish effect was exaggerated by her bright red dress, which sported a high white collar. Behind her, a dramatic, 20-foot-high blue curtain served as a backdrop. Flanking Heckler were Gallo and the agency's top brass: HHS assistant secretary Brandt, NIH director Wyngaarden, CDC director Mason, and NCI director DeVita.

Heckler opened the proceedings with a joke. "As I believe you anticipate, this press conference will be devoted to the subject of AIDS, in which area there is, of course, important news," she said from behind the thicket of microphones sprouting from the lectern. "But, unfortunately, we have not made similar breakthroughs in the field of laryngitis. So I apologize for the state of my voice today."

Heckler continued in an increasingly faltering voice. "Today, I am pleased to whisper"—the reporters laughed—"that the arrow of funds, medical expertise, research, and experimentation with the Department of Health and Human Services, and its allies around the world, have [*sic*] aimed and fired at the disease AIDS, and has hit the target only two or three rings from the bullseye itself."

It would be quickly forgotten, but at the time of this announcement, several far-flung theories about the cause of AIDS still competed with each other. Some researchers recently had argued that a fungus caused AIDS. Others maintained that AIDS was the result of using amyl nitrate inhalants, called poppers, that many gay men snorted to strengthen the power of orgasm. The list of suspected causes also included cytomegalovirus, Epstein-Barr virus, swine flu virus, herpesvirus, and the fanciful idea that several of these possible causes were assaulting the immune system in concert. Gallo himself for a time tried to pin the disease on one of the two known human T-cell leukemia viruses, HTLVs, obscure pathogens recently discovered by his lab.

Only a year before, Gallo and his coworkers had thrust HTLV-I and HTLV-II into the spotlight when they published, again in *Science*, two papers arguing that these viruses were the possible cause of the then two-year-old AIDS epidemic. Another paper in that 1983 issue from a group headed by Luc Montagnier—a researcher at the Pasteur Institute in Paris, France, who would later enter a protracted battle with Gallo about who deserved credit for discovering the cause of AIDS—bolstered the case. In the subsequent year, Gallo and his colleagues refined their argument, and now they were contending that the cause was a new member of this viral family, HTLV-III.

The Gallo lab's four 1984 *Science* papers showed that they had isolated HTLV-III from 48 patients with the disease and spelled out how to grow the virus continually in laboratory cultures, a critical feat because it routinely killed the cells it infected. Heckler emphasized to the journalists that the ability to grow the virus in mass quantities would allow scientists to characterize the agent in detail, understand its behavior, and develop a blood test in as little as six months. "Finally, we also believe that the new process will enable us to develop a vaccine to prevent AIDS in the future," she said. "We hope to have such a vaccine ready for testing in approximately two years."

Heckler proceeded to apportion credit for this discovery to the people sitting behind her. The next paragraph in her prepared statement, which had been distributed to the press, praised the work of researchers at the Pasteur Institute. But Heckler, her voice jumping into what sounded like a helium-induced register, abruptly abandoned the prepared statement, asking Gallo and Brandt to come forward.[2]

A hunch-shouldered Gallo moved into the spotlight. With his perpetually unkempt steel-wool hair, a too-casual camel-colored sport coat, and a slightly crooked tie, Gallo looked every bit the academic, which was in keeping with the NIH style. He proceeded to explain himself in a tortured English. "Many of you, or all of you, have seen discussions about work in Paris," he said. "There was. There is not. There has never been any fights or controversies between us and a group in France. . . . We have active collaboration in the coming month. If what they identified in *Science* a year ago is the same as what we now have produced more than 50 isolates of and in mass production, and in detailed characterization, if it turns out to be the same, I certainly will say so, and I will say so with them in collaboration."

Gallo fielded several questions, finally passing on this one: "How many years will it be before there is a marketable vaccine, based on previous experience?" Brandt, at Gallo's urging, came forward. "We're estimating a minimum of two years, probably more like three years," Brandt said. "In two years, we think it's possible to begin to start human trials. But I think we have a—one of the first steps that has to be accomplished is to mass-produce this virus in sufficient quantity to accomplish that. So I think we're talking about probably three years. We're going to hustle." There. Brandt had said it. Three years until a vaccine was on the market.

"Excuse me," the reporter said. "The French told me this morning that they would predict at least five years. And Luc Montagnier"—this was the first mention of Montagnier's name at the press conference by anyone—"said he thought it might be five to 10 years. Why are you more optimistic than the French are?"

"I'm more optimistic, I guess," said Brandt. "I don't know. I'm more optimistic. Except that I believe it can be done, that's all."

An astute reporter pointed out that there were a number of viruses, "known for as long as two decades," that scientists had yet to defeat with vaccines. "Why are you confident that you can get one for AIDS in two years?"

Gallo finally spoke his mind on the question, backpedaling mightily and, in effect, undermining what he had told Heckler and Haddow earlier. "I believe that the reason that there's a wide difference in what time it takes, you know, it obviously is not in the interest of the scientist to give a fast prediction," he said. "You'll only press us all, say, 'Remember what you said eight months ago, four months ago?' You'll be giving us a clockwatch. . . . Now your question gets tricky. How do you know you'll have one that really works in AIDS patients? Of course, the prediction for that is impossible. You have no hundred percent proof that it will be two years, a hundred years, or it may never come from vaccination. It depends on the subtleties of biology. But the principles are there. The likelihood, the best we can say in science, with the technology available, we should have things ready to be able to be tried by then."

Brandt and Gallo traded off answering questions, which ranged from uninformed to insightful, a reflection that the conference had attracted both general assignment reporters and those who regularly covered the science beat. The smartest questioners kept returning to the subject of vaccines. "Dr. Brandt," asked one, "in talking about approval of a vaccine, aren't you sort of blithely leaping over what could be a very difficult ethical problem, namely, the testing of this for safety and efficacy on healthy individuals?"

Brandt misunderstood the clairvoyant question. "Absolutely not. I mean, we would certainly test any vaccine," he said.

Another sharp reporter asked whether they intended to test a live- or a killed-virus vaccine. These are the traditional methods of making a viral vaccine: either killing the virus or weakening a live version of it. These modified versions of the virus then safely can teach the immune system how to defend itself should the real thing ever invade. "Can you comment on the difficulty of human clinical trials of such a vaccine?"

Brandt, an earnest and well-meaning man, did not seem to understand this question either. "Well, we're going to develop the vaccine first," he said. "Then we'll be able—it depends entirely on what kind of vaccine comes out of it. So we certainly are not going to test in humans any vaccine that would put those people at risk of the disease."

The reporter, obviously baffled, kept pressing Brandt, who accused the journalist of "presupposing the kind of vaccine" they intended to develop. "I'm not presupposing anything," the journalist snapped back.

"We don't have a vaccine at the moment," said Brandt. "And we would certainly take all . . ."

"You're talking about having one in a couple of years," the journalist interrupted, raising the question again about how they planned to stage tests of the vaccine in healthy volunteers.

"Well, as soon as we see the vaccine, then I'll be able to answer that question."

Had Brandt understood vaccine development better, this testy exchange need not have happened. Long before AIDS came around, scientists well recognized that the testing of a vaccine against a lethal disease raises difficult ethical questions. Inevitably, the issue comes down to weighing risks against benefits—and carefully describing the two sides of the scale to anyone who volunteers for the trial. What is more, Brandt's charge that the journalist was "presupposing" the type of AIDS vaccine that would be developed was particularly off-base because if anyone was presupposing anything about the formula for an AIDS vaccine, it was Gallo, as he revealed in a subsequent answer.

"You keep asking about the vaccine, when it comes up, what's going to be available in two years," Gallo said to a reporter who had *not* asked about the vaccine. "As best as a scientist can calculate. We can't predict anything 100%. But you've got to give some answers."

Gallo, who had never made a vaccine against anything, went on to detail a recipe for an AIDS vaccine that, instead of relying on the entire virus like the traditional live and killed approaches, only called for using a piece of the agent that had been produced with genetic engineering techniques. Just such a hepatitis B vaccine was then being engineered that contained the surface protein, or envelope, of that virus. When it came to AIDS, said Gallo, the same logic would apply. "In my view, you'd vaccinate with the envelope protein of the virus,"said Gallo. "We'll go to recombinant technology, we are doing that now, and we will produce the protein that way. That protein, I would certainly be willing to take if I were a high-risk person, and I'd do it voluntarily. . . . That's the direction we're kind of thinking about."

• • •

A COUPLE DAYS AFTER THE PRESS CONFERENCE, June Osborn, then a University of Wisconsin virologist who headed the NIH's AIDS Advisory Committee, received a phone call from an outraged gay researcher she knew in Los Angeles. The researcher explained to her how hard he and his colleagues had been working to convince the gay community in Los Angeles that sex clubs and bathhouses were spreading AIDS. Before the press conference, the researcher said these businesses were empty. But Heckler's enthusiasm about the development of a vaccine had created a euphoric sense that the end of the epidemic was near. Last night, he complained, he saw a bathhouse with a four-hour waiting line.

Media reports from the press conference did dutifully report Heckler's comment that an AIDS vaccine would be ready for *testing* in about two years. They also by and large left out Brandt's more declarative statement that, by three years, a vaccine probably would be on the market. Still, regardless of the specific language used in the distilled and packaged versions of what was said in the Hubert H. Humphrey auditorium on April 23, 1984, the public heard a simple message: An AIDS vaccine was just around the corner.

Looking back on the press conference nearly 14 years later, many of the participants had serious regrets about the vaccine predictions. Brandt said he still "winced" every time he heard reference to an AIDS vaccine being tested in two years, and he had completely forgotten about his own declaration of a marketed vaccine in three. "Certainly there wasn't much experience with the whole idea of a vaccine," acknowledged Brandt. "There was a general belief moving about at the time that this was really going to be a fairly easy nut to crack. It was just a matter of time until some quick breakthrough, or whatever the word is, until that happened. I suspect if we really understood the complexity of the organism at that time, things might have happened differently."

Gallo aknowledged that, at the time, he had "zero previous experience in my life on a vaccine" and only the crudest understanding of the basic immune mechanisms that protect people from infectious diseases. "You can't imagine the level of my own feeling of confidence at that time," he said, by way of explaining his role in the optimistic predictions. His lab had just played major roles in the discoveries of HTLV-I and HTLV-II, as well as uncovering a critical immune system messenger, interleukin-2, that tells cells to grow. With AIDS, they had convincing evidence for the first time that a virus isolated from ailing people caused the disease. Their ability to mass-produce this so-called HTLV-III—which later became known as human immunodeficiency virus, HIV, when it proved to have no relationship to the other HTLVs—laid the groundwork for a blood test, treatment, and a vaccine. They were hot on the trail of discovering which cells the AIDS virus preferred, the various ways it was transmitted, and what genes it had. "There was so much data pouring in that I, too, thought—probably believed—we could have solved the whole bloody thing, you know?"

Gallo's greatest regret is that the press conference was held at all, as it violated an agreement he had made with Montagnier and the other Pasteur researchers earlier that month—and seriously incensed them. "I was terribly, terribly nervous," remembered Gallo. "I had massive conflicting emotions. I felt this shouldn't have been done. I didn't act strong enough that we should have waited until we compared the virus with the French and made a joint announcement." The perception that Gallo made a credit grab was compounded by Heckler's laryngitis, which forced her to skip the parts of her prepared statement that acknowledged the Pasteur's contributions.

There is, of course, no way of knowing for certain whether an HIV vaccine would be any closer to market had the announcement of the discovery of the cause of AIDS been handled differently. But the April 23, 1984, press conference and the attendant backroom imbroglios communicated in bold type that the search for an AIDS vaccine was off to a bad start.

• • •

THE WHITE HOUSE
Office of the Press Secretary

PRESS BRIEFING BY
LARRY SPEAKES
December 11, 1984
The Briefing Room
12:03 P.M. EST

^^^^^^^^^^^^^^^^^^^^^^^^^^^^^^^^^^^^^

MR. SPEAKES: Lester's beginning to circle now. He's moving in front. (Laughter.) Go ahead.

Q: Since the Center for Disease Control in Atlanta—(laughter)—reports—

MR. SPEAKES: This is going to be an AIDS question.

Q: —that an estimated—

MR. SPEAKES: You were close.

Q: Well, look, could I ask the question, Larry?

MR. SPEAKES: You were close.

Q: An estimated 300,000 people have been exposed to AIDS, which can be transmitted through saliva. Will the President, as Commander-in-Chief, take steps to protect Armed Forces food and medical services from AIDS patients or those who run the risk of spreading AIDS in the same manner that they forbid typhoid fever people from being involved in the health or food services?[3]

MR. SPEAKES: I don't know.

Q: Could you—Is the President concerned about this subject, Larry—

MR. SPEAKES: I haven't heard him express—

Q: —that seems to have evoked so much jocular—

MR. SPEAKES: —concern.

Q: —reaction here? I—you know—

Q: It isn't only the jocks, Lester.

Q: Has he sworn off water faucets—

Q: No, but, I mean, is he going to do anything, Larry?

MR. SPEAKES: Lester, I have not heard him express anything on it. Sorry.

Q: You mean he has no—expressed no opinion about this epidemic?

MR. SPEAKES: No, but I must confess I haven't asked him about it. (Laughter.)

Q: Would you ask him Larry?

MR. SPEAKES: Have you been checked? (Laughter.)[4]

CHAPTER 2

The Next Breakthrough

Hang gliders from around the world flock to the Torrey Pines Glider Park in La Jolla, California, a caramel-colored bluff standing high above the expansive Pacific Ocean. On this same mesa sits a building that, from the road, looks like a slab of gray concrete. Inside, scientists explore these questions: how do humans age, reproduce, see, think, feel, use muscles, lose their minds, develop cancer and AIDS? It is called the Salk Institute for Biological Studies, and gold letters on the marble steps of the entryway explain that it was established by the National Foundation March of Dimes, led by Basil O'Connor, in 1960.

The steps lead into a rectangular, marble courtyard, which is flanked by two mirror-image, concrete-and-teak buildings that jut outward at milk-carton angles. A foot-wide gutter cuts through the courtyard's center; smaller, perpendicular gutters feed in, building an elaborate drainage system. The water courses down the big gutter's length and then splits into four waterfalls that appear to spill into the Pacific. At sunset, when amber and rose rays drench the marble, water, and concrete, architect Louis Kahn's mix of laboratories and offices looks like nothing less than a shrine to the gods of light and silence.

On April 15, 1986, Robert Gallo walked through the courtyard and climbed the five flights of stairs to Jonas Salk's office. Though the two men had spoken before, this was the first time they had met, as Gallo later put it, "on a personal level."

Robert Gallo and Jonas Salk at that moment in history had much in common. They both were members of the rarefied club of scientists who knew fame outside of science's halls. Salk's polio vaccine had made the world a safer place, as had the HIV blood test that resulted from the work done in Gallo's lab. And the great feelings of success that both of these men should have enjoyed at the peak of their careers had become marred by raging controversies engulfing the very work that had brought them to the top.

But besides having contentious, high-profile careers in biology, enjoying the company of admirers, and sporting leviathan egos, Robert Gallo and Jonas Salk had little else in common. Gallo was blunt, crass, frenetic, and competitive, a metallurgist's son from Waterbury, Connecticut, whose youth was centered around basketball. His mind worked too quickly for his own good, sometimes leading him to slop words together like "perfectly scientifically exactly." He was perpetually running behind and had a hard time recalling specific dates from the past. His crinkly hair defied every attempt to keep it in place. If he perceived so much as a hint of a threat from a colleague or journalist—and he courted both colleagues and journalists assiduously—he would dress down the person in any situation; often, he would charmingly apologize the next day, asking for forgiveness because of what he called his "Mediterranean temper." He treasured loyalty and had built a large family of colleagues, whom he hosted each year at his lab's prestigious invitation-only, weeklong conference. Some years, he coauthored hundreds of scientific papers. He hungered for a Nobel Prize more than a little. His public and private personae were indistinguishable, and he freely shared personal anecdotes, the most gripping being his explanation for choosing a career in cancer research: When he was 11, his little sister died of leukemia; their last visit together haunted him still, her wasted and bruised body trapped in a hospital bed, her teeth caked with dried blood.[1]

Salk's personality was less obvious, partly because it was obscured by his public image, partly because he hid it, and partly because it contained many contradictions. While Salk was reserved and philosophical, he enjoyed the spotlight and the image he cut—the dapper cravat around his neck; the author of thin, deep tracts (*Man Unfolding*, *The Survival of the Wisest*, and *Anatomy of Reality*). He selected each word as though he were a jeweler setting a gemstone, and he incessantly challenged the words that others used. But then his love of aphorisms could leave listeners with the sense that he spoke in riddles. He recalled the past lucidly, even tossing out the day of the week from several years before, but he sometimes gave opposite answers to the same question.[2]

Salk could move easily through a crowd of scientists and had an understated charm, yet he did not exhaust himself at winning friends or winning over enemies. Rather than competing with colleagues, he viewed his work as an attempt to understand nature. He avoided journalists, conflict, and impassioned rhetoric whenever possible. He rarely published scientific papers and joked that he didn't need to win the Nobel Prize because most people thought he already had one. He bristled at private questions and was calculated in public, explaining, "You wear different uniforms under different circumstances."

By 1986 the struggles of Jonas Salk had become legend, while Gallo's troubles were escalating. The outrage felt by Luc Montagnier and his colleagues at the Pasteur Institute two years before, when Gallo and the U.S. government officials stood before TV cameras and slighted the French contribution to the discovery of the AIDS virus, had led to a lawsuit. The law does not have a mechanism to punish a scientist for hogging credit, nor does it offer compensatory damages to the aggrieved party. But scientific credit feuds often end up in court under the guise of patent disputes. In this case, the Pasteur Institute sued the U.S. Department of Health and Human Services (HHS) over its patent claims for an AIDS blood test. The technical charge was that Gallo's lab had "misappropriated" from the Montagnier group the specific isolate of AIDS virus used to make the U.S. test.

The Pasteur suit rested on a question that surfaced at the 1984 Heckler press conference: how similar was the virus that Montagnier's lab had promoted as the cause of AIDS to the one spotlighted by Gallo's group? As had become clear in late 1984, not only were the viruses they had isolated similar; they were indistinguishable from each other. This was as unfathomable as the same person winning the lottery in Paris and Washington, D.C., on the same day. Like all viruses, HIV mutates when it copies itself. But it differs from many viruses because it replicates at a blinding speed, meaning the world is filled with an incredibly diverse array of HIV strains. So there were three main explanations for the similarity between the French and American viral isolates. The first, and least likely, was that the French and American isolates of the virus came from people who were having sex with each other. The Pasteur suit ignored this possibility.

The charge of misappropriation focused on a sample of the putative AIDS virus that Montagnier's lab had shared with Gallo's group during the summer of 1983. The French gave the Americans the virus, which they called LAV, to study it. But following the realization that the two labs had "discovered" identical viruses, the French arrived at the other two most plausible explanations: Gallo and his coworkers had either contaminated their own viral cultures with the French sample, or outright stolen LAV and renamed it HTLV-III.

Though Salk would later find himself standing between the combatants in what became a scientific *jihad* between France and the United States,

Gallo's visit in April 1986 had nothing to do with war or peace. In town on other business, Gallo simply wanted to meet Salk, who was then 71.

Salk didn't want to become involved with AIDS. "I was not looking for another career at that time in my life," Salk recalled. "This was something that needed a lot of attention and vision." Still, Salk had an idea of how AIDS vaccine development should proceed.

Gallo entered Salk's office, which could not have been more different from his own. Gallo then worked in an institutional structure with an institutional name—Building 37—on the campus of the National Institutes of Health (NIH) in Bethesda, Maryland. Gallo's office was a disaster, the floor littered with scientific journals and stacked boxes, the walls lousy with awards and snapshots of scientists, arms around each other's shoulders, sharing wine or boating in far-off countries during conference breaks. The linoleum-lined hallways outside of his office led to the labs of more than 50 scientists who carried out research under his direction. The Salk Institute, in contrast, was designed to be anti-institutional, the very antithesis of Building 37. Salk's office had the feel of an art gallery. Everything was in its place. Sculpture and contemporary paintings—including work by his wife, the painter Françoise Gilot—dressed the concrete walls. Bowing to space constraints and declining scientific interests, Salk had not had a laboratory at the institute since July 1984.

To Robert Gallo, the human body was a fascinating, infinitely complex organism, and the goal of science was to define in finer and finer detail every biological process. He loved mechanism. With HIV, he wanted to know how it breached the membrane of a cell, how it slipped its own genes into the cell's nucleus, how it copied those genes into proteins, how each of those proteins behaved, how new viral particles formed, and how they then killed the cell. The clutter in his life reflected a couple of simple truths: he had an insatiable intellectual curiosity; and intense debates surrounded the flood of data pouring out of labs in every corner of the globe that were competing with each other to answer these questions. Gallo relished the clubiness of science too—playing on a team, elbowing the competition, cutting slack for friends, tripping enemies.

To Jonas Salk, the human body was a fascinating, infinitely complex organism, and the goal of science was to understand what needed to be known to solve problems and improve life for humans. His interest in HIV did not then—and never would—reach beyond developing a way to stop it. The organization in his life reflected both his intense focus and, as critics would point out, the limits to his intellectual curiosity. Salk recoiled at the clubiness of science, and likely would not even have been hired at a scientific club to beat all clubs like the Salk Institute had he not founded the place.

The two men sat at a large teak table that April afternoon in 1986 and volleyed ideas. Time had curved the back and thinned the voice of Salk, and the hair surrounding his famous bald pate had long gone gray. But Salk still had a fire in his eyes. Gallo, who was a touch in awe of Salk—as many people, even scientists, were at first—appreciated the warm reception. "Jonas was extremely cordial, he was extremely open, and thanked me for coming to him and also encouraged me to think more optimistically about vaccines," remembered Gallo. "It was a punctuation mark that gave me a lasting memory, because it was Jonas, because it was the manner of the greeting, and because of the manner that he immediately made a statement that I would remember."

"You've had the vaccine since 1983," Salk told Gallo.

"What?" said Gallo.

Salk explained, drawing diagrams on a board. In November 1983, researchers in Gallo's lab had transferred the virus from an AIDS patient's blood cells into foreign cells living in a test tube. Then—and this was the great feat—the lab kept those infected foreign cells alive. Once this virus was thus "isolated," Salk reasoned a vaccine could be made the same way he made his polio vaccine: kill the virus with chemicals, heat, or light, and then mix it with an immune system booster known as an adjuvant. It was that simple.

"Jonas, if you're right, we could have had a vaccine three years ago," Gallo said.

By the end of the meeting, Salk thought Gallo was intrigued with the idea of a killed HIV vaccine. "I began to look at the problem from the point of view of how would I have approached it 40 years ago," said Salk. "So that's what set things in motion."

Salk may well have been overly enthusiastic about the prospects of an old-fashioned killed-virus vaccine thwarting HIV. But his grander point was right on target: Past successes offer powerful lessons that could speed modern vaccine development. And all too often, these lessons are given lip service and then blithely ignored.

• • •

AMONG THE GREEK, ETRUSCAN, AND ROMAN ART in Copenhagen's Ny Carlsberg Glyptotek museum resides an Egyptian stele of the Eighteenth Dynasty that depicts a young priest and a woman, offerings in hand. The stone carving appears typical of its era, 1580–1350 B.C., except for one detail: the priest's right leg is gnarled and shortened, telltale deformities of poliomyelitis.

Though this is the first recorded evidence of polio, no epidemics were charted until the mid-1800s. Hard-hit cities like Stockholm, Oslo, and Lyon were befuddled when the disease began its assault, much in the same way AIDS took San Francisco, New York, and Kinshasa by surprise. Many Americans probably first learned of polio in 1894, the year the disease killed 18 children in Rutland County, Vermont, and paralyzed 114 others, the largest outbreak to that date.

Nearly 15 years passed before scientists in Vienna showed that polio was spread by an infectious agent. In the signal experiment, the researchers caused monkeys to develop polio by inoculating their brains with spinal cord tissue taken from a human who had died of the disease. Just as the isolation of HIV spawned the AIDS vaccine search, so did the isolation of poliovirus. By 1910, investigators had fashioned experimental polio vaccines.[3]

The pioneers of polio vaccinology knew little about poliovirus—or, for that matter, any virus. Shaped like a minuscule soccer ball, poliovirus contains a single strand of ribonucleic acid (RNA), which holds the virus's genetic code. Poliovirus enters the body through the mouth and nose from droplets, such as saliva, or from microscopic pieces of feces. The virus then slides into the gut and reproduces. Normally the immune system can limit this infection before it causes serious disease, but on rare occasions, the virus travels through the blood and into the central nervous system, causing meningitis. Paralysis occurs only if the virus then enters nerve cells. Scientists now estimate that in an unvaccinated population, wild poliovirus paralyzes only 1% to 2% of the people it infects.

The first prototype polio vaccines worked, to a degree, in experiments with animals, though the results ultimately were inconclusive. Still, the fact that these scientists, for the most part working in the dark, had *any* success is instructive. Vaccinology at the turn of the 20th century was largely an empirical art. If the vaccine worked, it worked. How or why a vaccine protected a person from disease was relevant but not all-important. Vaccinologists aimed to stop human suffering and the loss of farm animals. If these investigators had an understanding of how a potion performed its magic, so much the better.

The drive to develop a polio vaccine paradoxically owes much to the success that late 19th century populations had at eliminating poliovirus from their environments. For centuries, polio kept a low profile because human behavior successfully exploited the immune system. The umbilical cord supplies fetuses with blood from their mothers. In addition to passing blood-borne germs—such as HIV—from mother to baby, this natural bridge helps newborns, too, by providing them with maternal antibodies that can ward off many diseases. Breast milk also can supply a baby with a mother's antibodies. But these "passively" acquired antibodies disappear within a few months, meaning the protection they offer is temporary. Still, if poliovirus infected an infant while a mother's antibodies were there, the baby not only would clear the infection unscathed; its immune system would "actively"—and safely—learn how to make antibodies against the bug. Owing to a phenomenon called immunologic memory, the child's immune system would never forget this active lesson. In essence, then, babies used to routinely receive what amounts to a natural polio vaccine.

But then came sanitation, which catapulted polio from obscurity to plague status. Indoor plumbing, the germ theory, and other advances of the industrial age reduced the exposure of infants to poliovirus, creating more and more people susceptible to infection and disease. Most often, the disease struck these "susceptibles" early in life, earning polio the moniker infantile paralysis.

By 1916 the pool of susceptibles in the United States had so deepened that the disease afflicted more than 27,000 Americans, killing 7,179. Panic hit especially hard in New York City, where more than 2,000 died. People fled Manhattan, only to meet unwelcoming committees of gun toters in the suburbs. Some hospitals refused to admit cases for fear of infecting others. Cities impounded cats and dogs, which authorities mistakenly thought could transmit the virus to humans. Parents sealed windows and doors and refused to let their children outside to play. Miracle cures like ox and frog blood, cedar shavings, and radium water abounded.

Poliovirus crippled the 39-year-old Franklin Delano Roosevelt in 1921. Just as the news in 1985 that Rock Hudson had AIDS would later stoke public concern about that disease, Roosevelt's condition changed the public's perception of polio and became a catalyst for increased scientific research.

Beginning in 1934, President Roosevelt staged charity balls on his birthday (January 30) "to dance so others may walk"—and, more to the point, to pay off the debts accrued at the Warm Springs Foundation, a polio center that he had established eight years before in the pine-studded hills of Warm Springs, Georgia. Under the direction of Basil O'Connor, Roosevelt's former law partner, the first ball became a countrywide "celebration." What has been described as "the biggest radio network in history" was patched together so that Roosevelt could broadcast a live message from the White House. "I thank you but lack the words to tell you how deeply I appreciate what you have done," Roosevelt told the nation's revelers, "and I bid you good night on what is to me the happiest birthday I have ever known."[4]

More than 4,000 communities staged balls that first year, raising a little over $1 million. Aside from Warm Springs, the money went to groups in towns that had held balls. Because health insurance was not yet common, city committees gave the money to polio victims. The Warm Springs Foundation, in turn, earmarked $100,000 to "stimulate and further the meritorious work being done in the field of infantile paralysis elsewhere," establishing the President's Birthday Ball Commission for Infantile Paralysis to distribute research grants. The next year, balls only took in $750,000, but $241,000 went to the new Commission for Infantile Paralysis.[5]

Maurice Brodie of New York University in 1935 received the first massive grant from the commission, $65,000. Brodie, who reported that he had protected monkeys with a killed poliovirus vaccine, distributed the experimental vaccine to hospitals and health departments, which then inoculated more than 9,000 children. At around the same time, Temple University's John Kolmer, who had not received any Birthday Ball dollars, had cooked up a vaccine made from live but weakened poliovirus and, with the help of private physicians, had inoculated nearly 11,000 kids.

Allegations soon flew that the vaccines had paralyzed and even killed a few vaccinees, the Birthday Ball Commission taking the heat for both disasters. In retrospect, the live vaccine that Kolmer made probably was the more damaging of the two, though both were ineffective, if not dangerous. Officials publicly lambasted Kolmer and Brodie for rushing their vaccines into humans. Brodie died four years later; many have written that he killed himself, though his death certificate lists coronary thrombosis as the cause.[6]

Stung by the vaccine failures, criticized by the chief surgeon at Warm Springs, charged with shamelessly politicizing polio, Roosevelt abandoned the birthday balls after the 1937 bash and announced the formation of a new group. Named the National Foundation for Infantile Paralysis and headquartered in O'Connor's New York law office, the group, announced Roosevelt, would "make every effort to ensure that every responsible research agency in the country is adequately financed to carry out investigations into the cause of infantile paralysis and the methods by which it may be prevented."

The first fund-raiser held by the National Foundation was a radio blitz, again coinciding with Roosevelt's birthday. Comedian Eddie Cantor supposedly hatched the idea of having radio stations donate time to ask for small contributions. "We could ask the people to send their dimes directly to the White House," Cantor enthused. "Think what a thrill people would get. . . . And we could call it the March of Dimes!" Two days after the appeal, only $17.50 had arrived at 1600 Pennsylvania Avenue. "You fellows have ruined the president," a White House aide groused.

A quarter of a million letters arrived during the following week stuffed with $1,823,045, including 2,680,000 dimes, all of which went to the National Foundation for Infantile Paralysis/March of Dimes. For the first time, a serious war chest existed to research and develop a polio vaccine.

When man does not intervene, epidemic diseases rise and fall in waves; the number of impacted people crests over several years as the susceptibles become infected, and then steadily drops into a trough, allowing another population of susceptibles to build. That year, 1938, the United States recorded only 1,705 cases of polio, the deepest trough the country would enjoy until 23 years later, when the vaccine developed by Jonas Salk was in wide use.

• • •

JONAS EDWARD SALK was born October 28, 1914, and raised by his Russian immigrant parents in New York City. His father, Daniel, designed women's scarves and blouses. According to Jonas's younger brother Lee, their father "was something of a Willy Loman character from *Death of a Salesman*, beaten down in business but still believing that success would soon be his."[7]

Jonas, who zealously shielded his private life over the years, once gave a speech to a group of educators in which he described the sway his mother, Dolly, had over him. When Jonas matriculated to the College of the City of New York as a precocious 15 year old, he registered as a prelaw student. "I had no notion of studying medicine ever," he told the educators. "Well, why did I suddenly make that change? Well, it was because my mother, who was a very powerful person [and] had enormous influence with me and everyone else with whom she was in contact, was able to put me into a state in which, when I had anything to say that was contrary to what seemed to be her wishes, I would stutter and stammer. And she didn't think I'd make a very good lawyer. She preferred that I become a teacher. . . . They got out of school in the midafternoon and they had long vacations, and she was one of those very overprotective mothers and didn't want her son to work very hard. And so, my form of protest was to study medicine. And that was the way I dropped out, so to speak. It was a discovery for me. But I didn't really want to study medicine. I decided that if I was going to do that then I'd do research."

In 1938–39, his final year of medical school at New York University, Salk met Thomas Francis Jr., chair of the Department of Bacteriology. Francis, a redoubtable influenza researcher, held the iconoclastic view that vaccines made from killed viruses had much to offer. Conventional wisdom held that only a live, attenuated vaccine could confer long-lasting immunity because it more closely mimicked a true infection. Salk, then 24, took an elective with Francis, who at the time was attempting to kill influenza virus with ultraviolet light. Salk, however modestly, helped isolate the virus from mice lungs and then irradiate it. The day after receiving his M.D., Salk married Donna Lindsay, a social worker. He stayed on with Francis until March 1940, whereupon he began an internship at Mount Sinai Hospital in New York.

In October 1941, Salk visited Francis at the University of Michigan in Ann Arbor, where Francis had moved a few months before to head the epidemiology department in the School of Public Health. Under contract with the Armed Forces Epidemiological Board, Francis was developing a killed-virus vaccine for influenza, which in the 1918–1919 epidemic had killed 43,000 doughboys—more than the number killed in combat during all of World War I.[8] Salk wanted in on the project. Even though the work did not involve polio, Francis encouraged Salk to apply for a fellowship from the National Foundation for Infantile Paralysis, which groomed up-and-coming virologists. "My ultimate intentions are to do investigative work in virus diseases," Salk wrote in a draft of his fellowship application. "At the present time, my chief interest lies in the fundamental aspects of this field relative to the problems of prophylactic immunization." Viruses. Vaccines. The 27-year-old Salk had charted his course.

The National Foundation awarded Salk the $2,100 fellowship, and he moved to Ann Arbor, beginning the yearlong project on April 12, 1942. Salk soon began publishing scientific papers, some of which revealed what his critics would call his essential flaw and his admirers would praise as his genius—his penchant for extrapolating from the data at hand. In *Breakthrough: The Saga of Jonas Salk*, a 1965 book by Richard Carter, Salk recounts how Francis once scolded him for attempting to publish a paper that was short on facts and long on inferences. "There may have been times when I made more of my data than might have been expected, but I was not functioning in the expected way," Salk said. "I was supplying *immunological* insights. I was attempting to elucidate the *interaction* of man and virus in a field which was accustomed to viewing the two separately. I engaged in extrapolation because I had always felt that it was a legitimate means of provoking scientific thought and discussion. I engaged in prediction because I felt it was the *essence* of scientific thought. The fact that neither extrapolation nor prediction was popular in virological circles seemed to me to be a shame."

Francis also grew impatient with Salk's independent streak. "Damn it, Salk," Francis chided his young assistant around that time, "why do you have to do things differently from other people?"[9]

Salk rose through the ranks at Ann Arbor, becoming an assistant professor. But when the University of Pittsburgh approached Salk in 1947 and offered him his own viral research program, he eagerly made the move.

With a grant from the Armed Forces Epidemiological Board, Salk continued his work on killed-virus influenza vaccines, focusing on adjuvants. Adjuvants serve as a fuel additive to killed-virus vaccines, boosting potency and durability, and can make or break the effectiveness of a preparation. Though adjuvants have taken on something of an alchemical mystique, they work because they are made of substances so foreign to the human body that the immune system cannot help but take notice. At the time, Salk was interested in an adjuvant known as Incomplete Freund's, an emulsion of mineral oil and water with the gooey consistency of white glue. He began testing this killed influenza vaccine on thousands of soldiers stationed at Fort Dix, New Jersey.

Salk worked hard, and luck came his way in the form of Harry Weaver. Appointed as director of research at the National Foundation a few years before, Weaver, with the blessing of Basil O'Connor and the damnation of many, wanted to do just that—direct research. Until Weaver came on board, the National Foundation simply funded projects that independent researchers selected, maintaining the age-old stance that yoking the minds of scientists would leave fields fallow. The U.S. government also shied away from directing polio research, but for different reasons. The NIH, impressed with the polio research funded by the nongovernmental National Foundation, reserved most of its grant money for other scourges.[10]

While researchers who are independent arguably unearth more truths about a disease, they often investigate questions that have little practical application. Soon after becoming research director at the National Foundation, Weaver began inviting leading polio researchers to roundtable conferences in luxury hotels, where stenographers recorded their remarks, which the foundation then published as progress reports. He quickly concluded, as he wrote O'Connor, that "only an appalling few . . . were really trying to solve the problem of poliomyelitis in man." In a different reflective note to O'Connor, Weaver wrote, "It became evident that, if real progress were to be made, more exact methods of research would have to be instituted, objectives would have to be clearly defined, procedures and techniques would have to be developed to permit attaining these objectives, and individual groups of workers would have to sacrifice to some extent their inherent right to 'roam the field,' so to speak, and concentrate their energies on one or, at most, a few of the objectives."[11]

At the behest of Weaver, the National Foundation spearheaded a research project in 1948 to identify strains of poliovirus, a critical step in the development of an effective vaccine. Researchers knew that there were at least three types of poliovirus, but no one knew how many existed. If a vaccine is not made from every type of the virus, people will remain susceptible to the disease. The National Foundation assembled an eminent advisory committee to lead the virus-typing project. Along with Thomas Francis Jr., Salk's old professor and boss, there was Albert Sabin from the Children's Hospital in Cincinnati. An M.D. who began studying poliovirus shortly after leaving Russia for the States in 1930, Sabin declared his interest in—though he had not yet begun devising—a weakened, live poliovirus vaccine.

The actual grunt work of typing viruses did not interest the members of the advisory committee, so Weaver went hunting for ambitious, younger researchers who had the necessary laboratory space. And that is what led Harry Weaver to Jonas Salk, and Jonas Salk to polio research.

That year, 1948, Isabel Morgan of Johns Hopkins University in Baltimore successfully protected monkeys from one type of poliovirus with a killed-virus vaccine. Before the year was out, the United States recorded 27,726 new cases of polio. American polio cases surged to 42,033 in 1949. Public panic was building.

Science advances nanometer by nanometer, but in 1949, March of Dimes–funded research yielded a bona fide breakthrough: The lab of John Enders at Children's Hospital in Boston discovered a way to grow poliovirus in nonnervous tissue cultures, a feat that later earned a Nobel Prize. Not only did the culturing of virus in nonnervous tissue remove a safety concern; it also dramatically reduced the need for monkeys, as the testicles or kidneys of one animal provided enough tissue culture for 200 test tubes. One monkey, then, did what used to require 200. Anxious to apply this technique to making a vaccine, Salk had his own tissue-culturing system functioning by 1950, and by June of that year, revealed plans to make a killed-virus vaccine and test it in children. "My orientation at the moment is to see if we can develop a satisfactory procedure for the prevention of poliomyelitis by immunologic means," Salk wrote Weaver. "The approaches to the realization of this solution are obvious enough. The problem at hand is that of devising ways and means whereby it can be accomplished."[12]

Problem. Solution. It was the language Weaver spoke, but he could not openly back the plan.

Developing a vaccine for humans clearly was the next step, but a cautious scientific mood prevailed, partly because the two vaccines in the 1930s had performed so pitifully. Scientific concerns and public ones, though, often differ greatly. The National Foundation was a cross-breed, with both a scientific and a public lineage. Part of its soul belonged to the very researchers who found comfort in slow progress. Yet the National Foundation also worked for every person terrified by poliovirus: The mothers wearing paper hats, who had begun arming themselves with tin cans, marching door-to-door to collect dimes during January fund-raising drives. The girl in the iron lung who clenched a mallet in her teeth and played the xylophone for a March of Dimes newsreel. The crutched-up poster boy wearing a wide cowboy hat and a smile to match. And the Hollywood stars who gave generously to the cause. Weaver and Salk were the type of scientists who moved judiciously, but with a palpable sense of urgency, the very sense that the panicked public felt.

Weaver counseled Salk that if he wanted the National Foundation to help him make a vaccine, he could not speak of using children in experiments. And so it was that when Salk submitted his proposal for a 1951 National Foundation grant, which he got, he left out the idea of vaccinating children. Salk even sidestepped the controversial notion of pursuing a killed-virus vaccine, stating that a preparation containing live, but weakened poliovirus might be the way to go.

During the typing project, Salk found weakened strains of polio that intrigued him, but he never seriously considered making a live vaccine. Indeed, he would become the standard-bearer for the noninfectious killed vaccine and for a scientific principle that would open the door for a new era of vaccinology.

Back then, common wisdom held that the only way to develop long-lasting protection against a disease was by "convalescent immunity"—in other words, getting an infection and recovering. This, of course, was how people naturally became immune to polio. A vaccine could only hope to improve upon nature by creating a weak infection that could not lead to disease, but could create durable immunity. This is what the live, "attenuated" vaccines for smallpox, rabies, and yellow fever aimed to do. Salk found the common wisdom unwise. Since the time he was a medical student, he questioned why infection was needed for viral vaccines but not for the bacterial vaccines that worked so well against diphtheria and tetanus. "From my experience in chemistry, logic suggested that if one were possible then so should be the other," Salk once wrote.[13]

Salk hardly thought up the idea of a killed-virus vaccine: the first was made in 1886 (for hog cholera), and France's Louis Pasteur had unwittingly made the first human one (for rabies) two years later. There also existed killed vaccines for human typhoid, cholera, plague, pertussis, and influenza before Salk began his polio work. Yet it was the diphtheria and tetanus vaccines that encouraged him to make a killed poliovirus vaccine. The bacteria responsible for diphtheria and tetanus cause disease by releasing toxins. In the mid-1920s, researchers found that chemically inactivating the toxins, creating what are now called "toxoids," made for effective vaccines. So then didn't it follow that infection and protection could be separated, that one was not necessary to create the other? That is a fundamental question Salk believed a killed poliovirus vaccine could answer. But the question tipped the dogma cart and, at least in the world of science, won him few friends and no shortage of detractors.

• • •

THE NATIONAL FOUNDATION HAD competition in the vaccine search in 1950, most notably the New Jersey–based Lederle Pharmaceuticals. One of the hotshot researchers at Lederle was Hilary Koprowski. A Polish physician who studied piano at an Italian conservatory in 1939, Koprowski sailed to Brazil once the war broke out and worked on an attenuated yellow fever vaccine. He came to Lederle in 1944 and by 1950 had developed a live vaccine that contained weakened virus, grown in the brains of cotton rats, to defeat one of what the typing project would later reveal were three main types of poliovirus. While a successful vaccine would need to be trivalent, this was an exciting start. In the spirit of self-experimentation, Koprowski and his boss ate their vaccine.

Koprowski subsequently inoculated 20 institutionalized children with his polio vaccine. The children developed antibodies to poliovirus without illness. Though their parents had put them up for the experiment, scientists branded Koprowski, two years Salk's junior, as brash and reckless. But as the first vaccine administered to humans since the fiascoes of the thirties, it proved an important point. Humans, like monkeys and chimpanzees, responded to weakened poliovirus.

That year, 33,000 Americans got polio. One of them was Basil O'Connor's 30-year-old daughter. "When, in 1949, Dr. John Enders grew poliovirus, O'Connor sniffed victory," wrote science reporter Victor Cohn in the 1955 *Minneapolis Morning Tribune*, "and when in 1950 his daughter got polio, his quiet controlled frenzy was doubled. Now he wanted in on the kill."[14]

Jonas Salk and Basil O'Connor became friends in 1951. Though they had met before, it was on the *Queen Mary*, carrying them home from the Second International Poliomyelitis Congress in Copenhagen, that their visions fused. Over the years, the friendship would draw charges of cronyism and, worse, that O'Connor exploited Salk or that Salk willingly let O'Connor exploit him. Time has rendered the once hot allegations inert. The robust Irish-American and the slightly built, reserved son of Jewish immigrants pushed each other higher than they might otherwise have risen and the world benefited. "We made a good team," Salk explained to me. "Whatever it was, he was hoping to succeed. That was his narrow outlook."

In 1951, 28,386 Americans got polio.

By the middle of 1952, Salk had developed a killed poliovirus vaccine designed to work against all three strains of the virus, had tested it in monkeys, and was ready to inject it into humans. In keeping with the trusting nature of the times, he did not have to clear this proposed human trial with a group of ethicists at his university, scientists at the Food and Drug Administration (which didn't exist), or even the scientific committee of the National Foundation. Rather, he just had to find people willing to participate, which he did at the D. T. Watson Home for Crippled Children, just outside of Pittsburgh.

Salk chose to test his vaccine on a few dozen sufferers of polio at the home, whose parents had agreed to let them participate in the trial. He chose this group for safety reasons: If something were dangerous about his vaccine—if, say, not all of the poliovirus was killed, as some critics of the approach feared—these kids would not contract polio anew. And his goal was simply to assess whether the vaccine could boost production of antibodies against the three distinct types of poliovirus. On July 2, Salk began vaccinating the children with three vaccines, each containing a different type of poliovirus that had been killed with the chemical Formalin, a formaldehyde solution. In all, 43 children received one of the three types of vaccines. Everyone involved swore to secrecy, and only the top brass at the National Foundation, which was funding the study, even knew of its existence.

The vaccine, just as Salk had hoped, goosed the immune system and led it to crank out antibodies that could "neutralize" the virus. Specifically, when Salk and his coworkers mixed fresh samples of the vaccinees' blood with poliovirus growing in tissue culture, the virus could no longer destroy the tissue. The vaccine, it seemed, worked. As Salk told writer Richard Carter, "It was the thrill of my life. Compared to the feeling I got seeing those results under the microscope, everything that followed was anticlimactic."[15]

That year, 57,879 Americans got polio. The United States had never experienced a higher crest in the epidemiological wave—and never would again.

At a National Foundation Committee on Immunization meeting in January 1953, Salk presented his results from the Watson Home and from another institution where he had tested his experimental polio vaccines on humans, the Polk State School, a home for the retarded. Both trials indicated that the killed vaccines were safe and triggered a strong immune response, stronger even than the one caused by natural infection. Some committee members immediately called for a larger field trial. Others, most notably Sabin and John Enders, cautioned against an expanded trial.

The next month, the National Foundation gathered luminaries, none of whom were directly involved with polio research, from the Rockefeller Foundation, the American Medical Association, academia, and the U.S. government to toss about the idea of launching a large clinical trial of a killed-virus vaccine made by Salk. One of the attendees, Thomas Rivers of the Rockefeller Institute, had long been a key adviser to the National Foundation, having come aboard during the days of the Birthday Ball Commission. After Salk told the group his findings, Rivers explained how Salk would soon publish his data in the *Journal of the American Medical Association* (*JAMA*). "Dr. Salk is over a barrel," Rivers said. "Terrific pressure is going to be put on him. Terrific pressure is going to be put on the foundation and there is always the danger of moving too fast. There is also the danger of going too slow, because if you have something that is good, the public should have it as soon as possible and at a safe time. That is the problem before the group this morning. . . .

"Wouldn't it be silly to wait 50 years or to wait 10 years to develop the ideal vaccine when there is the possibility of a vaccine being developed very rapidly that will last, say, for two or three years with one injection perhaps? We don't know anything about that, but have we the right to wait until the ideal vaccine comes along?"

The attendees concurred that, as Salk wished, he should move ahead cautiously, and that Rivers should write a letter to *JAMA* explaining their decision. The intent of these formalities was to inform professionals about the vaccine Salk was developing, while keeping the public enthusiasm in check. All of these careful maneuvers were undermined, however, when a syndicated New York columnist ran a story headlined "New Polio Vaccine! Big Hopes Seen."[16] Insiders have speculated that someone at the National Foundation, anxious to march more dimes into the coffers, leaked the story.

Salk, worried that the attention would slow him down and make him seem a publicity hound, forged a deal with O'Connor that he would speak directly to the public in a radio address the evening before the *JAMA* article came out. "I knew goddamn well the next day they [the media] would be exaggerating," Salk told me. "I wanted to modulate."

Entitled "The Scientist Speaks for Himself" and broadcast over CBS radio, the speech Salk gave, which followed a brief introduction by O'Connor, stressed patience and the fact that he was but one of many polio researchers. "We want to reach our goal as quickly as possible," Salk told listeners across the country, "and I am certain that you will understand that the actual accomplishment of our purposes cannot be achieved in a day."[17]

Salk had to duck behind this door and that over the next few weeks to avoid the torrent of press. It would be this way, off and on, for the rest of his life. Whether by design or default, Salk had walked into a hall that few scientists ever enter. The bold, tall letters of the daily press had spelled out his name. Sonorous announcers had repeated it over the wireless, on television and newsreels. The kind-faced, soft-spoken Dr. Jonas Edward Salk was famous.

Whatever the effect on Salk's personal life, the radio speech bridled any hopes the public had for a large-scale trial in 1953. With help from a pediatrician at the Watson Home and more than 500 volunteers, Salk and his little viral research lab spent the summer improving his vaccine.

During the summer of 1953, Harry Weaver left the National Foundation, in part because of a clash over the design of a large-scale field trial. The Foundation also lost the support of several high-powered members of its Committee on Immunization, including Albert Sabin, when it indelicately sprung the news at an October meeting that plans were far along for a large-scale test of Salk's killed-virus vaccine.

As one committee member later wrote, "Decks had been cleared for action on a vaccine, and if in the process some individuals had been hurt and some wrong decisions had been made, that was just too bad."[18]

That year, 35,592 Americans got polio.

In 1954, the National Foundation's polio field trials mobilized the country. One writer, looking back, compared organizing the experiment to teaching a large troupe of elephants to do the mambo. Trial officials, true to the gee-whiz fifties, compiled the largely meaningless but otherwise staggering tally of how many people pitched in: 20,000 doctors and public health officials, 40,000 registered nurses, 14,000 school principals, 50,000 teachers, and 200,000 lay volunteers.[19]

The one figure that was anything but meaningless was 1,830,000—the number of first-, second-, and third-grade children in 44 states who participated. They were the patients, though the National Foundation, fearful of having them labeled guinea pigs, dubbed them "Polio Pioneers."

An illustration of Salk appeared on the cover of *Time* magazine a month before the trial began, further lionizing him. The March 29, 1954, article itself, though, did not glorify Salk, recounting the contributions of many polio researchers and even quoting unnamed colleagues criticizing the young researcher. One particularly harsh complaint was that Salk had not published much of his data. Wrote *Time*, "A cautious Yankee with long years of experiences with viruses and vaccines objects: 'We want Salk to show us, not tell us.' "

From the outset of the trial, which formally began on April 26, 1954, Basil O'Connor foresaw that if the vaccine worked, there would be a stampede of parents demanding it for their children. So that summer, O'Connor decided to pay six pharmaceutical houses a total of $9 million to produce the vaccine. If the vaccine proved effective, warehoused vials would be at the ready. The only thing standing in the way would be the government licensing authority.

That year, 38,476 Americans got polio.

In February 1955, Salk appeared on *See It Now*, the popular television interview program hosted by Edward R. Murrow. Salk had a strong distaste for journalists. (He once described a press conference to me as "feeding the animals in the zoo.") But the no-nonsense Murrow, the exemplar of broadcast journalists who made his name dispatching clear-eyed radio reports while Nazi bombs crashed around him in London, much impressed Salk.[20]

Salk discussed the colossal trial underway, the criticism pelting him, and the contributions others made to the development of what he wished was not being called the "Salk vaccine." "You must remember that nothing happens quite by chance," he told Murrow, intonating in the high style that was his wont. "It's a question of the accretion of information and experience. You must see—it should be obvious that there is an enormous heritage into which

I was born, so to speak, and it's just chance that I happen to be here at this particular time when there was available and at my disposal the great experience of all the investigators who plodded along for a number of years. And essentially that's what we've been doing ourselves, but acknowledging fully that which has been contributed by everyone else. . . . So what I am really saying is that the . . . science of immunology is not a new one, and it's one that has been contributed to by many investigators ever since the time of Pasteur."

On April 12, 1955, Thomas Francis Jr., who oversaw data analysis of the trial, revealed the results. The announcement fittingly was made in Ann Arbor on the campus of the University of Michigan, 13 years to the day after Salk first arrived there to work with Francis. The 150 journalists who attended acted something like zoo animals being fed, climbing atop tables, fighting for copies of the report, elbowing each other as they rushed to phone in their copy. Having assembled and analyzed 144,000,000 separate pieces of data, Francis stated that the vaccine worked at least 60% of the time, and had not caused any significant harm to a single child.

Two hours later, the Department of Health, Education, and Welfare licensed the vaccine.

The country—the world—exploded with joy, dancing and singing and hugging, and blasting, ringing, and clinking anything that would make noise. Page 1 of the *New York Times* had a photo of Salk and Francis with this ebullient headline above it: SALK POLIO VACCINE PROVES SUCCESS; MILLIONS WILL BE IMMUNIZED SOON. Boxes filled with vaccine, RUSH printed on their sides, were whisked out of the warehouses.

That night, Salk and Francis appeared on *See It Now* with Murrow, live from the University of Michigan. "Who owns the patent on the vaccine?" Murrow asked Salk at one point.

"Well," said Salk, "the people I would say. There is no patent. Could you patent the sun?"[21]

• • •

DURING THE NEXT FEW WEEKS, letters, telegrams, phone calls, and in-the-flesh well-wishers descended on Salk. Elated folks suggested Salk make vaccines for cancer, multiple sclerosis, colds, Lou Gehrig's disease, Parkinson's, diabetes, herpes, colitis, and even anti-Semitism. There was no end to the awards, honorary degrees, and donations. It was a train of elation, with ever more cars hooking on.

And then the train crashed.

An infant arrived at a Chicago hospital on April 25, 1955, his legs flaccid, paralyzed from polio. Nine days earlier, the infant had received a shot of "killed" vaccine in the buttocks. On April 26, California had five more cases of paralysis in recently vaccinated children. Six manufacturers had made the Salk vaccine. All six of these children received vaccine made by the same company, Cutter Laboratories of Berkeley, California.

A superbly coordinated public health network pieced these facts together posthaste, and on April 27, Surgeon General Leonard A. Scheele asked Cutter to recall its vaccine. Fearing mass panic, Scheele assured the public that this was only a "safety precaution" and that there was "no cause for alarm." At the end of the day, two more cases, one fatal, were reported.

With the cases of vaccine-related polio mounting, Scheele suspended all polio vaccinations for nearly a week in May so that investigators could inspect all of the manufacturers' plants. They found live virus in the putatively killed-virus Cutter vaccine. In all, bad lots of Cutter vaccine caused at least 149 cases of polio, 10 fatal.[22]

The Cutter incident, as it became known, should not have cast a black cloud over Salk, whose polio vaccine went on to defeat the virus in this country and abroad without ever causing another case of disease. Nor should Cutter forever have tainted the reputation of killed-virus vaccines, which have been safely used for such severe viral diseases as rabies and Japanese B encephalitis. Nor should Cutter have found its way into the propagandist's toolbox, as it did in one Hungarian paper that blamed the incident on " 'free enterprise,' which did not hesitate to draw profits from a new discovery, and used living children for experimental animals. . . ."[23] But the Cutter incident, which began 13 days after the godsend dubbed the Salk vaccine was pronounced safe and effective, left a ghost that still flutters through the vaccine world.

That August, the Public Health Service fell on the sword and conceded that its own safety tests had to share blame for the industrial accident. Salk insisted that Cutter had not followed his precise instructions: specifically, Cutter had not properly filtered the virus before killing it with Formalin, allowing sediment to build up and make some live virus inaccessible to the deadly chemical; Cutter also had not checked for live virus frequently enough during the manufacturing process. Salk later helped Melvin Belli, the San Francisco-based king of tort law, when the flamboyant attorney represented people injured by the Cutter vaccine. Though the government essentially let Cutter off the hook, the courts held the company responsible for the accidents because of "implied warranty," a legal principle that holds a party liable even if it was not at fault. Cutter settled the 60 suits filed against it for more than $3 million.[24]

The Cutter incident scared many parents away from vaccinating their children that year. In 1955, 28,985 Americans got polio.

Only 15,140 cases of polio were reported in 1956. In 1957, the polio case count plunged to 5,485. That same year the USSR began a mass inoculation with a live vaccine made from Sabin's weakened strains of poliovirus.

The United States had wrangled with the idea of mass vaccinations. The concept invariably made free-marketeers bristle, as it implicitly meant that vaccination would have to be subsidized by the government and vaccine distributed without markup by physicians. The Eisenhower administration particularly did not cotton to the idea of bankrolling mass vaccinations. Like Ronald Reagan after him, General Dwight D. Eisenhower had come into office promising to trim government and find "things it can stop doing rather than new things for it to do." And the American Medical Association issued resolution after resolution aggressively discouraging the federal government from either purchasing or distributing vaccine for any group other than the indigent. The bottom line was that the United States had a vaccine, but millions of Americans were not using it. Warehoused boxes of unshipped product gathered dust.

In 1958, 5,787 Americans got polio. The National Foundation, convinced that it had done all it could in the battle, changed its focus to birth defects. The number of polio cases jumped to 8,425 the next year.

Millions of Americans were being vaccinated, so the rising numbers likely had something to do with the waves in which epidemics occur over time. The rising numbers also had something to do with the limits of the killed vaccine. Seven hundred and fifty of the paralyzed had received three doses of the killed vaccine. Salk's critics clung to this last factor. Albert Sabin, in particular, discounted everything but the shortcomings of a killed-virus approach. In 1959, with polio on the rise once again, Sabin's criticisms attracted the most serious attention. As for 1960, Sabin augured, the situation might even be worse.[25]

Sabin was wrong. Only 3,190 Americans got polio in 1960. A Soviet medical expert that year boasted that 50 million people had been vaccinated with Sabin's live preparation and not one had come down with polio (a claim later shown to be spurious). At the Fifth International Poliomyelitis Congress in Copenhagen, the Soviets were so confident of Sabin's live vaccine that they handed out candy laced with it. Salk vividly remembers Sabin walking up to him and saying, "I'm out to eliminate the use of inactivated vaccine." To which Salk replied, "More power to you."[26]

In 1961, the year the U.S. government first licensed Sabin's live vaccine, only 1,312 Americans got polio. The drop had little, if anything, to do with the live vaccine: the live vaccine, which of course needed to contain all three types of poliovirus, was licensed one type at a time; the first was approved in August (near the end of polio season), and the last type in March 1962. It was the killed vaccine, then, that had reduced the number of polio cases 96.6% since the plague years of 1950 to 1954.

Why was a new vaccine introduced when the old one was working so well? In part, because there were lingering cases of polio, even in some who had received all three of their Salk vaccine shots. In part, because free market forces were at work: pharmaceutical houses, notably Charles Pfizer and Lederle, stood to profit from the new product. In part, because experienced vaccinologists like Albert Sabin adamantly insisted that polio would never be eradicated by a killed-virus vaccine. Cold War competition also figured in. So did the American Medical Association's strong backing for the Sabin vaccine and mass vaccinations, a policy shift that Carter argued in *Breakthrough* was an attempt to make the AMA, which was opposing nascent Medicare legislation, appear less "Scroogelike." Though the AMA's realpoliticking did not scuttle Medicare, mass inoculations—called "Sabin Oral Sundays,"or S.O.S., and backed by local medical societies—began in 1963.

Many physicians, about to be crushed by the S.O.S. juggernaut rolling their way, sought Salk's help. At one such physician's behest, Salk wrote a 10-page letter to the Essex County Medical Society. "One has no choice but to question the motives of those who are the initiators and promoters of this movement," Salk wrote in one heavily marked-up draft of the letter. "Are they in any way benefiting from this promotional campaign to create the illusion of a need that does not exist? . . .

"The effort to blanket the nation with live-virus vaccine at a time when it is not necessary . . . is not only unscientific, it is anti-scientific. The reasons given to justify such action tamper with the truth and to a point that can be considered libelous and an insult to our intelligence."

Try as he might to avoid it, Salk was at the center of what had grown into a holy war. And like all *jihads*, this one would not end for years to come.

Albert Sabin is not what kept Salk returning to the battlefield, though that is what the caricatures of the fight routinely recounted in the press would assert. Nor was it the basic principle that lasting protection could be achieved without infection; Salk had already shown that. What kept Salk bound to the killed-versus-live debate his entire career was the fact that cases of polio apparently caused by the live vaccine began to mount, while the killed vaccine never caused a single case of polio after Cutter—and hundreds of millions of doses had since been used worldwide. Indeed, in 1996 the United States shifted gears again, and decided that to reduce the risk of vaccine-caused polio from the live vaccine, children should first receive the killed vaccine and then the live one; in 2000, official policy changed to exclusive use of the killed preparation. And Salk also insisted that because the live vaccine kept reintroducing live poliovirus into the environment, by definition it could never eradicate the virus itself. The killed vaccine could.

In a 1963 paper, Salk first spelled out the details of how only the killed vaccine could eliminate poliovirus—and other viruses, too. "Our knowledge of the disease and the agent would then become academic," postulated Salk in "Mechanisms of Immunity," part of the publication *Recent Progress in Microbiology*. "Man would then have one less active concern, but some new concern will then challenge his ingenuity."[27]

• • •

POLIO IS NOT AIDS, of course. Most people who become infected by poliovirus suffer no lasting damage and are left naturally immune for life. With HIV, in contrast, most people who become infected go on to develop AIDS. And if it is easier for nature to immunize a person against a virus, then, logically, it is easier for scientists to develop a vaccine that duplicates this natural protection. The mechanism behind the polio vaccine's success also is straightforward: antibodies can stop poliovirus, as researchers well knew for many years before a vaccine existed. With HIV, scientists to this day cannot agree whether antibodies alone can prevent an infection. Some even contend that antibodies *help* the virus dodge other, more effective immune weapons.

Making matters more complicated, even if antibodies can by themselves derail one strain of HIV, that says nothing about a vaccine's potential effectiveness, as the AIDS virus changes its shape most every time it copies itself. The problem is not, as common wisdom has it, that HIV mutates at a blinding speed: Indeed, some estimates suggest that poliovirus mutates more rapidly than the AIDS virus.[28] But HIV does *replicate* at a blinding speed, making 10 billion copies of itself each day. Researchers estimate that an average of one mutation occurs during each replication cycle. Although most of these genetic variants are defective and cannot themselves make new HIVs (that is, new copies, or virions, of a given strain), enough viable ones persist to give nightmares to vaccine developers who favor the antibody approach. The bottom line is that a vaccine made from one strain of HIV—or even, say, 20 strains—may work only against a small proportion of the many extant strains of the virus now circulating in the world.

The AIDS virus presents still other scientific obstacles that poliovirus does not. HIV weaves its genetic material into the human DNA inside a cell, effectively allowing this so-called retrovirus to hide beneath the immune system's radar system. Because HIV primarily enters humans through two dramatically distinct routes—needles and sex—a vaccine that works for gay men may not help injecting drug users. And HIV attacks the very immune system that is designed to stop it, creating something of a hall of mirrors for scientists, who could develop a vaccine that stops HIV but inadvertently leads critical immunologic weapons to malfunction.

On the sociological and political fronts, even more pronounced dissimilarities stand out. The search for a polio vaccine was fueled mightily by a collective sense of urgency to stop this crippler: The President had the disease, his former law partner headed the organization that ran the polio research show, and all humans were equally susceptible to poliovirus, regardless of their behavior, their income, or their race. AIDS, in contrast, has suffered mightily because, outside of the scientific world and gay communities, there is no collective sense of urgency to stop HIV with a vaccine. And politicians have dealt with AIDS awkwardly, even in the best of times.

A key reason why the public had such a keen sense of urgency to find a polio vaccine is that everyone—everyone—was connected to the disease by a relative, a friend, or an office mate. The world's subconscious was filled with terrifying, heart-wrenching images of toddlers in leg braces and hospital wards crammed with children in iron lungs. Mothers actually raised research money to spare their children from these horrors, deepening the trust that the public already had in medical science. With AIDS, most people in wealthier countries feel no real sense of fear—and rightly so: if they are heterosexuals who are having sex with condoms and not injecting drugs, their odds of becoming infected are infinitesimally low. Many, if not most, Americans, Europeans, Canadians, and Australians don't have a single strong connection to a person who is infected with HIV, and mothers who went door to door asking for money to support AIDS vaccine research likely would be chased off many a porch. As for the public's trust in medicine, people today who volunteer their bodies for medical research want—and modern ethics demand—many more assurances that they will not be harmed.

Finally, the culture of science has changed markedly. Salk, Sabin, Koprowski, and other scientists at the front of the polio vaccine search had made vaccines against other diseases. Gallo and his colleagues who launched the HIV vaccine search had no experience making vaccines. Forty years ago, vaccine makers didn't have as many high-tech tools at their disposal, either, making it easier to avoid the trap of asking fascinating research questions that have little to do with solving the vaccine problem. And funding for the polio vaccine work came from one nongovernmental body that actively sought out researchers like Salk, encouraged them to share their preliminary data, and tried to build on research successes as quickly as possible. No such organization exists for an AIDS vaccine.

But the polio success story also provided an invaluable road map for AIDS researchers because of the similar obstacles faced by both endeavors. Here is what the signs along the road to success said:

- ***Work together:*** Polio scientists organized themselves and figured out in short order how many types of the virus existed, how to exploit the finding that poliovirus could be grown in tissue culture, and how to stage a massive human trial of one vaccine and then another.
- ***Bold leadership pays:*** Basil O'Connor, Harry Weaver, and Tom Rivers made an unpopular decision to support moving forward with Salk's killed vaccine, alienating many scientists but ultimately advancing the field and saving lives. Weaver also encouraged directed research and, at his roundtable conferences, the sharing of unpublished data.
- ***Don't wait for the ideal vaccine:*** Even though the Salk vaccine was far from perfect and was only used by two-thirds of the U.S. population, it nearly wiped out polio from the entire country within six years of being introduced.
- ***An animal model can guide the way even if it does not closely mimic what happens in humans:*** Monkeys do not naturally become infected with poliovirus and can only be paralyzed by it if the virus is injected directly into their brains. Yet Isabel Morgan's successful experiment with a killed vaccine in monkeys paved the way for Salk to develop a killed vaccine for humans.
- ***The public will tolerate failure with a vaccine if they understand the***

potential risks and benefits of its use: Although the disastrous, slapdash Kolmer and Brodie polio vaccine trials in the 1930s likely set the entire field back by a decade, the Cutter fiasco did not lead to widescale abandonment of the killed vaccine because scientists carefully moved the product forward and aggressively addressed its weaknesses as soon as they surfaced.

- ***Beware of political and business aims masquerading as scientific ones:*** The behavior of the AMA and the pharmaceutical industry during the early 1960s wrongly steered polio vaccination policy in the United States, a mistake that led to hundreds of unnecessary vaccine-caused cases of polio and was not undone until 1996.
- ***Heroes can help:*** The March of Dimes's orchestrated glorification of Jonas Salk—distasteful as it was to scientists—galvanized the drive to bring forward a polio vaccine as quickly as possible.
- ***Empiricism works:*** While polio vaccine developers knew what they needed a vaccine to do, there was great doubt about whether Salk's killed vaccine could accomplish the feat—doubts that could not have been addressed in the laboratories and were only dismissed by a real-world trial in humans that rigorously assessed the preparation's worth.

Perhaps the most salient lesson from the polio vaccine era is that spectacular scientific advances do not always involve dramatic breakthroughs in understanding. Because of this, Jonas Salk won little respect from his peers, who viewed his work as applied research that added little to the fundamental knowledge about poliovirus, immunity, or vaccination. But these colleagues entirely missed the point of Salk's accomplishment.

Salk's brilliance was that he behaved as though he had a moral imperative to solve the problem as quickly as possible. He speedily exploited the knowledge gleaned from Morgan's monkey experiments, the March of Dimes's typing program that he participated in, and the tissue-culturing work done in Enders's lab. He knew that developing a vaccine from weakened live virus would take longer, and he doubted that it offered the advantages that others so fervently espoused. He moved into humans swiftly, and, with those positive data in hand, helped the March of Dimes rapidly stage the enormous polio field trials that proved the vaccine's worth.

Salk later explained his driving philosophy at a U.S. Senate subcommittee hearing on biomedical research. "I am frequently asked: 'What is the next breakthrough going to be?' " Salk testified to Senator Edward Kennedy's

Subcommittee on Health and Scientific Research on March 31, 1977. "My response usually is, 'To use the knowledge we already possess.' "[29]

When Robert Gallo visited Salk's office in April 1986, three decades had passed since the church bells rang and sirens blared to celebrate the arrival of the first polio vaccine. Salk, who was all but retired, knew little about HIV and even less about the structure of the nascent AIDS vaccine enterprise. What Salk did recognize, though, is that a critical experiment was not being done: The testing of the killed HIV approach. That oversight alone signaled that AIDS vaccine researchers in 1986 had little respect for the past and a haphazardly organized research effort, shortcomings that ultimately would waste much time and likely cost many lives.

CHAPTER 3

Empiricists Versus Reductionists

On May 12, 1986, Jonas Salk visited Robert Gallo's laboratory in Building 37, the six-story government-issue monolith set on the NIH's 70-acre campus a few miles north of Washington, D.C. "They put on a performance, so to speak, to show me all the things that they were doing," Salk later recalled. The show lasted an hour. Then Salk presented his thoughts for 45 minutes, describing his earlier work with influenza and polio vaccines and suggesting that the whole-, killed-virus approach should be attempted with AIDS.

Accompanying Salk on his visit to Gallo's lab was Joan Abrahamson, a right-hand woman to Salk who both studied him and helped him. Abrahamson, then 35, was ridiculously accomplished. A lawyer, artist, and winner of the MacArthur "genius" award, Abrahamson described her role in Salk's life alternately as "the string on the kite to bring him down to earth," a catalyst, and a red-tape cutter. Abrahamson had high hopes at the gathering's end. "It wasn't even low-tech he was proposing, it was no-tech," she recalled. "We went back to California thinking, Great, we've initiated something. We didn't think we'd have to do any more except monitor it and find out what was going on." Salk still was resolved not to get involved.[1]

But when Salk checked in a month later, Gallo's researchers had not made any headway.

Gallo, in fact, had no interest in developing a vaccine by taking whole HIV and killing it—even though he believed it would work. "What can be done with a killed whole can be done with a protein and a little brains," Gallo explained to me in 1990. "The country has enough brains to do it. Let people who can't do it with their own abilities or brains work with it as killed whole, and let the rest of the country figure out how do it with proteins, and don't stop either one."

The bottom line was that Gallo did not think a whole, killed HIV vaccine was worth the risk. "For me, it wasn't the right thing because I'm not going to take it," said Gallo. "And I'm not going to take it because a company is going to make a mistake one day on a lot, like they did with polio."

Gallo's lack of interest convinced Abrahamson that Salk, who was spending much of his time writing philosophical treatises, had to leave the stands and come down into the arena. "I wanted to get him back into science," explained Abrahamson. "I didn't feel that the most valuable course for him was to go on writing these things thinking he's going to make a big impact on the world just through the writing. Frankly I didn't believe that. But I felt there was a role for getting him involved in some problem solving, which he's very talented at, particularly in this AIDS area, where he's been trained and where his background is, that allowed him to use his philosophical way of thinking but to apply it."

Abrahamson was straight with Salk. "Look at it this way: If it works, you will have removed fear from the life of every living human being," Abrahamson told Salk. "Everyone has been worried about polio or AIDS." And if it doesn't work, she ventured, "you will have accelerated the development of a vaccine because the other scientists would rather kill themselves than see you do it again."

• • •

WITH FLASHBULBS BURSTING and boom microphones in his face, Robert Gallo attempted to carve through the press pack that mobbed him as he finished his presentation at the Second International Conference on AIDS. Jonas Salk also had traveled to Paris for the June 1986 gathering, and suddenly Gallo noticed his new friend at his side, a friend who knew a few things about the theatrics of scientific conferences and maneuvering around journalistic frenzies.

Gallo was scheduled to attend a press conference, but Salk steered him into the quiet of the men's room, where the two men stayed sequestered until the mob at the door lost interest. Salk then suggested they sit outside for a drink and a chat. Gallo was under the gun for his growing dispute with Pasteur's Montagnier, and reporters—especially in France, Montagnier's home turf—thrilled at the ever-quotable, fired-up Gallo. But Gallo's handler this day was doing his damnedest to deny their wishes. "Once I had calmed down, and only then, he took me into the press conference," remembered Gallo. It was a vintage Salkian move: meet the press on your terms.[2]

While helping Gallo through the jam was a simple act, it foreshadowed a major role Salk would begin to play in the AIDS world. "I was hit hard by the controversy between Gallo and Montagnier, from all sides," recalled Salk. "Everybody was very disturbed and upset about that.

"It had so contaminated the entire field, scientists and nonscientists, that I felt it had to be settled or it would do great harm to the scientific communities and national affairs. And it was inappropriate. It had to be cleaned up, it had to be stopped. I felt the need to do something about HIV itself, but I felt that before I got into that, this had to be taken care of. And I thought if I saw a way to bring it to an end, this should be done. I had no authority. I wasn't asked by anybody. It just occurred to me that there may be some way of mediating this so everybody gets something."

Salk enjoyed his outsider status at the three-day meeting, as well as the chance to learn a blossoming field. "All this was new to me because I didn't know who the characters were," remembered Salk. "I was not involved with this at all. Once I got into the zoo, I met all the animals and began to realize what the mosaic was, both on the [Gallo-Montagnier] controversy side and on the side of the subject matter itself. So I started my nursery school and kindergarten and very quickly went into first grade and began to have ideas of my own as to how the problem might be dealt with. As I kept being fed, I saw more and more clearly what the pattern was. So that was a very condensed period, and my only desire was to give away ideas, not to get involved myself. This was something that needed a lot of attention and vision."

Salk particularly noticed an odd pattern in the AIDS vaccine talks he attended. Everyone had staked out a similar approach to the one Gallo espoused: using a genetically engineered protein of HIV. No one, it seemed, wanted to make a vaccine from the whole virus.

A reasonable philosophy guided Gallo and like-minded researchers, who believed that they could design effective vaccines by coupling the tools of molecular biology with their understanding of the mechanism by which HIV infects human cells. Salk saw this as a hopelessly reductionist approach. He embraced a more empirical philosophy that said first test the whole, killed pathogen because science has shown the power of that approach against polio, influenza, pertussis, plague, typhoid, cholera, rabies, and other diseases of humans and animals. Who cares *why* something works, as long as it works?

Vaccinology began as a most empirical science, with Edward Jenner proving in 1800, before anyone knew that viruses even existed, that exposure to cowpox virus could safely protect a person from smallpox virus. Similarly, 79 years later, Louis Pasteur developed the first laboratory-made vaccine, a weakened form of chicken cholera bacilli, despite knowing little about the mechanism by which this pathogen killed chickens. Pasteur went on to concoct effective rabies vaccines for humans by both weakening and killing that virus, which to this day kills people by a mechanism that is not fully understood.[3]

Although Pasteur proved the biological irony that forms the foundation of vaccinology—the same organism that causes a disease can be modified to prevent it—his successors, as the polio story shows, would disagree passionately about how an offending organism should best be modified. And with the advent of biotechnology, the debate went far beyond whether a bug should be weakened or killed.

• • •

OLD-FASHIONED VACCINES ARE SHOTGUNS, while new-fashioned preparations are rifles equipped with high-powered scopes. The key difference between the two approaches is that killed and live vaccines present a modified version of the whole organism, which in the case of HIV contains a half-dozen proteins, fats, sugars, and even material from humans that the virus accumulates when it buds from a cell. New-generation vaccines only train the immune system using proteins from the organism or even pieces of the proteins. The supposed advantage of this approach is that new-generation vaccines cannot, under any circumstance, directly cause an infection.

The most dangerous part of a virus is the genetic material packed into its core. With HIV, this genetic material consists of two separate, serpentine strands of ribonucleic acid, RNA, the molecule that carries the genetic code and allows HIV to make copies of itself. (Some viruses have DNA, or deoxyribonucleic acid—RNA's more complicated sibling—which is made of two linked strands of nucleic acid that twist into a double helix.)

The AIDS virus is shaped like a sphere with mushrooms sprouting from its surface. The RNA resides at the sphere's center, protected by a capsule of core proteins. The sphere itself—commonly called the "surface" or "envelope membrane"—is made of fats. The mushrooms, which have their stems embedded in the envelope membrane, are proteins shrouded with sugars. Each of these glycoproteins weighs 160 kilodaltons and thus is called gp160. (One dalton equals the atomic weight of a hydrogen atom.)

HIV is an arsonist that targets firehouses: it selectively infects and destroys the immune system cells that are most critical in clearing the body of a viral infection. To accomplish this feat, the virus first must break into the firehouse—most commonly, a white blood cell that carries a set of molecules on its surface that scientists have designated the CD4 receptor. HIV's gp160 uses a secret handshake with this and other surface receptors, opening the cellular doors. The virus then pours its RNA into the cell, which effectively douses the firehouse with gasoline. A viral enzyme appropriately dubbed reverse transcriptase converts the RNA into DNA. This naked retrovirus then wiggles into the nucleus, where its DNA splices into the host cell's own DNA.

A cell can live unharmed with HIV DNA in its nucleus in perpetuity. The match that lights the cell on fire, ironically, often is the immune system's attack on the virus. When the body mounts an immune response, it relies on CD4 cells to wake up the firefighters and send out the trucks. To carry out their fire chief role, CD4s begin dividing, making copies of themselves. But in the very process of cell division, the CD4s grind out viral particles that self-assemble into whole HIVs. In the final stage of viral replication, the new viruses burn through the cell membrane, killing the cell and simultaneously incorporating human proteins into the viral membrane.

These HIVs next enter the bloodstream, racing like a prairie fire from one uninfected CD4 to another. The immune system at first contains the fire, but slowly, intractably, its fire engines seize, the hydrants run dry, and the chiefs die from exhaustion.

Genetically engineered AIDS vaccines can feature any part of the virus to serve as the immune-stimulating "antigen," although the most popular ones back then all banked on HIV's envelope protein, gp160, or some portion of it. Because such vaccines contain no genetic material, they can never cause the disease they are trying to prevent. Whole, killed vaccines contain genetic material that scientists have killed (or "inactivated") with chemicals, light, or heat. The genetic material in a whole, live vaccine is intact, but it codes for an organism that has been weakened (or "attenuated") and, theoretically, cannot cause disease.

In 1986, only one prominent vaccinologist, Stanley Plotkin, was interested in developing a live (or "attenuated") HIV vaccine. Then with Children's Hospital of Pennsylvania, Plotkin had made vaccines for rubella and rabies, a live polio vaccine that never came to market, and coauthored the authoritative medical book *Vaccines*. His proposals to make a live, attenuated HIV vaccine attracted little attention and, more importantly, no funding. "I have to tell you, I'm one of those who did not want to abandon a live-virus approach," said Plotkin, who later would take over the AIDS vaccine program at France's Pasteur Mérieux Connaught. "I have been turned down for grants five times with ideas for a live vaccine. This is ridiculous. It's just too risky they say. I think it's regrettable that we don't explore as many avenues as possible."[4]

Gallo, who had much clout as one of the world's leading retrovirologists, from the beginning was a strong critic of the live approach. "I think I would argue to put people in jail who did it, OK?" Gallo told me in 1990. "This would be so seriously nuts, beyond belief."

Gallo's main fear was that a vaccine made from a weakened strain of HIV could, like other retroviruses, cause cancer through a process called "insertional mutagenesis," in which supposedly innocuous genes cause mayhem simply by integrating with the DNA in the host cell. A second worry was that HIV copies itself quickly and makes many mistakes, or mutations, which could allow the RNA in the weakened virus to change back to a virulent, AIDS-causing killer. A third nightmare scenario was that the attenuated HIV might cause disease at a slower pace than the so-called wild type virus, with people developing fullblown AIDS, say, 30 years after becoming infected rather than 10.[5]

Gallo and many others dismissed the whole, killed HIV vaccine because of the risk that not all of the virus would be killed. Any time the idea was raised, in would flutter the ghost of the Cutter incident, the tragic 1955 mishap with Salk's whole, killed polio vaccine. Theoretically and esthetically, using a specific protein from the whole organism inherently makes more sense than using either an attenuated or a killed pathogen, especially with a lethal virus like HIV. But creating such a vaccine raises significantly more complicated hurdles than the old-fashioned approaches.

Just as people can identify Elizabeth Taylor simply by seeing her purple eyes, Kirk Douglas by his cleft chin, and Jack Nicholson by his mischievous smile, the immune system needs to see only a portion of a virus to identify it. This portion, this modified version of the whole virus, can then serve as a vaccine. The biggest obstacle in designing vaccines made from viral pieces is figuring out which pieces are the purple eyes, the cleft chin, and the mischievous smile. And there are more wrong pieces than right ones. How many people, after all, could distinguish Kirk Douglas's foot, knee, elbow, neck, or knuckle from Jack Nicholson's? Then, even once the right protein or pieces of it have been selected for a vaccine, there is the manufacturing problem of either cloning copies of the protein or synthesizing pieces of it in a machine. If the copy is just a little off, if that cloned piece of virus has magenta rather than purple eyes, the vaccine likely will fail.

A vaccine made from the whole virus does not face these obstacles. Yet even enthusiasts of the whole, killed or attenuated HIV strategies well recognized the difficulty of outwitting this virus with any vaccine approach. Ideally, as happened with poliovirus, scientists begin a vaccine search knowing that people who become infected routinely suffer a mild illness, develop an immune response, clear the virus, and then remain invulnerable to subsequent infections. The same sort of "adaptive" immunity protects people from measles, mumps, rubella, diphtheria, pertussis, smallpox, and many other diseases that vaccines can defeat. With HIV, there is scant evidence—and there was none in 1986—that anyone becomes infected, clears the virus, and forever remains invulnerable to it.[6] Scientists also realized early on that HIV both copies itself rapidly and mutates at a fast clip, creating many genetically distinct subtypes. This means that the immune system might learn to defeat one strain of HIV but remain vulnerable to all of the others—much in the way that immunity against an influenza virus circulating one year offers little protection against the strain circulating the next.

So it was that in 1986, Jonas Salk believed that getting hung up on mechanisms was a trap, while Robert Gallo and hundreds of other AIDS researchers around the world believed that a thorough understanding of HIV's mechanics—how it infects cells, what it does to the immune system, what the immune system does to it—would ultimately reveal the viral proteins needed for a vaccine.

• • •

AT THE PARIS CONFERENCE, reporters bird-dogged Laurence Lasky, a little-known researcher who was doing AIDS vaccine work at Genentech, the South San Francisco biotech *wunder*company.[7] "It was like I was a rock 'n' roll star," said Lasky. "I took a woman to the conference, who later became my wife, and she asked me, 'Is your life always like this?' "

By the time of the Paris meeting, Genentech was one of a handful of companies that had joined the AIDS vaccine race. It had not begun with a bang. The initial search was a clumsy, slapdash, anemic enterprise driven largely by sexy science, venture capitalists, and marquee scientists—not a knowledge of vaccine R&D and its history. And the speed with which companies and their scientists rose to the front of the pack and then disappeared completely underscored how the leading lights were not heading anywhere meaningful—or were beaten down by the flimsy support they received from their executives.

From the outset, major pharmaceutical companies stayed clear of AIDS vaccine work. As a page 1 story on the AIDS vaccine race in the September 4, 1984, *Wall Street Journal* rightly explained, big pharmaceutical companies were "avoiding vaccines like the plague" because they took a long time to develop, raised serious liability issues, and offered a paltry profit margin compared with drugs. Indeed the article quoted a spokesperson for Merck & Co., one of the most likely large drugmakers to become involved as it recently had brought a hepatitis B vaccine to market, saying it had "no interest in AIDS vaccine [*sic*]."[8]

Because established pharmaceuticals had decided to steer clear of the field, it became the purview of biotechnology companies. A few, like Genentech, had serious scientists on board and a growing research portfolio outside of AIDS. Other early entrants attempted to build entire companies around their AIDS vaccine work.

While Merck had a distinguished history of making vaccines and total revenues in 1984 of $3.6 billion, Genentech had never marketed a vaccine and had total revenues that year of $69.8 million. But Genentech wanted a cut of the AIDS vaccine business, and it had a more solid vaccine foundation than might have been suspected.

Genentech's scientists, not management, launched the company's AIDS vaccine project. "It didn't start from the top down at Genentech," remembered Phillip Berman, a molecular biologist who joined the AIDS vaccine project. "We had a big vaccine effort. . . . It was obvious to do it."

Berman at the time worked with Lasky on a celebrated Genentech project to genetically engineer a herpes simplex vaccine. Just as the Merck hepatitis B vaccine simply contained one viral surface protein, the herpes project focused on the envelope protein of that virus. The logic was straightforward. Surface proteins on viruses play a critical role in helping them infect cells. Vaccines made from surface proteins thus, theoretically, would trigger production of antibodies that, theoretically, could attach to the virus and "neutralize" it, derailing the infection process.

Berman, Lasky, and coworkers decided to construct an AIDS vaccine with the same game plan. Rather than making a vaccine that contained the entire surface protein from HIV, the mushroom-shaped gp160, the scientists set their sights on the mushroom cap, gp120. "It was a guess," acknowledged Lasky.

To "clone" gp120, the Genentech team stitched the gene that carries the code for the protein into Chinese hamster ovary cells. The scientists harvested newly minted gp120 made by the ovary cells and injected the "recombinant" molecule into guinea pigs and rabbits. They then drew blood from the animals to see whether the protein had triggered the production of antibodies that could neutralize the AIDS virus.

In early 1986, Genentech did not have its own test for neutralization, so it sent the guinea pig blood to Jerome Groopman at Harvard Medical School's New England Deaconess Hospital. Years later, Lasky remembered clearly when the results returned. "I remember the moment Groopman called," said Lasky. "Groopman read the numbers and Phil [Berman] had a big smile on his face. That's when I knew we had it."

That evening, they cracked open the champagne. "It was nothing good: Cordon Negro, Freixenet," said Lasky. "In those days we were poor as church mice." At the June international AIDS meeting in Paris, Lasky announced Genentech's results. "It was the first real evidence that you could do it, that you could make a vaccine out of [recombinant gp120]," he said.

A few years later, Genentech's AIDS vaccine developers would be toasting their successes with Dom Perignon.

• • •

ACROSS THE SAN FRANCISCO BAY in Emeryville, researchers at a young biotech firm called Chiron had begun R&D on a strikingly similar AIDS vaccine. Chiron came to HIV vaccinology more purposefully than any other player, as the company saw itself as a vaccine maker. And it had recently had great success cloning the envelope protein for the hepatitis B vaccine.

For many biotech firms flirting with making an HIV vaccine, Chiron's success with the hepatitis B vaccine was a potent liquor. And no one was drunker on the success than William Rutter, one of Chiron's cofounders.

In the late seventies, Rutter, chair of the biochemistry department at the University of California, San Francisco, was a front-runner in the race to clone a human gene, which had never been done.[9] The gene was for insulin, the hormone that regulates blood sugar and must be injected each day by millions of people who have diabetes. But back then, the public had terrific qualms about genetic engineering, which often conjured images of Pandora, and the cloning of insulin didn't seem like a great boon to the masses.[10] "I became very concerned about developing a model that demonstrated the benefit over risk [of recombinant DNA technology]," Rutter told me. "I thought vaccines would be a tremendous model."

In collaboration with an academic colleague, Rutter made a deal with the pharmaceutical giant Merck & Co. to genetically engineer a hepatitis B vaccine. Merck already knew that a vaccine made from the surface protein of the hepatitis B virus would protect people from that disease. But Merck's method of fishing the virus from the blood of infected people, isolating the surface protein, and finally purifying it was laborious. Cloning, if it could be done, offered an elegant solution.

To clone the hepatitis B virus, Rutter and his colleagues decided to hatch a company of their own. They named it Chiron, after the wise Centaur whose tutoring led Aesculapius, the man famed for his serpent-entwined staff, to become the god of medicine and healing. By 1984, a short six years after the research began, scientists had inserted the gene for the hepatitis B surface virus into baker's yeast, made a vaccine from the recombinant protein, and gathered evidence that it worked.[11] (A month after the Paris conference, the vaccine became the first genetically engineered vaccine approved by the Food and Drug Administration for human use.[12])

Like Genentech, Chiron fashioned an HIV vaccine from the gp120 gene. Chiron never considered a traditional whole, killed or attenuated HIV vaccine, explained Paul Luciw, a retrovirologist who launched the project there. "Chiron is not interested in those approaches because Chiron is a biotech company," said Luciw. "Look, we had made a subunit in yeast for a deadly virus. The attenuated or whole inactivated approaches held no interest for us. It's like asking a car maker to make an airplane because they both have wheels and go places.

"It was the one success, right there: hepatitis B surface antigen. Maybe we could do it again."

Chiron's AIDS vaccine program, which did not really lift off until the middle of 1985, was distinguished by a crack team of scientists who carefully proceeded from A to B to C. But Chiron had not arrived at a vaccine approach by comparing several divergent strategies and selecting the best one. Rather, when it came down to it, Chiron, like every other AIDS vaccine developer, had simply decided to invest in its best guess.

One of Chiron's early snags was that it cloned the HIV protein inside of baker's yeast. Yeast, unlike the mammalian ovary cells that Genentech used to clone gp120, refuses to sugar-coat the HIV envelope protein, yielding Chiron a nude "p120" that was as different from gp120 as a fresh apple is from a candied one. This so-called p120 violated a basic principle of vaccinology, which held that a vaccine should, as closely as possible, mimic the "native" look of the pathogen in order to trigger the strongest immune response. Still, Chiron went forward with its p120, which the company believed—with a logic that escaped outsiders—would make neutralizing antibodies capable of thwarting HIV.

• • •

A FEW MONTHS BEFORE THE PARIS MEETING, Shiu-Lok Hu at Oncogen, a Seattle biotech company, reported in the prestigious journal *Nature* that he, too, had made a potential AIDS vaccine from HIV's surface protein, but Hu introduced a novel historical twist.[13] Hu spliced HIV's gp160 gene into vaccinia, a derivative of Jenner's cowpox virus used for nearly two centuries as the smallpox vaccine.[14] The scientific rationale for this approach had much merit, and many researchers subsequently would try variations on the theme. Yet Hu had to wrestle with an extra scientific factor, too: his corporate sponsors had a flimsy commitment to the work, and what little interest they had was dwindling.

Shiu-Lok Hu was one of the few early HIV vaccine developers who actually had experience making vaccines. A Hong Kong native whose shoulder-length black hair and relaxed posture gave him the look of someone who attended Berkeley during the sixties (which he did), Hu had genetically engineered vaccines before coming to Oncogen in 1985. He concocted a vaccine for a sheep disease, Rift Valley fever. Again, his vaccine contained vaccinia, which he modified to carry the gene for the surface protein of the Rift Valley fever virus.

Engineering vaccinia to hold genes from another virus lured researchers like Hu (who did not originate the idea) for both practical and theoretical reasons.[15] To appreciate the practical advantages, compare the process of making an engineered HIV protein using vaccinia with the yeast system that Chiron mastered or the hamster ovary cells employed at Genentech. At Chiron and Genentech, the researchers had to stitch the HIV gene into the yeast or ovary cells, grow the engineered yeast or cells in test tubes, fish the HIV protein from the concoction, and then purify the protein. The engineered vaccinia, in contrast, could be injected into the body, which would make the HIV protein inside of human cells. No need to fish it from anything. No need to purify. And all the sugars would be attached in their native configuration.

On the theoretical front, researchers like Hu favored engineered vaccinia over purified proteins for the same reason that polio vaccine developers favored the live vaccine over the killed: the engineered vaccinia more closely mimics a natural infection. During the polio days, Salk challenged the very idea that a vaccine had to mimic natural infection, arguing that a killed vaccine could train the immune system just as robustly as the live approach. This indeed may be the case for polio and some other diseases, but in the intervening years, researchers revealed the mechanisms that elicit specific immune responses, and it became clear that the two vaccine approaches could teach the immune system critically different lessons.

When a virus enters a person's bloodstream, a race begins between the bug and the white blood cells that make up the immune system. The virus quickly tries to enter cells and copy itself. The immune system rushes to prevent the invader from infecting cells, and to eliminate any cells that do become infected.

To prevent entry into cells, the immune system's B cells secrete antibodies, Y-shaped proteins that can identify a specific invader in the blood and lock onto it—much in the way that a Y could lock its two arms onto the point of a V. With antibodies latched onto it, an invader cannot infect cells and is rendered inert. Killed-virus vaccines excel at training B cells to produce antibodies.

Live-virus vaccines, on the other hand, specialize in prodding the immune system's T cells into action. This arm of the immune system is known as "cell-mediated immunity," which, unlike the antibody arm of the immune system, relies on living cells to engage in what amounts to hand-to-hand combat with the enemy. And the Green Berets of cell-mediated immunity are the aptly named killer cells, which clear already infected cells.

Live vaccines exploit the natural way T cells learn to search and destroy cells that harbor dangerous foreigners. After an invader, say HIV, infects a cell, it copies itself. In the process of manufacturing new HIV proteins, extra pieces of the viral proteins are cranked out and move to the cell's surface. When faced with these enemy flags stuck in the cellular topsoil, T cells note which foreigner has invaded and then furiously make copies of killer T cells that can patrol the body for other HIV-infected cells and then annihilate them. Vaccinia that contains a gene for HIV's gp160, then, aims to infect cells and copy itself, producing gp160s along the way. Pieces of these gp160s that make it to the cell surface create a mock boot camp for HIV-specific killer T cells, instructing them how to dispatch troops quickly should a real HIV invasion occur.

It would be ludicrous to suggest that Shiu-Lok Hu or anyone else in 1986 knew what it would take to stop HIV, be it antibodies alone, antibodies plus killer cells, killer cells alone, or some altogether different immunologic warriors. The immunologic mechanisms behind the various vaccine approaches also were not hard and fast: live vaccines do stimulate antibodies, and killed vaccines can trigger killer cell production. Still, Hu had an intriguing concept that merited testing, but, unfortunately, the scientific community had much more enthusiasm for it than did his corporate sponsors.

Oncogen executives, like those at Genentech, did not launch an AIDS vaccine project. Rather, when Hu in early 1985 moved to Oncogen, it was a two-year-old biotech firm in Seattle that hoped to develop diagnostics and therapies for cancer. "I started in AIDS I wouldn't say by accident, but sort of a combination of the circumstances," explained Hu in 1990. "No project was ready for me to get into at that stage, and I needed something to set up the lab, using vaccinia virus as an expression system." The most interesting something close at hand was HIV.

Oncogen was an offshoot of Genetic Systems, a biotech company launched in 1980, the year Wall Street first came down with a serious case of biomania. Genetic Systems specialized in making diagnostics for sexually transmitted diseases, including HIV. In Robert Teitelman's 1989 book *Gene Dreams*, he described how brothers Isaac and David Blech, nonscientists both, imagined the company into existence. "Genetic Systems . . . was a company that raised money with no hard assets, no products, no history to speak of," wrote Teitelman. "It was pure concept; it was the biotech everyman. . . . The Blechs created an abstraction existing only on paper and named it."[16]

France's Pasteur Institute selected Genetic Systems in early 1984 to develop and market its version of the HIV blood test. To explore genetically engineering a blood test by expressing HIV proteins, Genetic Systems logically enough turned to Oncogen, its cancer-oriented branch downstairs. And that is how Shiu-Lok Hu, the budding cancer researcher, became Shiu-Lok Hu the budding AIDS vaccine developer. "There was virus and they had all the reagents for me to set it up, and if we were really lucky we might even start something that can be used as a candidate vaccine," recalls Hu. "The implications were there—but clearly I did not go into it thinking that is my project, that is my future."

Soon Hu showed that he could integrate HIV's gp160 gene into vaccinia. So Oncogen entered the AIDS vaccine business. Not because the company had experience making vaccines. Not because the company focused on AIDS.

Oncogen got into the business because one of its researchers, on his own volition, had journeyed into an unexplored biotechnological mine and found a speck of gold.

At first, the biotech company—run by George Todaro, an accomplished tumor virologist recruited from the National Cancer Institute—encouraged Hu. But in February 1986, Bristol-Myers purchased Oncogen and its parent company, Genetic Systems, for $294 million. The biotech company's founders and executives scooped up millions for themselves. Scientists like Shiu-Lok Hu carried on as before. At least they tried to.

As a maker of blockbuster products like Bufferin, Excedrin, No-Doz, and Ban Roll-On, Bristol had the deep pockets to fund the research and development of an AIDS vaccine, which some estimates then predicted might end up costing more than $200 million.[17] But from the get-go, Bristol had reservations about Hu's project. "Bristol is interested in Oncogen because Bristol is the number one cancer drug company," said Hu, "and partly because we're doing the biotechnology stuff and Bristol really does not have the expertise in-house.

"Bristol has all along taken a semi-interested [position], tolerated more than anything else. We're just a bunch of guys fooling around. We clearly generate publicity based on our work, but does Bristol need that? I doubt it. Number two, it might not want it. But really it comes down to a bottom line with vaccine: because of the liability and because of the business of vaccines in general, big companies like Bristol, for that matter most of the companies, shy away from it."

Yet Bristol did not unplug Hu's AIDS vaccine project, which, a few weeks after the ink had dried on the takeover deal, netted a paper in *Nature* describing his success at cloning gp160 in vaccinia. At the Paris meeting, Hu described how the vaccine, when given to chimpanzees, stimulated various immune responses, as hoped. For a time, Bristol let Hu and his coworkers keep fooling around with their AIDS vaccine, which moved into a front-runner slot and doggedly stayed there for years.

• • •

ALTHOUGH ONCOGEN, GENENTECH, AND CHIRON HAD but the vaguest idea in 1986 about how to make an AIDS vaccine, each company had accomplished scientists in management to steer the research efforts. The same does not hold true for MicroGeneSys, a nascent biotech based in West Haven, Connecticut.

Franklin Volvovitz, a businessman who had some scientific training but no advanced degree, started MicroGeneSys in 1983 to develop genetically engineered insecticides and other agricultural products. Volvovitz originally hoped to build his business around baculovirus, an insect virus that infects Lepidoptera—the order that includes moths and butterflies—and is often on lettuce and other common vegetables. Like vaccinia, yeast, and Chinese hamster ovary cells, baculovirus readily accepted genes from other organisms.

To head the lab, Volvovitz hired Mark Cochran, a young researcher working at the National Institute of Allergy and Infectious Diseases (NIAID). After building a lab from scratch, Cochran soon began to play with engineering baculovirus to hold genes from hepatitis B and then HIV. "I didn't want to do insecticide stuff," explained Cochran, years later.

Volvovitz knew the sound of opportunity's knock, and soon after Cochran began expressing interesting genes in baculovirus, Volvovitz refashioned MicroGeneSys as a vaccine company. In addition to AIDS, R&D began on vaccines for hepatitis B, malaria, Japanese encephalitis, dengue, and yellow fever. Volvovitz won a handsome Small Business Innovation Research Program contract with the Department of Defense to develop several of these vaccines, which would pump more than a million dollars into the company's coffers over a few years time. Best of all, AIDS would put the company on the map.

A mention of the name Franklin Volvovitz to AIDS scientists who crossed paths with him triggered a remarkably consistent response. First they would say, "Frank." Their eyebrows would reach for the ceiling. Their nostrils would inhale a large quantity of air, which would exit slowly while the eyebrows retook their seats. "He's got chutzpah," said Cochran, years after he had left the company. "And he's got no understanding of human nature."

Brad Ericson, who worked as a staff scientist under Cochran, described Volvovitz as an overly aggressive businessman with a dangerous amount of scientific knowledge. "Frank had had some graduate school," said Ericson. "As a result, Frank thought he knew immunology better than anybody in the company. . . . He didn't know what he was talking about."

In 1979, Volvovitz dropped out of a doctoral program at New York University, where he was studying interferons (natural proteins that fight viruses), and founded BioTechnologies Inc., which hoped to market these natural proteins. Interferons were a Wall Street darling back then, as many proclaimed that they might—and this was just for starters—cure cancer. But neither interferons nor BioTechnologies Inc. lived up to expectations. "After about a year of operations, the investors simply liquidated the company," Volvovitz later explained to the *Wall Street Journal*.[18]

Volvovitz was only 34 when he started MicroGeneSys in May 1983, and though he still had a pudgy, boyish face, he had the harried demeanor and pinched smile of someone 15 years older. He alienated employees like Cochran and Ericson because of what they described as his paranoia, refusal to share scientific information, and micromanagement. But he knew how to take advantage of connections and move his company into the limelight.

At NIAID, Cochran had worked in a lab run by Bernard Moss, one of the researchers who pioneered genetic engineering experiments with vaccinia. Shortly after the Paris AIDS conference, Moss organized a meeting between the MicroGeneSys team and other leading NIAID AIDS researchers, including Malcom Martin and Cliff Lane.

Cochran described his attempts to clone gp160 in baculovirus, which still had yielded precious little. "It was very ethereal to me," said Cochran. "We had worked on recombinants but hadn't seen any protein yet. Cliff Lane is a clinical scientist who wanted to look at immune responses in individuals. He saw Frank as the person who'd give him what he needed.

"Frank pissed a lot of people off. Just by talking to them. Malcolm Martin went crazy—these are very important guys and Frank was wasting their time. They thought he was using them. Then again, they were using Frank."

By winter 1987, MicroGeneSys had produced gp160 from the insect cells and mixed it with alum, an immune system booster, or adjuvant, commonly used in many vaccines. The company quickly asked the Food and Drug Administration to approve a trial in humans.

• • •

IN 1986, BOB GALLO WAS THE HEAD OF A sprawling scientific family and if you were in, knowing him would make things happen. He would connect you with the best scientists. You would collaborate well and publish prominently. You would present your work at large conferences. Your projects would receive ample funding. You would publish and present more and more frequently, becoming an authority. Your name would become familiar to many people who otherwise would never know you. You would, in short, succeed. At least that's what happened to Scott Putney, a would-be AIDS vaccine developer from a young biotech company named Repligen.

Launched in 1981 and based in Cambridge, Massachusetts, Repligen hoped to develop genetically engineered drugs, but one of its longest-running projects in 1986 was a contract job developing "personal care" items for Gillette, the shaving giant. Potential products included deodorant, hair dyes, and hair-waving treatments. Like MicroGeneSys, the company also had a project to genetically engineer insecticides.

Putney had much expertise engineering proteins using *Escherichia coli*, a bacteria found in the human gut. Rather than constructing a vaccine from HIV's gp160 or gp120, Putney, in his spare time, decided to hunt for the very part—or "epitope"—of the surface protein that could best stimulate production of neutralizing antibodies. This hunt for *the* neutralizing antibody epitope was the ultimate reductionist exercise, with Putney using *E. coli* to manufacture different fragments of HIV's gp120. Putney next injected these fragments into rabbits and goats to see whether the animals would produce antibodies that could neutralize HIV. "I had never worked on a virology project prior to HIV," Putney acknowledged. "I had to learn a lot about viruses, retroviruses, and so forth, but I had an *entrée*: I knew how to make proteins, which is something that virologists didn't know how to do."

Gallo's lab then had one of the few reliable assays to test the power of antibodies against HIV. Repligen's cofounder, Alexander Rich of the Massachusetts Institute of Technology, was an old friend of Gallo's and so the company and the NCI researcher decided to collaborate. Gallo also hooked up Putney with Dani Bolognesi, a friend of his at Duke University who had developed a similar assay and was quickly emerging as the unofficial dean of AIDS vaccines.

In July 1986, a few weeks after the Paris meeting, a researcher from Gallo's lab called Putney with the verdict. Antibodies from one of his fragments powerfully neutralized HIV. The next morning, Duke University called with similar positive results. Putney had found a remarkable epitope. "I was so happy," recalled Putney. "I was shaking on the phone. It was really something. Here it was."

Gallo called the next day, urging Putney to write up the paper. "That really started the AIDS vaccine project at Repligen," said Putney. "The whole company had never published a really important scientific paper. . . . At Repligen, many of the projects were not going that well, so this was a time we could really put a fair amount of the resources onto the project. With these guys involved—Bob Gallo's an international AIDS researcher—that was great for Repligen, a little company with 60 people, struggling at the time."

On December 12, 1986, a paper appeared in *Science*, with Putney's name getting top billing on a list that featured Gallo, Bolognesi, and 11 other researchers.[19] Putney's 15 minutes of fame began—his face on TV, his voice on radio, and his quotes in newspapers. All of which was great news for Repligen, which soon would attract Merck, the pharmaceutical behemoth, as a collaborator. But ultimately, the Repligen story would amount to little more than a sorry reminder of how far off track the AIDS vaccine search had strayed.

• • •

BEFORE THE 1986 AIDS CONFERENCE in Paris ended, Jonas Salk realized that as much as he wanted to stay out of the AIDS vaccine search, his resolve was crumbling. "I wasn't looking for another career, so to speak," Salk later explained to me. "But, when I sensed or saw clearly no one was going to follow this path, then my sense of responsibility as a physician and human being, knowing what I knew, having the visions that I have, imagining what needed to be done and that there was no one to do it, I was moved. I had no choice in the matter. I was seeing a drowning person or somebody who's about to be hit by a great passenger train. If you can see that far ahead and you can avoid it, then you should do something about it even if it's self-sacrificing. So I suppose that kind of force forced me to do it."

Salk spent much of his time at the conference talking shop with people he already knew, the small fellowship of scientists who actually had experience designing and testing vaccines. Donald Francis and John Petricciani were two such vaccinologists.

Francis had become well known while head of the AIDS lab at the CDC, a job he recently had relinquished, moving from Atlanta to his old home of Berkeley to be the CDC's liaison for the state of California. He had a sleeve full of stripes. After he finished his residency in pediatrics at the University of Southern California Medical Center in 1970, the 28-year-old M.D. joined the CDC, and within five years he had worked on the campaigns to eradicate smallpox in Yugoslavia, Sudan, India, and Bangladesh. At the Harvard School of Public Health, he earned a doctorate in microbiology, focusing on feline leukemia virus, a retrovirus that causes an AIDS-like disease in cats. (He took time off from his doctoral studies for a brief stint with the World Health Organization, where he and three others stanched the first outbreak of African hemorrhagic fever, a devastating disease caused by Ebola virus.)

He was at the front in the testing of the vaccine for hepatitis B, a sexually transmitted virus that had a high prevalence in gay men. When AIDS cases started cropping up, Francis clicked in early—before scientists even had identified HIV (coincidentally, a distant cousin of feline leukemia virus)—with the CDC team that first unraveled how and where AIDS had spread.

Francis had a keen sense of urgency, just what AIDS needed—and just what the U.S. government, his boss, lacked. This impassioned northern Californian with longish hair publicly whipsawed the Reagan administration for not spending enough money on AIDS and doing everything it could to ignore the burgeoning pandemic. He assailed the people controlling the blood supply for failing to take basic precautions early on. He broke the only-speak-when-spoken-to rule of government charges, directly courting the media to spread the word about what was then considered a mysterious new disease. In the process, he became the hero of Randy Shilts's 1987 bestselling screed, *And the Band Played On*, which helped open the eyes of the world to the epidemic and the difficulty leading scientists were having with the disease and each other. Francis also collaborated with Montagnier and his Pasteur colleagues, helping them in their unsuccessful race against Gallo's lab to be the first to prove that HIV causes AIDS. And along the way, Francis, who had once given a pint of his own blood to Gallo for an experiment, became Gallo's arch enemy. "The ethics of science are that you've got to build on the shoulders of predecessors," said Francis. "You've got to give credit where credit is due. Here's a man who's powerful and receives major funding. This guy violated all the standards of virology I know. I had to watch this man for two years after the press conference [announcing that HIV causes AIDS] never mentioning the French. What 'I, I, I did.' "

Petricciani, who earned a master's degree in chemistry before becoming an M.D., had been in vaccinology the better part of his career too. In the sixties, he began working on the regulatory end, evaluating vaccines for the NIH and later, when the division was transferred, the FDA. He moved on to the WHO, where he headed their division of biologicals, the class that vaccines fall under as they, unlike the chemicals used in drugs, are based on living organisms. A gentle, soft-spoken man with a broom mustache and jade-green contact lenses, Petricciani had a beatific aura—a perfect foil for "St. Francis," the rebel with a cause.

The preceding December, Francis and Petricciani had published the first medical journal article devoted to AIDS vaccines. "The Prospects for and Pathways Toward a Vaccine for AIDS," published in the *New England*

Journal of Medicine, mapped out the course for developing an AIDS vaccine, marking the obstacles and opportunities. Unlike writers of similar review articles over the next decade, Francis and Petricciani were not dismissive about the whole, killed approach, but they stacked their chips on genetic engineering. In keeping with the experimental hepatitis B vaccine then being developed with recombinant DNA technology, they lined up behind the high-tech approach, writing that it was perhaps "the best option at this time." But, all in all, their farsighted meditation had a decidedly Salkian feel.[20]

After rendezvousing at the Paris home once owned by Françoise Gilot's mother, Salk and the two researchers went out for dinner and talked about the possibilities that a whole, killed HIV vaccine held. Said Petricciani, "It was a consideration of the big picture and a discussion of the global need and strategy that might be developed to get to a usable product in the shortest possible time."

The Big Picture. Jonas Salk loved this phrase, and it surfaced often when colleagues and friends spoke about him. Because science favors a reductionist understanding of mechanism, researchers forever are probing into smaller and smaller pictures, from the body to the cell, cell to proteins, proteins to amino acids, amino acids to DNA, DNA to nucleotides, nucleotides to sugars and bases, sugars and bases to carbon and hydrogen, carbon and hydrogen to electrons, neutrons, positrons, quarks, and whatever else could be found. This focus helps explain why scientists often have no more concern for the application of their science than poets have for the application of their words. The aim is basic science: to accrue knowledge, to reveal, to shed light; utilitarian ends are secondary, tertiary, or antithetical. Medicine, on the other hand, is an applied science, where mistakes can kill and successes can save. Salk, Petricciani, and Francis were physicians.

Vaccinology goes one better still, applying science to whole populations rather than individuals, and it holds the potential to save entire communities, entire countries—even, as in the case of smallpox, the entire planet. Let scientists view the world through the lens of an electron microscope. Let doctors view the world one patient at a time. Vaccinologists had a public health duty to view the world on a seven-story-high IMAX screen, which forces the eye to constantly scan in every direction, and forces the mind to acknowledge that whatever you are focusing on is a snippet.

Over the course of the evening, Salk spoke with Petricciani and Francis about the possibility of first vaccinating already infected people—called "seropositives" because their blood tests positive for HIV antibodies—with a whole, killed HIV preparation. Although these people were already infected with the virus, Salk suspected that a vaccine could boost their immunity to it, helping them to contain the fire. A century before, Pasteur had first shown that a similar postinfection strategy with a rabies vaccine prevented disease in a nine-year-old boy, Joseph Meister, who had been attacked by a crazed dog on his way to school. "The idea of immunization of seropositives became obvious to me immediately in June when I was in Paris for this scientific meeting," said Salk. "I realized the information gave the obvious analogy to the Pasteur immunization of the young boy who was already bitten. . . . You have the immune response giving the immune system an advantage before the virus gets the upper hand."[21]

During the remaining months in 1986, Salk's role in the AIDS vaccine field steadily grew, and many colleagues began to wonder out loud whether his contribution was self-sacrificing or a misguided quest by a passionate old man. Joseph Melnick, a veteran polio virologist, called Salk "the Rip Van Winkle of virology." Said Melnick: "He came back after a long sleep. And when he came back, he thought science was exactly where he left it."[22] Salk just didn't get it—and never had. "No one's ever accused Jonas Salk of being a scientist," once chided epidemiologist D. A. Henderson, who ran the successful smallpox vaccine program for the WHO until that disease was eradicated.[23]

Whatever Salk's motivations and skills, his entry into the arena signaled a division in fundamental philosophies about how to proceed. And although some reductionists would later switch to being empiricists when decisions had to be made about whether to stage real-world tests of their vaccines in thousands of people, the division between these two camps would grow wider and wider, eventually bringing the entire enterprise to what looked like a standstill.

CHAPTER 4

Moving into Humans

During a rousing Sunday morning game in September 1984, Daniel Zagury volleyed both tennis balls and ideas with his friend and sometime collaborator, Bob Gallo, who was visiting Paris. The main topic: where did the AIDS virus come from? Maybe, they speculated, the virus harmlessly lived in people for many years, woven into their chromosomes but otherwise invisible. Only when infected by a protozoa like the one responsible for malaria, a tuberculosis-causing bacteria, or some other bug, would this "endogenous" virus surface and start whittling away at the immune system. "It would be good to go to central Africa," Zagury said as he drove to the Paris airport for Gallo's noon flight.

It made sense to test this hypothesis in the center of Africa, as it is home to most every infectious agent that has ever plagued humans. But had Zagury not been friends with Gallo, their discussion about the origin of AIDS would have been little more than tennis-court chatter.

The Moroccan-born Zagury was an M.D. who in the early sixties traded in his stethoscope for the microscope, specializing in the T-cell actions of the immune system. He met Gallo, whose lab discovered a critical T-cell growth factor, in the mid-1970s. Despite a wide range of medical and laboratory experience, Zagury, a professor at Paris's Université Pierre et Marie Curie, was far from expert on the evolution of viruses. And notable as his own work was, he mostly published in obscure journals and was not in the same league as his American friend. Yet Zagury, who with Gallo had just completed a paper for *Science* that detailed how they isolated the AIDS virus from seminal fluid, was on a career trajectory that was heading straight up. The peripatetic doctor-scientist, then 57, also had worked outside of France many times, conducting research in French Indochina, Algeria, Israel, and the

United States. If Daniel Zagury wanted to go to central Africa and investigate the origin of AIDS, well then, why not?

Zagury soon met a Zairian surgeon at his university, who helped him arrange a trip to Zaire's capital city, Kinshasa, at the end of December 1984. While there, Zagury gave a talk to a group of physicians at the University of Kinshasa's hospital. Lurhuma Zirimwabagabo, a well-connected Zairian doctor, was intrigued by the impassioned Zagury, whose jungle of black eyebrow hair give him a brilliant, if somewhat crazed, look. Lurhuma offered to collaborate. By December 28, the researchers had cobbled together a description of the work they hoped to do and, with Lurhuma pulling strings at the Ministry of Health, had received the go-ahead from the Zairian government. As part of their agreement, the Université Pierre et Marie Curie would pay all the costs, including building the infrastructure at the University of Kinshasa. In addition to attempting to unravel the epidemiological links between malaria and AIDS, the work would also investigate how the immune system behaved when assaulted by both diseases.

Zagury carted the first samples home with him in early January and then, every two weeks, Lurhuma shipped more. The samples interested Zagury, but the only light they shed on the origin of AIDS was that the virus was not, as he had hypothesized, an endogenous patch of genetic material that other infections brought to life. When Zagury returned to Kinshasa in mid-1985, he began investigating the origin question by hunting for the virus in different monkey species. So Zagury, in a move that further emphasized his throwback, independent personality, went to the marketplace there and began to buy monkeys.

But the ultimate evidence that Zagury was driven by an old-fashioned, just-do-it approach to science surfaced in his pursuit of an AIDS vaccine. As early as 1985, Zagury had raised the idea of testing an AIDS vaccine in Zaire with Jonathan Mann, an epidemiologist who then headed Projet SIDA, a Kinshasa-based, collaborative AIDS research program between CDC (his employer), the NIH, the Zairian Ministry of Health, and the Belgian Institute of Tropical Medicine. Mann was less than enthusiastic, and encouraged Zagury to work closely with the Zairian government. "I wanted to make sure he didn't just round up people and vaccinate them," said Mann in 1998, by which time he had become the dean of the School of Public Health at Allegheny University of the Health Sciences in Pennsylvania. "A fair amount of research had been done that way in Zaire."

Mann soon learned that his concerns about Zagury working independent of the government were misplaced. Before Mann left Projet SIDA in June 1986 to take the job as director of the WHO's newly formed Global Programme on AIDS, an official from the Ministry of Health called him in to explain why his project wasn't doing more for the people of Zaire—and the official held Zagury's work up as an example of the way things should be done. Although no one had yet vaccinated anyone, Mann noted that Zagury's collaborator, Lurhuma, recently had held a press conference to announce the great progress he was making in developing a vaccine, a treatment, and a diagnostic for HIV infection.

The speed at which Daniel Zagury progressed during the next few months was compelling and demonstrated what a driven, unshackled researcher could accomplish, much in the way that Salk and other polio vaccine researchers made great strides in years rather than decades. Yet Zagury's vaccine studies also raised serious ethical issues that would come back to haunt him. They would highlight, too, the uncomfortable reality that the nascent search for an AIDS vaccine urgently needed strong leadership if it was to become a collaborative, collective effort of a scientific community—rather than a scattershot, anything-goes free-for-all that resisted organization, cherished secrecy, and often defied logic.

Many leading researchers would attempt to organize the field, and they wisely mapped out the difficult terrain in front of them. But these leaders, who all the while paid close attention to their own position in the pack, would prove much more adept at identifying the obstacles than at charting ways around them.

• • •

THE FIRST HUMAN EVER TO RECEIVE AN AIDS VACCINE is lost to history. Daniel Zagury, who made the vaccine, did not document for his records the person's name, the exact date of the vaccination, or who gave the shot. Indeed, he was not even in Zaire for this historic event.

Zagury, however, was hardly in the dark about this vaccination. In July 1986, he visited Kinshasa and decided with his Zairian collaborators to begin tests of an AIDS vaccine. By then, Zagury had collaborations with both Lurhuma and Jean-Jacques Salaun, a colonel in the Zairian army who headed the French-funded National Institute of Biomedical Research (INRB). Zagury had also started a monkey colony at INRB, which soon would receive AIDS research money from the U.S. Department of Defense.[1]

Although Zagury wanted to make a conventional vaccine to protect uninfected people from HIV, he, like Jonas Salk, also wondered whether a vaccine might somehow supercharge the immune systems of people who already had become infected with HIV. "The patients were dying like hell," Zagury remembered. "You had to do something you think could help."

So-called therapeutic vaccination has been attempted for more than a century, but scant data support the idea that a vaccine can thwart an infection once it has a foothold in the body. One partial exception to this rule, Louis Pasteur's rabies vaccine, must be given within weeks of a person becoming infected by that virus. Doctors also routinely use hepatitis B vaccine to abort infections in newborns of mothers who carry that virus, and again, timing is critical. Outside of those two examples, vaccine therapy has been a much ridiculed idea, which is not to say it has no chance against HIV but, in science, extraordinary claims demand extraordinary proof. Zagury's studies would not provide that proof.

It is difficult, even in retrospect, to make sense of Zagury's earliest studies. Traditionally, researchers begin with small studies that involve, say, a dozen patients. These "phase I" tests simply aim to show that a vaccine is safe and can stimulate the immune system. If the vaccine meets those standards, it moves on to a larger, "phase II," study with maybe 100 patients. Researchers evaluate the same parameters, but now also start looking for hints that the preparation works. Researchers advance vaccines that clear these hurdles into "phase III" efficacy trials, which attempt to determine conclusively in a few thousand patients whether the vaccine actually works. Zagury did not proceed in this stepwise fashion. He also never published detailed results of these first tests and, more than a decade later, remained reluctant to discuss them in any detail.

What Zagury would say is that before he left Zaire that July, Lurhuma took him to the university hospital to meet two women with AIDS. The researchers drew blood samples from these women, which Zagury then carted back to his Paris lab, where he made a most unusual vaccine.

Zagury's first vaccine consisted of white blood cells that he had isolated from these patients and then, in the test tube, purposefully infected with their own HIV. He next mixed formaldehyde onto these cells to render the HIV noninfectious. This *gammish* of dead, HIV-infected cells was the vaccine.

The unusual approach had a sound mechanistic rationale behind it. An HIV-infected person has many cells that the virus has not penetrated. Theoretically, this vaccine would fool the body into thinking that a massive infection was underway, requiring a massive counterattack by the immune system. And Zagury specifically wanted his vaccine to boost cell-mediated immunity by tricking the immune system into flooding the blood with the warriors called killer cells. That is why he designed the vaccine to contain dead, HIV-infected *cells*—as opposed to simply whole, killed HIV: These infected cells presumably would have pieces of HIV on their surfaces, the enemy flags that would tell the immune system to deploy battalions of killer cells.

In July or August, said Zagury, either Lurhuma or Salaun injected these tailor-made vaccines into the women. Zagury distinctly remembered a phone call to Lurhuma near the summer's end. Zagury feared that one of the women, who was hospitalized, might have died.

"No, no," Lurhuma said. "She's back at work."

According to Zagury, this woman had seen a jump in the number of CD4 cells in her blood. The hallmark of AIDS is the steady destruction of CD4s, T cells that control the baton for the immune system's orchestrated actions. Normally, people have between 600 and 1,200 CD4 cells per milliliter of blood. Before receiving the vaccine, this woman had 250 CD4s, said Zagury, and, a month later, she had 380. And the news that she had returned to work was further proof, as far as Zagury was concerned, that the vaccine was helping. "When I heard that she left the hospital, I said bravo to me," said Zagury. "That prompted me to be very optimistic." The trial was soon expanded to include eight more HIV-infected people. No results from this trial were ever published.[2]

• • •

ACCORDING TO ZAGURY, the first HIV-negative person to receive an AIDS vaccine is not lost to history: it was Daniel Zagury himself.

Zagury repeatedly refused my requests to discuss this self-inoculation, which took place sometime in early November 1986. "It's embarrassing to me," said Zagury. "Any scientist had to do so. Otherwise, you cannot convince others."

Zagury inoculated himself with a vaccine that closely resembled the one that Shiu-Lok Hu was developing at Seattle's Oncogen: it contained vaccinia (the smallpox vaccine) that had stitched into it a gene for HIV's surface protein, gp160. Zagury hoped that the vaccinia would infect his cells and, once inside, produce gp160s that would pop up on the cell's surface, staging that mock boot camp for killer cells. The vaccine, he postulated, might also trigger production of antibodies that could neutralize HIV.

Before Zagury put the vaccinia-gp160 into his own body, he tested the preparation in monkeys, baboons, and one chimpanzee to assess whether it could stimulate immune responses and whether it was safe. He did not, however, conduct the formal "challenge" experiment, in which researchers vaccinate chimps and then inject them with HIV to see whether they can resist the infection.

Zagury staunchly defended his decision to proceed without first proving that the vaccine worked in chimpanzee experiments. "You don't know the situation in Zaire," he said. "It's like you're in the desert and you're talking about the level of calcium of the water. We were sure that it was absolutely harmless. We couldn't have more evidence." He also had little faith in the chimpanzee model. Although chimps were similar enough to humans to be infected with HIV, no infected animal had yet developed AIDS. "I felt that there was no animal model at all," said Zagury. "And if you could protect some people, you will learn."

Sometime in November 1986—it's unclear whether it was before or after he vaccinated himself[3]—trials of Zagury's vaccine began in Zaire on nine uninfected children, who ranged in age from 2 to 12. According to Zagury, the children's fathers all had died from AIDS and most of the mothers, who were infected, were taking part in his therapeutic vaccine studies. "They begged us to do something for their children," Zagury told me. That decision would later come under close scrutiny.

But Daniel Zagury's first vaccine studies did not receive any outside attention at the time because hardly anyone besides his collaborators and the participants knew anything about them—until a page 1 *New York Times* story on December 17, 1986. The story, by reporter Lawrence Altman, relied heavily on unnamed "medical sources," and only mentioned the therapeutic immunizations. Zagury said he had "no comment," stating that he was "under oath to the Zairian government" not to discuss the work until it was published in a scientific journal. When Altman contacted INRB's Salaun in Kinshasa, Salaun referred all questions to Zagury. Jonathan Mann, then head of WHO's Global Programme on AIDS, said he had asked the Zairian government about the studies but had not received a reply.

Altman closed his story with a quote from Gallo:

> "If anybody stumbles upon a way to open the door it might be this guy because he has a good smell for what is going on," Dr. Gallo said.

Nevertheless, he said, he had told Dr. Zagury he would not conduct similar experiments "because I would be afraid to do it, afraid of having troubles."

Le Monde, the Parisian daily, ran a follow-up story that included a sidebar entitled "Le danger des essais 'sauvages' "—The danger of wildcat trials:

> "The worst thing of all," Dr. Mann confided to us on this occasion, "would be the realization of 'wildcat' trials, outside of all international framework and reference. Far from advancing the situation, such initiatives would have the immediate consequence of delaying the whole process."
>
> . . . All medical or scientific initiatives giving the impression that black Africa would be a field for human experimentation on AIDS would not escape having formidable political, diplomatic, and, of course, medical consequences.[4]

• • •

WHILE DANIEL ZAGURY was forging ahead on his quixotic quest, the forces that for years to come would most powerfully shape and direct AIDS vaccine tests in humans began to emerge. WHO would provide the moral voice. NIH would pony up the bulk of the funding for the research, which mainly would be carried out at universities around the United States and Europe. The few companies seriously interested in AIDS vaccine R&D would court each of these parties, as well as politicians, whom they hoped would enact legislation to reduce their risks. It was a fragmented venture with many competing interests that excelled at spelling out what needed to be done but had a tremendously difficult time building on those insights. Still, enough activity was taking place in enough quarters to create the impression that an AIDS vaccine was moving closer to reality every day. All of this played out at a January 16, 1987, hearing held by the Senate's Committee on Labor and Human Resources that focused largely on AIDS vaccines.

Massachusetts Senator Edward Kennedy, chair of the committee, first directed AIDS vaccine questions to WHO's Mann, and the focus was Zagury's work. "Given the need for international coordination, is it not unusual for a secret vaccine trial to be under way?" Kennedy asked.

"Well, I think it's extremely unfortunate," intoned Mann. He went on to explain that WHO learned of the trials through the press and knew next to no details. He worried that the trial might not be well designed or conducted, which could create false hopes. "Our best interests lie in [the] sharing of the information as rapidly as possible," said Mann.

Kennedy asked whether it was "realistic" for the world to coordinate the development and testing of AIDS vaccines. Mann said it would be "difficult," but he stressed that WHO had a track record of establishing standards and guidelines for vaccine development. "I think we must make this effort, which ultimately boils down to an effort of voluntary participation by vaccine manufacturers, scientists, and others throughout the world to realize the benefits of collaboration benefit the whole world and that this problem is too important to leave to any one laboratory or to any one research facility," said Mann. "However, we may not succeed, but the alternative to trying is potential chaos in this area."

Next came Anthony Fauci, who effectively established that his institute, NIAID, was running the AIDS research show. Since 1984, when Gallo's lab proved that HIV causes AIDS, NIH positioned itself as the center of the AIDS vaccine universe. At first, Gallo's lab was in charge, forming HIVAC, what he has described as an "informal collaborative network" that meshed NCI researchers with groups from around the world working in academia and industry.[5] But by 1987, it was clear that the power base had shifted across the NIH's campus to NIAID and Fauci.

By then, Fauci had been designated the coordinator of AIDS research at the NIH, which that year had a $261 million budget to fight the disease. A smart dresser who explained science eloquently in his street-smart Brooklyn twang, Fauci was a darling of politicians, and had perfected many of their techniques, including the no-answer answer to tough questions. Fauci commanded wide respect from scientists, too, as he was a serious immunologist and clinician who had published widely on AIDS, helping to elucidate the finer aspects of how HIV destroyed the immune system and also keeping colleagues abreast of the latest treatment strategies.

After detailing the NIH's AIDS vaccine budget—which amounted to about 10% of the annual AIDS allocation, a percentage that wouldn't change for the next decade—Fauci described NIAID's plan to pull together industry and academia through "multi-institutional, multi-disciplinary research teams," then called National Cooperative Vaccine Development Groups, which would "develop creative and targeted approaches to vaccine development."[6] Six vaccine evaluation units at universities around the country would run human tests of promising candidates, one of which he said might begin by the end of the year. Everything sounded on track.

The most difficult issue Fauci tackled that day was the relationship between primate vaccine studies and human trials. "What is your opinion about whether it is going to be essential to test it in primates or humans first?" asked Kennedy.

Fauci said there was "no question" that before going into humans, a vaccine should be tested in animals for safety and its ability to stimulate an immune response. But he explained that the best animal model, the chimp, did not appear to develop AIDS when infected with HIV. "So it could be argued by some that even though you inject a vaccine into the chimp and either show efficacy or not show efficacy, this may not be directly extrapolated to man," said Fauci. "So it is an arguable point. It is not an open-and-shut case."

The hearing closed with testimony from a scientist and a lawyer representing Genentech, who urged the senators to wrestle with such staggering issues as guaranteeing a market for a vaccine and reforming tort laws to protect manufacturers being sued if their preparation harmed someone. "We urge that these steps be taken now before the biotechnology companies developing HIV vaccines have to face the decision of whether they will commercialize their inventions or not," declared Genentech VP David Martin in his statement.

There was bravado in the testimony, but Martin spoke sensibly. Companies feared entering the AIDS vaccine search because they had such a shaky handle on who would actually take a vaccine if it existed. Would the government make it compulsory, as it does with polio, diphtheria, pertussis, and other vaccines? Would the government buy vaccine in bulk? "We do not know what the market size is, and therefore when it comes to managing essentially a small-venture company such as Genentech, we always have to concern ourselves with whether we are betting the company on what is, in effect, a very risky project," said Martin.

The liability issue also frightened most every company thinking of making an AIDS vaccine. Vaccines have a long history of creating liability nightmares for manufacturers, because even when properly made, they often harm a small percentage of the people who receive them; Sabin's oral polio vaccine, for example, causes polio every 2.6 million doses.[7] Many pharmaceutical companies indeed ran from the business by the early 1980s, leading to fears of vaccine shortages, which compelled Congress in 1986 to pass the National

Childhood Vaccine Injury Compensation Act. This no-fault system collects a tax from vaccines and then uses the money to compensate people who agree not to litigate and can provide some evidence that they were harmed.

Rather than adding an AIDS vaccine to the list of preparations covered by the National Childhood Vaccine Injury Act, the Genentech representatives suggested that the federal government pass tort reform, similar to what the company had just pushed through the California legislature. This legislation protected manufacturers unless they blatantly failed to inform people of the risks or if the injury was caused by their negligence or misconduct.

Kennedy liked what he heard. "What we are talking about here is doing this in a logical, thoughtful way on something which we know we are going to face over a long period of time and giving some degree of predictability to the private sector and some reliability," said Kennedy. "That is the kind of thinking that I would hope we would have. I am not sure that we have got it right now, but it is an important point, and I am glad you made it."

The discussion that took place between the senators and the scientists that Thursday morning in 1987 cut to the bone, and carefully detailed some of the most pressing ethical, scientific, organizational, and legal conflicts surrounding the testing of AIDS vaccines in humans. Unfortunately, so little progress was made resolving any of these issues that a nearly identical hearing could have occurred more than a decade later.

• • •

DANIEL ZAGURY DID NOT SHARE ANY DETAILS about his human trials until March 19, 1987, when *Nature* ran a letter—not a full-fledged scientific report that had withstood the review of peers—from Zagury, Lurhuma, Salaun, and their collaborators. The letter focused on the results of Zagury's self-immunization. He suffered no complications, and, after two months, his immune system produced both killer T cells and antibodies against the virus.

From the outset of the AIDS vaccine search, a major issue has been whether a vaccine can protect against the many strains of HIV that exist. The *Nature* letter addressed this point. The gp160 in Zagury's vaccine came from one strain of HIV, dubbed 3B, which Gallo's lab first isolated. In a test-tube experiment, explained Zagury and his collaborators in the letter, they had shown that antibodies taken from his blood after the vaccination could neutralize HIV_{3B}, preventing it from infecting cells, but the antibodies did not work against a closely related, but different strain of HIV. Killer T cells isolated from his body after vaccination, however, responded to both HIV_{3B} and the other strain, indicating that he might be able to mount a killer cell response against distantly related strains of HIV. "Preliminary confirmation of this result has come from immunization of a small group of Zairians," the letter read, failing to note that the small group consisted of small children.

Reactions to the news were intense, both for and against. "This is a crazy experiment," Genentech AIDS vaccine researcher Laurence Lasky told the *Wall Street Journal*. "It's like a B-movie. . . . It's brave but ultimately it's meaningless." The same article quoted Jonas Salk, who had inoculated himself with his polio vaccine before testing it widely, supporting Zagury's self-inoculation. "I look upon it as ritual and symbolic," Salk said. In the *Washington Post*, NIAID's Fauci suggested that Zagury's experiment might help shorten the time it takes to find a vaccine because he showed that animal challenge experiments were not necessary before moving into humans. The same article had Dani Bolognesi from Duke University, the Gallo intimate and budding guru of the AIDS vaccine field, outright showering Zagury with praise. "The principle of an individual trying to do something in humans and being first to try it himself is exciting, daring and probably of considerable value," said Bolognesi, a bald and bespectacled man who had an air of wisdom about him and specialized in assessing the field's zeitgeist.[8]

That June, the third international AIDS conference attracted 6,350 researchers and 800 reporters to Washington, D.C. One of the biggest headline grabbers was Daniel Zagury, who gave his first public presentation about his vaccine trials.

After his talk, Zagury found himself deluged by the media. "Zagury was besieged by a horde of television camera crews and photographers, who pursued him through the auditorium knocking over chairs and shoving spectators aside," reported *Newsday*. "Zagury, surrounded, pushed through the crowd, left the hall, was forced out into the street, turned back into the hotel, sped through the auditorium and finally reached an elevator, where he stood cornered by glaring television lights until the doors slid closed."[9]

The absurdity of all of this is that, aside from showing that injecting someone with pieces of HIV was not inherently dangerous, Daniel Zagury's vaccine studies, in retrospect, did not teach the AIDS vaccine field much at all. Rather, these experiments became most noted for, as Gallo predicted, the troubles they caused Daniel Zagury—and what they taught the field about how *not* to stage vaccine trials. "There is such a thing as excessive empiricism," said Jonathan Mann, years later. "He could have accelerated the process to develop a vaccine, but instead, he sabotaged it. To this day, people say, 'You're being like Zagury,' which means you're being too aggressive."

• • •

AS A STARK CONTRAST TO the secrecy and dramatic derring-do that surrounded Daniel Zagury's initial trials, consider how the National Institute of Allergy and Infectious Diseases staged the first trials of an AIDS vaccine in the United States. NIAID advanced the vaccine made by MicroGeneSys and, scientifically speaking, it rested on as shaky a foundation as the one that Zagury tested. Yet unlike Zagury, NIAID all but advertised that fact.

As with Zagury's vaccine, the MicroGeneSys preparation consisted of HIV's gp160, which the company had genetically engineered with help from baculovirus, the microorganism that normally attacks moths and butterflies. This baculovirus-derived gp160, when coupled with the immune-boosting adjuvant called alum, stimulated animals (particularly guinea pigs) to produce "high" levels of antibodies that could neutralize the same strain of HIV. Although tests in monkeys and chimps did not lead to any serious side effects, no vaccinated chimps were challenged with live virus. "That was an era when anything was better than what we had," explained NIAID's Fauci.

On August 18, 1987, NIAID and MicroGeneSys held a press conference and confirmed stories, which had first leaked the preceding weekend, that said the FDA had given them the green light to begin the trials.[10] NIAID that day also distributed to the press seven pages of questions and answers, explaining in great detail why this vaccine was being tested, what it contained, what the risks were, and how it would move into larger trials, if warranted. In all, NIAID explained, they would recruit 81 healthy, homosexual men at low risk of HIV infection to participate in the trial.[11] Some of the questions, which were written by NIAID staff and scientists, were remarkable in their candor. "Are you cutting corners in the FDA process because of political pressure to begin vaccine trials?" NIAID asked itself. "We are not cutting corners, and in fact the FDA and our own Institutional Review Board [a committee that must approve human experiments] have been especially scrupulous in this regard," NIAID answered itself. "The pressure we feel is in response to the gravity of a worldwide public health emergency."

The media jumped all over the story, focusing mainly on the underdog status of MicroGeneSys, a startup biotech company that had 30 employees, no products, and no big-name researchers on board. Many newspapers played the story on page 1, and so many TV camera crews descended on the company's headquarters in West Haven, Connecticut, that employees could not pull their cars out for lunch. "Moving to the Fore in AIDS Research" read a *Wall Street Journal* headline on an article that profiled MicroGeneSys founder and CEO Frank Volvovitz. "Tiny Company Joins Front Ranks of AIDS Fight," announced the *New York Times*. "A Dark Horse in the Race Against AIDS," chimed the *Los Angeles Times*.[12] NIAID's choice of MicroGeneSys even surprised the company's top scientist, Mark Cochran. "I remember thinking this is crazy because we're up against Genentech, Merck, Chiron," said Cochran. "I always thought, Geez, why did we come first? This makes no sense to me." (A key reason they came first, ironically, is because of Cochran's connections at NIAID, where he used to work.)

The attention was all for the good. By selecting a virtually unknown company to start with, NIAID created the impression that it was interested in helping anyone and everyone interested in moving an AIDS vaccine closer to reality. The institution's openness and the attendant publicity also defused the type of criticism that had clouded Zagury's trials. Just as the March of Dimes did with the polio vaccine trials, NIAID set a standard for how to conduct and publicize AIDS vaccine trials that the rest of the world would attempt to emulate.

But NIAID had a serious shortcoming, and it would undermine its ability to lead with the authority that the March of Dimes once had: As eager as the institution was to find an AIDS vaccine, it did not move products forward the way industry did—namely, following a "critical path" that streamlined the discovery process by identifying milestones up front and advancing or scuttling projects accordingly. Rather, NIAID partnered with MicroGeneSys on human tests because it believed the vaccine could help elucidate how the immune system worked in relationship to HIV. This clinical research project in essence, then, addressed fundamental questions about the biology of HIV. That was certainly a worthy goal, but it was not a reliable plan to find a working vaccine as quickly as possible, and it ultimately would lose NIAID many friends in industry, who would come to see the institution as an unreliable partner.

• • •

BEFORE THE END OF 1987, the most visionary public meeting yet held about AIDS vaccine development took place at the National Academy of Sciences in Washington, D.C. Sponsored by the NAS's Institute of Medicine, the

December 14 and 15 gathering mixed old-time vaccine makers with the younger crew that steered the AIDS vaccine world. Several of the elders prodded their colleagues to look backward.

Rubella, polio, and rabies vaccine developer Stanley Plotkin sung the merits of live, but weakened-virus vaccines. But the younger set, enthralled with genetic engineering, had dismissed the approach out of hand.

David Nalin, an experienced developer of vaccines against influenza, hepatitis A, and cholera who worked at Merck, Sharpe & Dohme (a subsidiary of Merck & Co.), questioned why no one was pursuing "something that we used to like to know in setting out for vaccine development, namely, what is the natural mode of protection in man?" Exposed but uninfected people could be found, Nalin suggested. "I am urging people to pursue negatives," said Nalin.

Jonas Salk, in addition to promoting the whole-, killed-virus idea, pushed the merits of the old-fashioned empirical approach. "Perhaps it is useful at this time when we are about to examine the AIDS question to look at the successes of the past and not merely the failures, and see perhaps whether or not that could offer us some inspiration," said Salk.

Maurice Hilleman, a cantankerous Merck & Co. researcher who had a hand in making more vaccines than anyone alive, pointed out how little vaccine developers actually understood about their successes. "I think the big problem of trying to get up here and talk about how vaccines work is that we don't know a damn thing about how they work, and in the old days, you know, we used to try to solve problems without understanding them, and it was great," said Hilleman, a tall, can-do man with a bloodhound's eyes who was raised on a farm in the high plains of Montana and had a Wild West streak. "This is the first time we have ever had need to understand anything, and maybe what we should do now is to go back and try to figure out how those old vaccines worked, and maybe we would learn a lot about AIDS. I am sorry I couldn't talk about how vaccines work because I don't know."

"Maybe that's the point," said Salk. "You did it by the seat of your pants before, and you didn't know, but somehow you succeeded."

There was much wisdom in what each of these men had to say. But their words did not lead to actions. Human trials of AIDS vaccines already were underway, and it would take another five years before researchers began to realize that their first paths set them on a road that was leading nowhere.

CHAPTER 5

Animal Illogic

When Edward Jenner wanted to test the worth of his smallpox vaccine, he injected smallpox virus into his vaccinated nephew. If Jenner lived today and ran a similar test, he would lose his medical license. Ethical regulations, spelled out by the Nuremberg Code in 1949 following the discovery of Nazi experiments on concentration camp prisoners, explicitly require that human studies not only avoid all unnecessary physical suffering but be "designed and based on the results of animal experimentation." The Declaration of Helsinki in 1964 echoed these principles, stating that research involving humans "should be based on adequately performed laboratory and animal experimentation."

No AIDS vaccine researcher has ever proposed vaccinating a human and then challenging the person with live HIV. But animal experiments have played a confusing role in the hunt for the best AIDS vaccine formulations, highlighting that the Nuremberg Code and the Helsinki Declaration have offered little guidance in this realm. The essential dilemma is that no animal perfectly mimics a human HIV infection. Chimpanzees can readily be infected with HIV, but the virus typically does not damage their immune systems, and they only appear to develop AIDS in rare circumstances. Asian monkeys develop AIDS from simian immunodeficiency virus, a relative of HIV—but, in the end, SIV is not HIV. So AIDS vaccine researchers could either ignore or celebrate animal results, depending on whether they liked the message.

By the time of the December 1987 Institute of Medicine (IOM) meeting, human experiments had begun with vaccines made by Daniel Zagury and MicroGeneSys, and the FDA had approved tests of another preparation made by Oncogen, the Seattle-based Bristol-Myers Squibb subsidiary. Each of these vaccines contained different formulations of HIV's gp160 protein. None had yet proved that they could protect an immunized chimpanzee from a subsequent challenge with live HIV, which then was considered the gold standard. Indeed, only the Oncogen vaccine had even been subjected to a chimp challenge experiment. As Oncogen scientists first reported in June 1987 at the international AIDS meeting held in Washington, D.C., they challenged six vaccinated chimps and three "control" animals. All of the vaccinated animals became infected by the challenge virus, meaning the vaccine soundly failed.[1]

The issue of how animal tests should influence AIDS vaccine experiments in humans sparked intense debates at the IOM meeting. "If there is essentially, as some people would interpret from this conference, no basis for believing that [some] of the things that people want to put into phase I and phase II trials could actually be protective in the future, putting things into I and II trials raises public expectations to a considerable extent," said Roy Widdus, an IOM staffer who specialized in vaccines. "You are then faced with a situation of either turning around and explaining that there was, in fact, no basis for putting into the preliminary trials what was called a vaccine candidate or letting something go forward into phase III trials on very dubious grounds."

Further compounding the dilemma was the possibility that injecting pieces of HIV into a person could do harm. Although HIV gp160 by itself could not cause an HIV infection, Merck's Maurice Hilleman stressed that pieces of other viruses can lead to immune responses that make subsequent infections by that agent more dangerous. The classic example of "immunologic enhancement" is dengue virus, which typically causes a benign fever. But when people who have antibodies against one strain of dengue become infected with another strain, the antibodies can combine with the incoming virus and lead to severe hemorrhaging and death.

Hilleman well knew the risks of moving a new vaccine into humans: In November 1975, he had first tested the hepatitis B vaccine he had developed on nine Merck executives—"we picked the most worthless people we could find in the world who would be the least likely to sue," he cracked. With AIDS, he contended, the rationale for human tests was much weaker. "I would be reluctant to ask any company executives to take any vaccine that I have heard anything about to the present time," he said.

Hilleman detailed a specific plan for using animal experiments to guide decisions about human trials. "My feeling is that there is an ethical need to do everything you can prior to going into man," he said. "At least this was the policy that was in force for decades." First, he said a vaccine should be made from SIV and tested in monkeys.

Discovered in 1984, SIV is an HIV cousin that naturally is found in

African monkeys, to which it does no harm—presumably because they have been infected for so many generations that their immune systems have learned how to defeat it—but quickly causes AIDS in Asian monkeys, none of which are infected in the wild.[2] HIV and SIV, both of which are slow-acting "lentiviruses," each have nine genes, eight of which are so similar that scientists consider them "equivalent," and they cause nearly identical diseases.

Yet there are significant differences between these virological cousins. Although the genes for the two viruses code for similar proteins, the genes themselves, on an RNA level, differ by about 50%. Giving even more researchers pause, SIV typically causes AIDS in a couple of years, while humans carry HIV for an average of 10 years before they start to develop symptoms. Still, many researchers had come to see SIV in Asian monkeys as the best animal model for AIDS. Chimpanzees, after all, not only suffered less damage from HIV than did humans; each chimp infected with the virus cost around $45,000—$15,000 or so for the animal and another $30,000 to set up an endowment fund required by primate centers to cover the costs of housing the infected chimp separately for the rest of its life. Chimps, a rare species that an international treaty dictates can only be used in research if they are bred in captivity, also attracted much more attention from prominent animal rights activists like Jane Goodall. The abundant rhesus macaques, in contrast, cost about $2,000 each and, once infected, died from AIDS.

If a vaccine worked in an SIV challenge experiment and did not cause any harm, said Hilleman, researchers should then subject a similar HIV formulation to a chimp challenge. "This would show public responsibility, and from there, then you go into man, and at least you would do it with a body of data and perhaps a consensus in the scientific community," said Hilleman. "You cannot have a vaccine for which one half of the scientific community expresses alarm and the other half doesn't. I don't think you can do that because when it becomes a public issue, you are going to have too many voices, and the scientific community isn't speaking with one voice." These words turned out to be prophetic.

Gerald Quinnan, second in charge at the division of the FDA that approved the MicroGeneSys and Oncogen tests, defended the decision to move into humans despite the absence of animal protection. Quinnan argued that scientists who insisted on seeing protection in animals before moving into humans had set too high a standard, as the animal data might be misleading. Carefully staged human tests, he contended, made more sense.

When the IOM committee that organized the conference wrote their report about the meeting, they sided with Hilleman. "The committee recommends that a vaccine candidate enter into clinical trials only when the vaccine has demonstrated protective efficacy in a suitable animal model (HIV in chimpanzees, SIV in macaques, or a new system) or its design is based on a fundamental new understanding of relevant human immune responses that cannot adequately be modeled in animals," concluded the committee.[3] "The IOM conference committee felt that the current human trials should not necessarily establish a precedent for future approval of vaccine candidates of similar design." The committee also addressed the criteria for efficacy trials of the vaccines that already had gone into humans. "The conference committee felt that such subsequent trials of vaccine efficacy should only be conducted after protection has been shown against HIV infection in chimpanzees or after demonstration of efficacy in a closely related lentivirus model such as SIV."

As it turned out, the committee's recommendations had little impact on how the field proceeded. The philosophy elucidated by Quinnan and used to justify the first human AIDS vaccine trials would continue to undermine proceeding in the more logical fashion advocated by the IOM. As long as animal tests did not show harm, developers of just about any type of HIV vaccine could win regulatory approval to conduct studies in humans. Adding to the confusion, the scientific community did not favor vaccines that performed well in animal experiments.

The root problem had less to do with different philosophies about the value of the various animal models than it did with the inability of the primate research world, the NIH, and industry to work together. Labs, which used different reagents and experimental conditions, could not reliably compare their challenge experiments with one another. The exorbitant cost of chimp experiments, coupled with the intense scrutiny that such work attracted from the animal rights community, forced people who liked the model to use so few animals that the resultant data had little scientific value. Promising leads from successful challenges often languished. And a bias against old-fashioned approaches, which widely were written off as too risky and too crude, kept what clearly were the most powerful vaccines from ever moving into human tests.

• • •

ON FEBRUARY 19, 1988, Peter Fischinger, the AIDS coordinator for the U.S. Department of Health and Human Services, took part in a briefing on Capitol Hill of congressional staffers about chimpanzees and AIDS research.[4] "We don't have a perfect animal model, but the chimpanzee is an excellent model," said Fischinger, who was on leave from his job as deputy director of the National Cancer Institute, where he headed the National AIDS Vaccine Development Program. After explaining how the chimp had already provided critical information about the safety of candidate AIDS vaccines, Fischinger directly addressed the question of whether chimpanzee challenge experiments should steer vaccine development. "Let me say unequivocally that the chimpanzee is an excellent model," said Fischinger. "We are talking about the question of an efficacy model, and I feel very comfortable that the chimpanzee is it."

Merck's Hilleman came next. Hilleman explained that he knew of only three examples in which a vaccine had been developed without an animal model: measles, mumps, and rubella. "We were able to get away without animal models there, first of all, because there were none, but more importantly, because even if the patient came down with the full-blown disease, it was not all that serious, so that studies could be done with man," he noted. "You can't do that with AIDS." Hilleman then seconded Fischinger's assertion that chimps provided an excellent model. The only problem was supply.

In 1973, the United States became a signatory to the Convention on International Trade and Endangered Species. CITES had deemed the chimp an endangered species in the wild, which meant that no member country could import a wild-caught chimp. This led the United States in 1986 to set up a $5-million-a-year breeding program of chimps already in the country, but, suggested Hilleman to the congressional staffers, there was another approach that should be tried too: setting up breeding colonies in Africa, as Alfred Prince of the New York Blood Center had done in Liberia, and then importing the babies.

Unmentioned by Hilleman was that Immuno AG, an Austrian vaccine maker, had explored the idea of setting up just such a breeding program in Sierra Leone, which led to an international fracas that ended up entangling Robert Gallo, Jane Goodall, England's Prince Philip, Prince Sadruddin Aga Khan (the former UN High Commissioner for Refugees), and even the U.S. Supreme Court.

Immuno's attempts to acquire chimps began in 1982, before HIV was identified as the cause of AIDS. The company initially sought chimps to do hepatitis research, and proposed establishing a center in Freetown, Sierra Leone. This plan won widespread condemnation, including a sharply critical article in *New Scientist* magazine.[5] That December, the *Journal of Medical*

Primatology ran a letter to the editor from the head of the International Primate Protection League that similarly criticized Immuno's plans. Immuno promptly filed libel suits over both of the articles.

As the libel suits wound their way through the courts, opposition to Immuno's plans steadily built in the animal rights community. Princes Philip and Sadruddin emerged as two prominent critics, both writing to Sierra Leone's president and asking him not to allow the building of a chimp research lab in his country or the exportation of wild-caught chimps.

By 1986, Immuno had decided it needed chimps not for hepatitis research, but to support its nascent AIDS vaccine program. That March, Robert Gallo visited the company in Vienna and, when asked about criticism of Immuno's drive to obtain chimps, the local press reported, he replied that "I would lock up or send into psychiatric treatment those people who want to hinder these experiments." NIH Director James Wyngaarden, when apprised of Gallo's comments by animal rights activists, said they were "exaggerated and intemperate."[6]

Immuno never did build the Sierra Leone lab, but the pleas of the princes went unheeded: On July 30, 1986, Immuno imported 20 chimps from the country to its headquarters in Vienna. The next month, Gallo, who wanted to collaborate with Immuno, joined company officials in Vienna for a meeting with government officials. As Gallo later explained to a reporter from the Austrian magazine *Profil*, Immuno's chimps were "definitely very important" in his decision to link with the company. He also told *American Medical News*: "To do the studies I would like to do over the next two years, I need 75 chimps, but I will be lucky to get two or three."[7]

Gallo's links to Immuno's AIDS vaccine program, which centered on a recombinant gp160 vaccine similar to the one made by MicroGeneSys, led Jane Goodall and other members of the Committee for Conservation and Care of Chimpanzees—the four C's—to stage a vocal campaign criticizing the NIH for steering around CITES.[8] Immuno fought back with its own Committee for the Conservation of Chimpanzees, a name, critics charged, meant to cause confusion with the CCCC. The CCC attended CITES meetings and insisted that the criticism against Immuno was "not only irrelevant but irresponsible."[9]

The secretariat of CITES determined that "the chimpanzees should not have been imported into Austria, as the transaction was in violation of CITES, and, probably, of Sierra Leone law."[10] NIH officials replied that Immuno had sent them convincing documentation that the company had done nothing wrong, and they contended that CITES had not officially made a decision on the matter. The officials did stress, however, that the NIH would "not support research by, or otherwise collaborate with, organizations that have obtained chimpanzees improperly."[11]

At the same time as Immuno was fending off attacks for abrogating CITES, it was filing libel suits right and left against the press and animal rights activists. Many of the suits were settled out of court, but one defendant, the editor of the *Journal of Medical Primatology*, refused to concede wrongdoing in the $4 million suit against him. At stake was a scientist's right to express an opinion. The case received intense media coverage, and was slammed as "a top candidate for the outrageous litigation prize" in a *New York Times* op-ed piece.[12] After an appellate court ruled resoundingly in the editor's favor, Immuno doggedly took the case to the U.S. Supreme Court, which kicked it back for another appellate review. Again, Immuno lost.[13]

Immuno's Martha Eibl, who headed the company's AIDS vaccine program, said the dispute about the chimpanzees seriously hurt the company's efforts. "We were very upset that we were doing something that we considered the right thing and that didn't get appreciated," said Eibl. "From the earliest times on, we had been spending for AIDS research really as much as people possibly can, and that has not been properly acknowledged. We were really attacked by almost everybody for many years."

Even if Eibl's defense of Immuno's actions is less than compelling, the company's experience sent a loud message to researchers developing AIDS vaccines: if you use chimpanzees to test your preparations, animal rights activists are going to be watching you closely. And, despite the eventual success of the NIH's breeding program, the supply of chimps would remain so tight that no AIDS researcher ever would have access to anything like the 75 animals that Gallo once said he needed to complete a mere two years' worth of experiments. So chimpanzees, regardless of whether they were an excellent model for AIDS vaccine research or a misleading one, would remain a luxury item that never would provide much scientifically persuasive guidance to the field.

• • •

LIKE MOST OTHER PRIMATE RESEARCHERS testing AIDS vaccines in 1988, Ronald Desrosiers knew the taste of failure.

Desrosiers worked at the New England Regional Primate Research

Center, an affiliate of Harvard Medical School that housed 1,600 monkeys in the Boston suburb of Southborough, Massachusetts. NERPRC had made its name in AIDS research in 1985 when Desrosiers, who then studied herpesviruses, and other researchers there first reported the discovery of simian immunodeficiency virus, which had infected three of the 800 rhesus macaques in its colony and was killing the animals with AIDS-like diseases. In 1986, Desrosiers and his coworkers attempted to make their first vaccines against SIV. "We went in very wide-eyed and optimistic that we were going to be able get protection," recalled Desrosiers, a mustached, compact man who, true to his roots, would develop a reputation as anything but a wide-eyed optimist.

The son of a shoe factory worker, Desrosiers hailed from a rough neighborhood in Manchester, New Hampshire, and was the first member of his family to attend college. He put himself through school with money he made working, like his father, in a Manchester shoe factory; he also earned money digging graves and pouring steel at a mill, a job his parents made him quit the second time his clothes caught on fire. As his approach to developing an AIDS vaccine would show, Ron Desrosiers was a practical, die-hard realist.

Instead of joining the crowd and genetically engineering a vaccine from a piece of the AIDS virus, Desrosiers and his coworkers, as Jonas Salk had been advocating, began at the beginning: They injected four monkeys with an old-fashioned concoction of whole, killed SIV. Each animal then received two booster doses of the vaccine, spaced three weeks apart, and, as expected, developed antibodies to SIV. On June 27, 1986, the NERPRC researchers injected live SIV into the four monkeys to challenge their vaccine-induced immunity. As a control, they also injected live SIV into four unvaccinated monkeys. All eight of the animals became infected and, within two years, seven of the monkeys were dead or so sick that the researchers "sacrificed" them. The vaccine clearly did not work.

One year later they tried again, stacking the decks more in their favor. This time, the researchers made a purer version of the killed SIV (the earlier vaccine had lost much of its surface protein during its preparation), upped the dose in each injection, gave more booster shots, and challenged one week after the last shot, when SIV antibodies levels were near their peak. For six months, it appeared that the two vaccinated animals had been protected, while the health of the two controls steadily declined until they died from AIDS. But in early 1988, SIV popped up in one of the two vaccinated monkeys.

On February 16 of that year, the *New York Times* ran a page 1, above-the-fold story, "Recent Setbacks Stirring Doubts About Search for AIDS Vaccine," that revealed the failures occurring in Desrosiers's monkey experiments and, separately, in chimpanzee studies. The chimp results were especially sobering because they directly assailed the prevailing notion that an AIDS vaccine could work simply by stimulating production of anti-HIV antibodies. In what is known as a "passive immunity" experiment, the researchers inoculated two chimpanzees not with a vaccine but with anti-HIV antibodies that had been pooled from infected humans. They then challenged the animals with live HIV, and all of them became infected. A repeated test in which two other chimps received 10 times as much of the antibody also failed.[14] The finding, wrote *New York Times* reporter Gina Kolata in the same article, "had disturbed experts." The lengthy, detailed article only breezily mentioned cell-mediated immunity and killer cells.

Desrosiers's group also had a ray of hope that received no mention: the one vaccinated, challenged animal that had not become infected. The scientists certainly could not reach any conclusion based on this one healthy monkey: it may have been protected by the vaccine, or it may have just been chance. Still, it was intriguing, and they decided to do the experiment again, further stacking the decks by only using one-fifth as much live SIV in the actual challenge.

• • •

THE 1988 INTERNATIONAL AIDS CONFERENCE, which took place in Stockholm from June 12 through June 16, offered five days straight of depressing vaccine news. Save for a little-noticed talk by Finnish researchers that described evidence of cell-mediated immunity to HIV in *un*infected partners of HIV-infected people—suggesting that they may have developed some natural protection—basic researchers appeared to have uncovered few clues about how to proceed.[15] Chimpanzee and monkey tests of potential AIDS vaccines had all come up empty-handed. Zagury described results from his human studies, yet all they basically showed, again, was that vaccinating people with gp160 did not cause harm and could stimulate the immune system.

Zagury's talk did elicit an interesting exchange that well summed up the confusion sowed by conducting human AIDS vaccine trials without first demonstrating efficacy in an animal model. During his presentation, Zagury announced that he and his Zairian collaborators were "preparing a clinical trial which will be done in a large number of individuals with high risk of

HIV infection." This news was met with many raised eyebrows, including those of one Swedish vaccine researcher who went to the microphone at the talk's completion. "Considering the failure with primates, isn't it premature?" asked the man. "It seems like you're doing trial-and-error experiments." Duke University's Dani Bolognesi, co-chair of the session and by then the establishment voice on AIDS vaccines, fielded the question with a most curious reply. "It is important to understand that these are not vaccine trials that we are speaking about now," said Bolognesi. "These are experiments that try to understand safety and immunogenicity of these materials to lead us to what an eventual vaccine candidate might be."

Zagury's experiments, of course, were vaccine trials. They simply were not resting on a solid scientific foundation, which meant they offered no tangible information that readily could be applied to developing an eventual candidate and were, in a word, aimless.

The bleakness became even more apparent the next day when Gordon Ada, a prominent Australian immunologist then with the WHO, spoke on "Prospects for HIV Vaccines." Ada joked that he had not wanted to give this talk. "The usual thing under circumstances like this is to thank the people who invited me here, but I've come to realize more and more that I've really put my head into a noose," he said. "We're some way yet off from making a vaccine which may really help to control infection by HIV."

The week after the Stockholm meeting, the *Washington Post* ran one of the most in-depth stories that had yet appeared about AIDS vaccines. The story, which captured the prevailing mood in the field, was headlined "Why an AIDS Vaccine May Never Work."[16]

• • •

SIX MONTHS AFTER THE STOCKHOLM CONFERENCE, Desrosiers and the other researchers at the New England Regional Primate Research Center had the first compelling evidence that a whole, killed SIV vaccine could protect monkeys from that virus. Of the four monkeys they challenged with a lower dose of SIV a few weeks before the Stockholm meeting, one of them had resisted infection altogether and the three others, though infected, were faring better than the two control animals, one of which had by then died. Combined with their preceding experiment, then, the NERPRC group had protected two of six vaccinated monkeys. When presented with the data in June 1989 at the fifth international AIDS conference in Montreal, the media paid scant attention[17]—they were too enthralled by odd results from a chimpanzee AIDS vaccine study spearheaded by Jonas Salk.

Jonas Salk made scientists and journalists alike go goofy. As one of the only living scientists whose face was known the world over, Salk, in the public's eye, had a superstar aura. Airplane pilots would announce that he was on board, and passengers would burst into applause. Hotels routinely would upgrade him into their penthouse suites. A meal at a restaurant inevitably meant an interruption from an admirer. Scientists and journalists who regularly dealt with Salk would come to see him in more human terms, but many still initially approached him with the same drop-jawed wonder, as though some of the stardust might rub off. Even the toughest critical thinkers, including officials at regulatory agencies like the FDA, would trip over themselves to pump his hand after a talk, and publicly criticizing Salk was a risky venture. "Fighting with somebody like him, you don't win," explained James Carlson, a pathologist at the University of California at Davis who collaborated with Salk on his early AIDS vaccine work. "He demeans you and you demean yourself. Even if he speaks something unintelligible, nobody's going to say the king has no clothes. People want to please him. If you have data that support his point of view, that's fine. But if you have questions, it's the turd in the punchbowl."

Salk's chimpanzee data were peculiar for several reasons, not the least of which was the way in which he attempted to use animal studies to gain support for human experiments that he already had quietly begun. It would become an all too common gambit: Rather than having animal experiments guide them in developing an AIDS vaccine, Salk and other researchers would use experiments *ex post facto* to justify the paths they already had chosen. Salk exacerbated the suspicions and alienation many of his colleagues felt toward him by rejecting the scientific ethos of openness and wrapping much of his work in secrecy.

In 1987, when Salk first proposed in *Nature* that an HIV vaccine might help people already infected with HIV, he was eager to test the idea in humans. Although no animal data supported the concept, he cofounded a company, San Diego's Immune Response Corporation (IRC), to aggressively push forward the idea. The vaccine itself would contain whole, killed HIV, minus gp120, as the surface protein fell off during the manufacturing process. As an adjuvant, they would use Incomplete Freund's, a potent oil-and-water emulsion that Salk had used in the flu vaccine he had developed for the military decades before.

In the original plan, Carlson and others at UC Davis would help test the material and provide patients for the trial. "I always thought I was part of a team that was going to make a vaccine," remembered Carlson. "I saw power with Salk. Everybody had certain advantages working with someone who has an international reputation."

Salk and IRC rightly argued that deciding whether to embark on human trials of a therapeutic vaccine raised fewer issues than a similar trial of a preventive vaccine. The risk/benefit ratio was different: If a potentially therapeutic vaccine harmed a person already infected with HIV, it was, given the possible benefit, a much more acceptable risk than a preventive vaccine injuring a perfectly healthy person. Given this logic, then, little debate surrounded Salk and IRC's moving into humans with no animal evidence that the strategy would work. The debate also never erupted in public because Salk and IRC kept quiet about their actions—so quiet, in fact, that they even kept UC Davis researchers out of the loop.

Salk initially hooked up with UC Davis at the Paris AIDS conference through Murray Gardner, an ex–cancer researcher who reported at the 1986 meeting that his lab had successfully vaccinated monkeys with a whole, killed vaccine against a retrovirus only distantly related to SIV and HIV. But Gardner and the other UC Davis researchers early on discovered that the marriage with Salk and IRC had serious shortcomings. The first sign of trouble surfaced when Carlson reported that his lab had found live HIV in a supposedly killed batch of vaccine material that they were preparing for the chimpanzee experiments. "Salk wasn't excited," said Carlson. "It wasn't what he wanted to hear. It was sort of, well, Davis made a mistake there. We're going to cut them off and let them float away."[18]

The debate about the worthiness of that batch of vaccine—which never was intended for human use—was not what ultimately scuttled the collaboration, though. UC Davis, rather, insisted on conducting their work in the open. "Here, people put their cards on the table," explained Don Martensen, head of public affairs at the time for UC Davis. "It really wasn't something that could be kept a secret. I frankly felt that we should be on the record that it was happening. I knew any vaccine project, especially one bearing Salk's name, when word got out to local AIDS organizations, there's no way it wouldn't get out." So it was that Martensen fed the story about the human trials to the *Sacramento Bee*, incensing Salk and the IRC.

Salk's and IRC's intense interest in secrecy grew out of their strategy for keeping criticism to a minimum as they navigated through the regulatory process as quickly as possible. Typically, investigators developing new treatments ask the U.S. Food and Drug Administration to grant them an investigational new drug, or IND, approval before human tests can begin. But Salk and the chief scientific officer at IRC, Dennis Carlo, wary of negative reactions from the FDA about the whole, killed approach—and all too familiar with the agency's snail pace—by the summer of 1987 had been seriously exploring another regulatory avenue: the state of California's Food and Drug Branch. As long as Salk and Carlo made and tested their vaccine within the state, they only needed the FDB's blessing.

The shift to the FDB coincided with an emergency measure racing through the California legislature. The bill specifically earmarked $500,000 for the FDB to speed the development of AIDS treatments by beefing up its staff and by taking advantage of a 1939 regulation that gave the agency authority to approve INDs. Despite FDA director Frank Young making it clear that he was none too happy about what widely was perceived as an end-run around his agency's authority, the bill had widespread, bipartisan support.[19]

Carlo and Salk saw publicity as both a nuisance and an invitation for public scrutiny that could sap, if not destroy, their momentum. The UC Davis researchers, who knew nothing about these delicate backroom negotiations, had no such concerns. And there was another important detail the UC Davis group knew nothing of: On August 29, IRC—which had shown that the vaccine could stimulate immune responses in rabbits, guinea pigs, and chimps and appeared safe—had received the FDB's approval to begin trials of the vaccine. But the trials were to take place at the University of Southern California, not UC Davis, where Carlson and his coworkers had no idea that their collaborators had begun working with others.

As Carlo noted in his private log on September 22, the day after the *Bee* article, the UC Davis group wanted to apply for an IND with the FDA.[20] "Well, that could take a long time," wrote Carlo. "Jonas and I agree to keep them on the string and let them go when the time's right. Don't tell them of their fate."

UC Davis went ahead and recruited patients for the trial of the vaccine. In all, 95 people volunteered for the test and agreed not to take any other medications.

Salk and Carlo went to the University of Southern California's Kenneth Norris Jr. Cancer Center on the morning of November 3, 1987, to witness the first human injected with their vaccine. USC, which had agreed to keep the injection mum, was not about to let this potential historic event slip away undocumented: A public relations man from the university, Gordon Cohn, showed up with a photographer in tow. "Dr. Salk maintained that what was needed was absolute silence," said Cohn. "But in my opinion, their desire for secrecy exceeded the need."

The *Los Angeles Times* blew their cover on February 11, 1988, in a page 1 story that declared, "USC Secretly Testing Salk Approach to AIDS Therapy." The article quoted Salk defending the stealth trial. "My preference is not to engage in making promissory notes but rather to deliver results when they become available," Salk said. "I did the work on polio for quite a long time before there was any notice or attention."

Researchers at UC Davis responded that they knew nothing of the trials, and said they still had 95 patients ready to test the preparation. All they needed, they said, was FDA approval (the conservative university never considered going through the California FDB) and the material itself.

By that July, UC Davis researchers had become disgusted with the whole idea of testing the IRC vaccine and formally withdrew plans to participate in the trials. "Given the already significant delay in obtaining the material, and without having any expectation that it will soon arrive at UC Davis, we believe it is in the best interests of our volunteers and the AIDS research effort at UCD to pursue other options," said the dean of UC Davis's school of medicine.

Two weeks after UC Davis pulled the plug on the collaboration, Joseph Gibbs at the National Institute of Neurological Disorders and Stroke, who had access to chimps and began working with Salk at the request of Don Francis, challenged three animals that he earlier had immunized with the IRC vaccine. It was an unusual experiment—and the results were odder still.

At the time of the challenge, two of the three chimps had already been infected with HIV for more than four years. This part of the challenge aimed to test the therapeutic vaccine strategy. The third chimp, which was "naïve" to HIV, would test the traditional preventive approach. Gibbs infected a fourth, unimmunized animal as a control.

As Gibbs reported on June 8, 1989, at the Montreal conference and Salk amplified at a press conference there, the vaccine appeared to have cleared HIV from the two previously infected chimps and prevented them from becoming reinfected upon challenge. The evidence was squishy at best. After vaccinating these two animals, the researchers could not prod cells isolated from the animals to produce HIV. This "culture" test did not prove that the animals had eliminated the virus; indeed, chimps, unlike humans, routinely control HIV so effectively that it doesn't register on this crude assay. Gibbs then challenged the animals, and the culture test still came up negative. Again, this did not surprise chimp researchers, who already had well established that it was difficult to "superinfect" an HIV-infected chimp with a new strain of the virus. As for the naïve, vaccinated animal, it readily became infected upon challenge (which convinced Don Francis not to pursue this approach).

All in all, it wasn't much to write home about, but the journalists covering the conference spun the story as the first upbeat AIDS vaccine news in a long time. AIDS vaccine dean Dani Bolognesi was widely quoted as praising the experiment. "We are talking about virus being there and then not being there after the treatment," said Bolognesi. "I take it as a very positive step for vaccine development."[21]

In October 1989, Gibbs, Salk, and coauthors submitted a paper describing their results to *Science*, which sent it to two outside peers for review. Neither recommended publication, with one critic writing, "the data presented in the manuscript do not unequivocally support these definitively presented conclusions and interpretations."[22]

The paper finally appeared a year and a half later in the *Proceedings of the National Academy of Sciences*, which has much lower peer-review standards and, at the behest of academy members, publishes many studies that other publications have rejected.[23]

Desrosiers's reaction to the experiment summed up the sentiment of many primate researchers. "It's bizarre," he said. "We've probably infected 100 or more animals. We've never seen a single case of an animal being infected and the virus disappearing. It does not seem consistent with the natural course of events."

• • •

WHEN DESROSIERS'S LAB protected two of six monkeys with a whole, killed vaccine, it, too, only merited a *PNAS* paper, which was published in the August 14, 1989, issue. A vaccine that protected a mere 33% of the animals that received it obviously had significant shortcomings. Yet it showed what was possible, and in the fall of 1989, a group from Tulane University's Delta Regional Primate Research Center made the whole, killed SIV seem a more viable approach still. Led by a gregarious, enthusiastic woman named Michael ("Mickey") Murphey-Corb, the Tulane group revealed in the December 8 issue of *Science* that its vaccine protected eight of nine challenged monkeys. Five of seven control animals became infected and developed disease.

The numbers left little room for equivocation. "This is a giant leap forward for AIDS vaccine research," NIAID's Anthony Fauci told the *Washington Post*. The same page 1 article quoted Bolognesi saying, "This is making me feel better than I have in years."[24] Murphey-Corb appeared on *Good Morning America*.

Groundbreaking as the Tulane study was, it left open many Big Questions. Murphey-Corb had challenged the monkeys with 20 times less virus than the amount Desrosiers had used in his experiments. How sturdy was the vaccine, and did the protection reflect the amount of virus that a person would encounter in a real-world situation? The researchers also challenged the animals by injecting the virus into their veins. Could the vaccine protect against virus smeared onto their rectal or vaginal mucosa? How long would the immunity last? Another critical limitation was the strain of virus used in the challenge: it matched the strain used in the vaccine. What would happen if the researchers subjected the animals to a challenge with a widely divergent strain of SIV? Skeptics noted, too, that the SIV used in the challenge was cell-free virus—not virus inside a cell, a likely mode of transmission. Would the same vaccine protect against such a cell-associated SIV challenge? And most perplexing of all, the Tulane group did not identify the immune responses that explained the protection. Was it anti-SIV antibodies? Killer cells? Both? Neither?

The answers to these Big Questions ultimately would make or break the worth of an AIDS vaccine, and many observers stressed that declarations of success were premature. Indeed, NBC science reporter Robert Bazell went so far as to write in the *New Republic* that Murphey-Corb's work did not even reveal "significant research progress" and that "many prominent scientists . . . saw it stretched far beyond its significance."[25]

But rather than detracting from the Tulane group's advance, these questions provided a road map for SIV vaccine researchers. The field finally had a foundation upon which to build. From the outset, however, the people leading AIDS vaccine research in government, academia, and industry had ordained that, for safety reasons, a whole, killed approach would never fly in humans. So the whole, killed strategy—even if it protected against a high-dose, mucosal, cell-associated viral challenge with a highly divergent strain from the vaccine virus—would always mean more for what it represented than for what it did. It proved, in short, that a vaccine was possible, and the researchers pursuing sexier, genetically engineered approaches that featured pieces of HIV used this news to bolster support for their own projects.

Murphey-Corb, on the other hand, believed that the whole, killed approach could safely be produced for humans. "I'm running into a lot of resistance to my work," she confided to me in 1989, a few months before the *Science* paper appeared. And Murphey-Corb told me about a recent visit that Salk, her new ally, had made to her lab. Salk recounted for her how he had struggled to convince his colleagues that a whole, killed poliovirus vaccine could safely provide long-lasting, powerful immunity. "He patted me on the back and kissed me on the cheek and said, 'I have paid dearly for proving them wrong.' "

• • •

WAYNE KOFF'S OFFICE featured a magnet board that in 1990 charted the progress of more than two dozen different potential AIDS vaccine strategies as they moved through animal and human tests. The 38-year-old Koff then headed NIAID's AIDS vaccine program, and although he had a fire in his belly to organize the field and push it forward, the chart testified that his job had more to do with helping researchers and companies accomplish their own goals. "I've heard people say if polio came along now, we wouldn't have a vaccine," mused Koff, who had a firebrand's heart disguised by the nerdy beard-and-glasses look he favored. "People would be doing all of these epitope maps and gene sequences instead of plugging away at it."

As Koff's chart made plain, HIV vaccine research in 1990 was driven by molecular biologists who could exploit fancy technologies to hunt for epitopes—regions of the virus that trigger specific immune responses—and gene sequences, which they then used to engineer small pieces of the virus. Of the five preventive vaccines that had moved into humans by then, four featured variations on the same HIV envelope theme: Zagury's vaccinia-derived gp160, MicroGeneSys's baculovirus-derived gp160, Oncogen's vaccinia-vector gp160, and Chiron's yeast-derived p120. (The last of these lacked the "g" because yeast did not put the sugars—the glycosylates—on the engineered protein.) The fifth vaccine to enter human trials, designed by George Washington University's Allan Goldstein, contained a synthesized piece of 30 amino acids from HIV's core protein, p17 (and had so underwhelmed the FDA in early studies that the U.S.-based biotech company backing it, Viral Technologies, launched its clinical studies in the United Kingdom).[26] The true darling of the field was Repligen's ultrareductionist approach, which focused on synthesizing 24 amino acids at the tip of gp120 known as the "variable third region" ("V3 loop" for short), a putative neutralizing antibody epitope extraordinaire. Of the remaining preparations, more than half also limited themselves to envelope proteins, ignoring HIV's other eight genes.

The chart did list two whole, killed formulations. One, Salk's, not only was aimed at treatment rather than prevention; the manufacturing process had altogether stripped off the envelope protein from HIV. The other—the SIV whole, killed vaccine, which did retain the envelope protein—was limited to the domain of primate researchers like Murphey-Corb and Desrosiers.

So Murphey-Corb's success, combined with the studies from Desrosiers's lab, presented a problem for Koff and others at NIAID. Essentially, the data from the whole, killed vaccine experiments were pointing in one direction—use all viral proteins and don't rule out old-fashioned vaccine techniques—and the vast majority of vaccine developers were heading in another. (The live, attenuated approach, notably, did not even earn a slot on the board.)

Since questions about safety remained the main objection to the whole, killed (or inactivated) approach, Koff and others in NIAID's Division of AIDS organized a daylong meeting on February 9, 1990, to confront the issue. The discussion focused on several formidable challenges, and there was "strong sentiment that pressing ahead at this time for an inactivated HIV vaccine was premature, until there was more evidence from animal model studies of the full efficacy of this approach," explained the minutes from the conference, which NIAID closed to the public. They then listed the Big Questions.

But the surprising conclusion of the two dozen experts gathered by NIAID that day—including such vaccine veterans as Hilleman, Salk, Plotkin, Merck's Gordon Douglas, the NIH's Robert Chanock, CDC's Walter Dowdle, and Vanderbilt University's David Karzon—was that science could effectively deal with the safety issue. "There was general agreement that it should be possible to successfully inactivate HIV so that there would be a virtually zero risk of acquiring HIV infection," the minutes noted. "In summary, it was felt that there was no insurmountable barrier to the eventual use of a classic, whole-virus, inactivated vaccine for AIDS." Even Gerald Quinnan from the FDA gave what amounted to a green light, "stating that a well-rationalized approach to inactivation can be made if the virus and the process is studied well enough, and that there were no regulatory roadblocks to the development and licensing of an inactivated HIV vaccine."

NIAID's marching orders were clear: build on this input and attempt to steer the field toward answering the Big Questions as quickly as possible, and, if the data remained positive, address the safety issues and move the whole, killed HIV vaccine into humans. NIAID, however, as it would prove, did not

have the influence, the will, or the Basil O'Connor type of leadership needed to carry out those orders.

Industry, fearful of liability, also shied away from translating the monkey successes into human trials—although Salk's company, IRC, gingerly had begun to explore the possibility. IRC initially planned to ask the California Food and Drug Branch to let the company inject 10 uninfected people with their whole, killed vaccine. The company, working with its collaborators at USC, intended to recruit elderly nuns as volunteers, as they had no risk of becoming HIV infected and had devoted their lives to altruistic causes. While this experiment would not prove that the vaccine was safe—and indeed had the far-fetched goal of harvesting antibodies from the vaccinees that could then be used to treat infected people—it would break the ice for human testing of a whole, killed preparation.

In March, the Archbishop of Los Angeles, Roger Mahony, sent a letter to 3,500 priests and nuns that explained the need for 10 volunteers who were at least 65 years old. The quiet appeal made national news on March 11, when the *Chicago Tribune* ran the letter as its lead story. Written by Pulitzer Prize–winning investigative reporter John Crewdson, the article stated that "[m]any scientists who have followed Salk's work on AIDS are privately concerned about the proposed experiments and the safety of his prototype vaccine." Salk, as Crewdson reported, had urged the *Tribune* not to publish the story until after California's FDB had approved the trial. "If you do I'll tell you what's going to happen," Salk told Crewdson. "It's going to arouse the interest of people who are interested in slowing us down. It stirs up the animals, not only the state regulators but colleagues."

Crewdson's lengthy article made no mention of the recent successes in the monkeys, the liability issues, or the recent NIH meeting about the feasibility of safely inactivating HIV. Crewdson also did not quote a single scientist by name who had criticisms of the proposed experiment. Still, the story—to which even *Nightline* devoted a segment—had the precise impact Salk predicted, and IRC scotched its plans for the trial.

There was, of course, no guarantee that a whole, killed HIV vaccine would work. The IRC preparation also lacked gp120, the single most important antigen in the minds of many leading vaccine researchers. The hush-hush scheme with the elderly nuns seemed overly dramatic and scientifically unnecessary too. Still, if NIAID or some other influential organization wanted to build on the monkey success and launch a carefully designed trial that attempted to test a similar approach in humans, it could have happened. And the lack of action to assess the possibility as quickly as possible spoke volumes about the disorganization plaguing the field, the liability fears paralyzing industry, the media's shallow understanding of the issues, and, ultimately, the disdain researchers had for animal data that did not jibe with their biases.

• • •

ON JUNE 1, 1990, Genentech issued a curious, one-page press release about a vaccine study done by their scientists that showed protection of chimpanzees, which the company said it would not discuss in detail until the findings appeared in the June 14 issue of *Nature*. The release curiously mixed business and science to create the impression that this news was momentous—and it suddenly thrust the firm to the front of what then looked like an AIDS vaccine race.

One of the first companies to declare itself an AIDS vaccine developer, Genentech disbanded its 25-member AIDS vaccine team in 1987 after its lead candidate, a genetically engineered version of gp120, failed a chimp challenge and, simultaneously, its marketing people got cold feet about the entire vaccine business. But two of the biotech company's scientists kept engineering new HIV proteins, hoping to make purer products.[27] By 1989, they had upped the purity of the gp120 from 50% to greater than 99% and had also engineered a gp160 that had a purity of 95%. They convinced their higher-ups to let them do another chimp challenge. This time, the two gp120-immunized chimps remained uninfected after an injection of HIV. The two animals that received the gp160 vaccine and an unimmunized control all became infected.

The press release appeared odd, however, because *Nature* and most other scientific journals have strict embargo policies that forbid researchers from discussing upcoming papers with the media. But as Genentech explained in their release, *Nature* had let the company off the hook "due to the proximity of the June 14 publication date to Genentech's annual stockholder meeting on June 8."

The chimp study, which received widespread coverage—including a coveted page 1, above-the-fold *New York Times* story—helped reinforce the company message that Genentech was on a roll. And that message took on an added significance because at the meeting, the company planned to ask its shareholders to approve a $2.1 billion merger with the goliath Swiss pharmaceutical house, Roche Holding Ltd (which the shareholders did, by 97%).

The protection of the chimps *was* newsworthy, but the story was inflated. For one thing, the Genentech scientists used so few animals that they could not reach any statistically significant conclusions, which means that luck, rather than the vaccine, might explain why the challenged animals remained uninfected.[28] Another qualifier is that the paper was not, as the company stated in its press release and the scientists claimed in *Nature*, the first report of an HIV vaccine working in chimps: At a meeting in Vienna that March, Immuno had reported that its gp160 vaccine appeared to have protected one of two immunized chimps; additionally, that April a French researcher had revealed at a small AIDS meeting in Keystone, Colorado, that he had protected chimps with a series of vaccines that contained several different HIV proteins.[29] The Genentech scientists finally did not have a solid immunologic explanation for the protection, although they did make the trendy speculation that it was due to the gp120 preparation stimulating higher levels of antibodies to the V3 loop.

The logical follow-up to these experiments would have been to run the same test on, say, four chimps with two controls, which would have led to data that a statistician could evaluate.[30] But that was an extraordinarily expensive undertaking. Alternatively, a similar large-scale challenge could have been run in monkeys. But then, SIV's envelope protein did not exactly match HIV's gp120. Whatever. Genentech did not need to worry about these details. Chimpanzee experiments had the impact of propaganda, and these two protected animals had provided all the evidence that the company needed to convince the field of its gp120's great worth. The Genentech scientists broke out the Dom Perignon. Human trials would soon begin, and NIAID—thrilled to have an envelope product that had *any* evidence of efficacy in animals—would use its considerable might to help the company establish itself as the leader of the pack.

• • •

SAN FRANCISCO, ONE of the most AIDS-ravaged cities in the world, hosted the Sixth International Conference on AIDS in June 1990, and the actions of activists shared the spotlight with—and often overshadowed—the scientific reports.

The highest-profile group, the AIDS Coalition to Unleash Power, or ACT UP, had perfected the art of using the media to push their agenda. With a slogan of "SILENCE = DEATH," ACT UP stoked the sense of urgency, its signature method of acting up being the "zap," a Yippie-ish tactic that ideally mixed nonviolent protest with wit. The group had had a history of success. After an ACT UP protest shut down the Food and Drug Administration's headquarters, the FDA streamlined its drug-approval process. After ACT UPers invaded the New York Stock Exchange one morning, cuffed themselves to a handrail, and unfurled the banner SELL BURROUGHS, the pharmaceutical house Burroughs Wellcome lowered the price of its anti-HIV drug AZT (azidothymidine). That May, hundreds of ACT UPers deluged the National Institutes of Health in a demonstration called "Storm the NIH," which led researchers to work more closely with HIV-infected people when designing and running trials of HIV treatments.

In San Francisco, ACT UP made its presence known inside and outside the conference halls. One afternoon, ACT UP lobbied for women with AIDS by staging a play in the middle of a busy downtown street and stopping traffic, braving riot-outfitted officers mounted on horses and Yamaha 350s. When Dan Hoth, Wayne Koff's boss, presented a paper on human trials of AIDS drugs, 26 ACT UPers silently stood before the podium and displayed placards emblazoned with the Hoth quote, "We have not devoted much attention to how we actually *do* these trials." In their most notorious conference zap, they went after U.S. Department of Health and Human Services Secretary Louis Sullivan during his keynote address. Ostensibly a protest of the U.S. government's decision to bar the country's doors to HIV-infected immigrants, the ACT UPers abandoned the Zen theatrics, hurling wads of paper and drowning out Sullivan's speech with air horns, whistles, and shouts.

Yet anxious as ACT UP was to "kick the shit" out of AIDS, as one member put it when he spoke at the conference's opening ceremony, the group focused on treatment—and did not devote any attention to vaccines. In keeping with this treatment-centric agenda, ACT UP did not zap a single vaccine session held at the five-day meeting. Had the activists known more about the workings of the vaccine world, though, the scene might have unfolded differently.

Adding fuel to the small band of researchers who believed in the whole, killed approach, Murray Gardner and coworkers from UC Davis announced at the meeting that they too had protected monkeys from SIV with the old-fashioned vaccine. The UC Davis group tested three different formulations, one of which protected three of three challenged animals from a low dose of SIV injected into the muscle, while the other two vaccines each protected one of three. All told, then, their various whole, killed vaccines had worked in five of nine animals. The 14 unvaccinated monkeys used as controls in these experiments all became readily infected when challenged with the same dose of virus given via the same route.

A physician who left general practice decades before and joined the NIH-sponsored crusade to find viruses that caused cancer, Gardner had a spunky, cheerleader attitude about the AIDS vaccine search, and, in particular, about using the SIV model to guide the field. "I'm convinced that the place to develop the vaccine is the SIV vaccine model," the fast-paced, ebullient Gardner told me a few days before he presented at the San Francisco meeting. "I think it's excellent." And unlike many of his peers, Gardner believed the whole, killed approach was viable.

"The Ph.D.s I work with who are the great brains and intellects of this field are much more cautious about things than I am," he said. "I'm more willing to say I don't care how these molecules make disease any more than I care how polio kills neurons if you can prevent it. I would put this stuff into people. I mean, shit, we learn by climbing over the bodies of humans, they're dying anyway, it's a terrible thing. I'd go out and try anything if it had a chance and if it would stop it. Everybody now is gun-shy. It's a different world lawsuit-wise than it was years ago."

As with Desrosiers and other primate researchers, Gardner badly wanted the field to become better organized and to systematically address the Big Questions. "What's needed now is some degree of cooperation, instead of all of us guys out there cutting each other's throats and competing to beat each other out and hide knowledge, lock up our ice boxes—which is the basic instinct of scientists to get ahead, because that's the system we're in that promotes you only if you do that, write papers, gotta be first," said Gardner. "We need to have some people do this and some people do that and put the results together and more quickly get on with the answers."

Murray Gardner, for all the world, sounded like a member of ACT UP. If only he had had as much influence.

At the conference itself, great enthusiasm surrounded the talks on AIDS vaccine research, which made it appear as though everything was on track. "We've seen now in a variety of animal systems that there's a protective response against the AIDS virus," Wayne Koff said at a jam-packed press conference. "I think what you've seen in the past year or so is that we've cracked open the door on the optimism of a vaccine. In the next two or three years, we're going to knock it down."

• • •

THE FOOD AND DRUG ADMINISTRATION puts much weight on the input from its advisory committees, which, like tribunals, listen to researchers describe their work and then vote on whether the given product should move forward in human tests and, ultimately, receive licensure. On August 21, 1990, the FDA's Vaccines and Related Biological Products Advisory Committee asked Genentech and MicroGeneSys for an update about their studies.

A Genentech scientist explained in great detail how the company's recombinant gp120 vaccine had protected two chimpanzees. A MicroGeneSys scientist, in contrast, described how its recombinant gp160 vaccine had triggered immune responses in humans.[31]

At the end of the MicroGeneSys presentation, two advisory members asked whether a chimpanzee challenge had been done or was in the works. No, said the MicroGeneSys scientist, who explained that the company had no plans to do such a test. "I am not sure the chimpanzee model is a good one," he said. "The response in humans is quite different from the chimpanzees. There are many epitopes that we see responding in humans that you do not see in chimpanzees."

One company, Genentech, builds its program around a successful chimp challenge. Another company, MicroGeneSys, says the chimp model is so unconvincing that they have no plans to even do a challenge. These were two of the leading AIDS vaccine developers in 1990. The FDA advisory committee adjourned this public meeting without discussing the discrepancy. It is one that the field never adequately would address either.

If the AIDS vaccine field had true leadership, a decision could have been made about the relative importance of challenge experiments in monkeys and chimpanzees. Yes, it would have been a gamble to conclude that both were relevant and, as Maurice Hilleman had suggested in 1987, that successful SIV vaccines should then be converted into HIV vaccines and vigorously tested in chimps. Alternatively, it would have been a gamble to declare that the models were not that important. But leaders make educated guesses, convince their charges to carry them out, and then face down their critics. The March of Dimes so believed in the imperfect monkey model for polio vaccine research that it funded the purchase of tens of thousands of animals, and Basil O'Connor even started a monkey farm.

As it happened, no one with authority in the AIDS vaccine search offered any conclusions about the worth of animal models, which, in the long run, would keep the door of optimism cracked open but would prevent it from being knocked down.

CHAPTER 6

Market Forces

We are raising our opinion to a Buy on Repligen, the most exciting—and virtually the only—AIDS vaccine play among biotechnology and pharmaceutical companies.

—AMY BERNHARD BERLER, First Boston report, November 20, 1989

Although it is speculative to guess which of the various vaccine candidates is most promising, we have long been especially impressed with the Repligen/Merck effort.

—ROBERT KUPOR, PH.D., Kidder, Peabody report, December 8, 1989

Despite the long time horizon, we think Repligen stock is the best vehicle for maximizing a return on an investment in an HIV vaccine.

—JACQUELINE G. SIEGEL, PH.D., Hambrecht & Quist report, March 7, 1990

We continue to believe that Repligen-Merck is the leading candidate to produce an effective AIDS vaccine.

—R. BRANDON FRADD, M.D., Shearson Lehman Hutton report, May 2, 1990

We believe the Repligen/Merck team represents the best investment in this highly competitive arena.

—DAVID K. STONE, Cowen & Company report, September 14, 1990

By 1990, most any outsider to the scientific world surely would have concluded that a fierce race was on to develop an AIDS vaccine, which would reap the winner hundreds of millions of dollars in annual revenue. And at the front of the pack of this supposed race was Repligen, the Cambridge, Massachusetts, biotech firm that had made a name for itself by identifying a supposedly remarkable piece of HIV. According to Repligen scientists, the V3 loop at the tip of HIV's surface protein, gp120, was key to a vaccine, as it had an amazing ability to stimulate production of antibodies that could neutralize the virus. Repligen called the loop "the Achilles heel of HIV."[1] The nascent company's scientists had won the support of such opinion makers as Robert Gallo and Duke University's Dani Bolognesi. This high-powered linkage helped explain how Repligen's Scott Putney had quickly shot from obscurity to publishing a half dozen papers in the best journals. Merck & Co., the pharmaceutical giant, also had agreed to pump millions into Repligen's AIDS vaccine program in exchange for an exclusive worldwide license to any resultant product.

But the AIDS vaccine "race," as Repligen's fate ultimately would demonstrate, more closely resembled a crawl. Large pharmaceutical companies like Merck had little interest in developing an AIDS vaccine; the flimsy relationships they had formed with biotech companies like Repligen simply represented small bets on dark horses. Basically, big companies deemed the risk too high. In addition to the long list of scientific unknowns, the market size remained a mystery, the liability issues staggered corporate attorneys, and developing drugs—which people need to take repeatedly for a lifetime—appeared a better financial gamble. As a result, the AIDS vaccine field mainly consisted of cash-poor biotech companies that perpetually tried to lure new investors with each tidbit of data they could produce about their "product." This gave stock analysts a tremendous power. No matter that these people often traded in the very companies they "analyzed." No matter that their predictions often fell wide of the mark. No matter that they often made great hay about supposed scientific advances that to scientists meant next to nothing. No matter that their reports had no insights about vaccine development history. Like novelists or magicians, the analysts, many with Ph.D.s and M.D.s after their names, defined reality.

In time, it became clear that the stock analysts knew little about what it would take to develop an effective AIDS vaccine, and they stopped writing their reports. The desperation of the companies in the "race" to find a preventive vaccine also became obvious as one after another followed the lead of Jonas Salk and the Immune Response Corporation and began testing their vaccines as potential therapies for already infected people. From a marketing vantage, the therapeutic vaccine had many attractive features. Liability did not present a major issue: if a vaccine harms an already infected person, the manufacturer risks drawing a suit for making an ailing person sicker—a much less vulnerable situation than injuring a perfectly healthy person with a faulty preventive vaccine. A therapeutic AIDS vaccine, like drugs, presumably would require repeated doses for a lifetime. And determining whether a therapeutic vaccine actually worked likely would not cost as much as staging an efficacy trial of a preventive vaccine, nor would it take as long to bring a product to market.

All this activity, combined with the confusion sown by the stock analysts, unfortunately obscured a serious problem. The National Institutes of Health, the main funder of AIDS research in the world, does not make vaccines. The pharmaceutical industry does. Yet the pharmaceutical industry largely decided to sit on the sidelines. The result: many promising vaccine strategies never made it out of academic laboratories, and the biotech companies that did commit their resources to the effort did not receive the support they needed to determine whether their preparations worked.

Many scientists at the front of course recognized this problem, and some even forcefully argued to create more incentives for industry by addressing their liability concerns and establishing a more predictable market. These proposals, however, never moved forward. In a setting where companies appeared committed to winning the supposedly intense race underway, researchers who realized the magnitude of the problem could not build political momentum. Also working against these reformers were their colleagues who insisted that industry would come rushing into the field as soon as basic research had unraveled the fundamental immunologic mechanisms necessary to make an effective vaccine. This self-serving logic—these were, after all, the basic researchers who would receive government funds to do the basic research—suggested that publicly funded scientists should chase the prey from the bushes and then allow industry to hunt by helicopter. It was a viable strategy—as long as time was not of the essence.

• • •

IN 1989 AND 1990, R. Brandon Fradd, a young M.D. working as an analyst at Shearson Lehman Hutton, wrote the first comprehensive reports for investors about the AIDS vaccine field, and Repligen in particular. The reports boldly defined the size of the AIDS vaccine market, handicapped the companies in the race, and even predicted time lines for a product winning FDA approval. "It was very naïve," Fradd said years later of his aggressive arguments. "But it wasn't like I was getting guidance from thin air. I talked to people in the companies. They tended to be optimistic, too."

Fradd went to Shearson Lehman in 1988 as a 26 year old fresh out of Harvard Medical School, and he made a name for himself with these reports. He also influenced how other analysts viewed the field. "There's a Wall Street version of reality that may or may not coincide with results in the world," said Fradd, who in 1995 left the investor analyst business to start his own investment fund. The analysts, in turn, influence more than just investors. "Journalists pick it up from Wall Street," noted Fradd. "And it goes and it goes and it goes."

As Fradd emphasized, analysts rely heavily on input from company scientists who, naturally enough, thrill to the slightest evidence that their products work. This helps explain why Fradd and at least four other analysts loudly celebrated a marginally interesting chimpanzee study done with a V3 loop–based vaccine. In the study conducted by Marc Girard of the Pasteur Institute and first reported at a small scientific meeting in October 1989, one chimp received a whole, killed HIV vaccine and then a preparation that contained gp160.[2] The animal did not develop a high level of neutralizing antibodies, so Girard injected it with a V3 loop vaccine, which did the trick. Girard gave the V3 loop vaccine to another chimp that similarly had failed to develop potent neutralizing antibodies after receiving injections of several HIV pieces (including surface and core proteins). It, too, developed the special antibodies. After challenging both animals with live HIV, Girard showed that they resisted infection. "Repligen states that the AIDS research community has been galvanized by this result," wrote Robert Kupor, Ph.D., an analyst with Kidder, Peabody, in his December 9, 1989, Repligen report.

Kupor listed the caveats that appeared in other analyst reports. To begin with, scientists did not consider this a "clean" challenge experiment; as Girard himself noted when he eventually published the data, "it is difficult to determine which of the many antigens were instrumental in eliciting protection."[3] In other words, Girard used so many different vaccines in these chimps that he could not assess which ones triggered the immune responses responsible for protection. Girard's data also had no statistical significance whatsoever: He did not have an unvaccinated control nor did he have enough vaccinated chimps to reach a meaningful conclusion. And, as it turned out, one of the chimps that initially appeared to be completely "protected" later showed evidence of having been infected. Still, this "exciting study," in the words of First Boston's Amy Bernhard Berler, demonstrated for the first time that an AIDS vaccine could protect a chimp—if, that is, the protection was real and not due to chance. The scientific insights offered by the analysts thus amounted to a house of cards.

The marketplace analyses they offered looked shakier still.

According to Fradd's 1989 analysis, the U.S. AIDS vaccine market in 1993 would total at least $974 million. Fradd spelled out how he had arrived at this number. Fradd assumed that the main market would consist of 30% of the 15 million male homosexuals in the country. The next largest consumer group would consist of the 5 million health workers in the United States, 30% of whom would line up for shots. Add in 30% of the 2 million prisoners, 30% of the 1 million injecting drug users, 10% of the 12 million college students, and 100% of the 15 thousand hemophiliacs, and you come up with a total of 8.11 million people. Now price the vaccine at $120 for the series of shots, then the current price of Merck's hepatitis B vaccine, and you arrive at the $974 million figure. Leading business newspapers and magazines repeated Fradd's estimates, which had the imprimatur of Shearson Lehman, helping to create the general notion that the AIDS vaccine market would amount to about $1 billion annually.[4]

Repligen had done its own market analysis, which investor analyst Berler detailed in a report for First Boston.[5] Berler wrote that Repligen, using figures from the CDC, projected that in 1993 the United States would have 13 million people at high risk of HIV infection. Assuming the vaccine would sell for the same $120, "a 50 percent market penetration is worth $780 million," she wrote. Repligen, she noted also, factored in "medium risk" populations, which included health care workers, the military, prison inmates, and college students. Assuming that 20% of these 19.6 million medium-risk people took advantage of the vaccine, that would add $477 million more to the pot, which meant a U.S. market of $1.26 billion.

Jacqueline G. Siegel, Ph.D., of Hambrecht & Quist offered her own estimate of the marketplace—"at least $500 million"—without giving any explanation for her number, other than to suggest the vaccine likely would cost $50 to $100. In other words, her low end of the vaccine price was about 40% of the $120 assumed by Fradd and Repligen. Multiply 40% by $1.26 billion. It equals $504 million.

It does not take a great analytical effort to see the loopiness of these predictions. No one knows with precision how many male homosexuals live in the United States, nor does anyone know how many of these men are sexually active. Why would 10% of college students take an AIDS vaccine? Why not

2% or 90%? And why only 30% of injecting drug users? Fradd's own 1990 reports underscore the squishiness of these numbers. Fradd estimated the size of the fiscal year 1997 market in the United States and Europe at $1.655 billion. For these calculations, he suggested that 20%, not 30%, of homosexuals (who would then number 18 million, not 15 million) would provide the main U.S. market. A scant 10%, rather than 30%, of injecting drug users would take advantage of the vaccine. Only 25% of health care workers and 10% of prisoners would take advantage of the vaccine (both down from 30%) and the 12 million college students were replaced by a mere 8 million "urban, ages 20–30," only 5% (not 10%) of whom would be vaccinated. Fradd cautioned that his estimated market size might be "much too big." In the next breath, he noted that his numbers "may be too conservative," noting that AIDS had greatly impacted gay men and health care workers, and that the government might mandate vaccination for some populations.[6] Fradd, who in addition to his Harvard M.D. had a bachelor's degree from Princeton in math and a master's in biology from Oxford, might as well have consulted a Ouija board to arrive at these numbers.

To understand the dilemma faced by pharmaceutical companies that entertain the idea of entering the search for an AIDS vaccine, consider how simply they can run the same calculations for most other vaccines, which have a defined population: children. To enter public schools in the United States, children must have received more than a dozen vaccines. Because of this, after the age of 5, more than 95% of the kids have taken the full series of these shots.[7] Knowing the number of newborns in this population, then, one accurately can predict the market for these vaccines.

For even more evidence that the analyses done by these stock analysts are hooey, look at Fradd's 1990 predictions about the Repligen/Merck vaccine time line to market. Using an "aggressive time period," he wrote, the vaccine could receive FDA approval by mid-1994, though it was more likely that this would occur in late 1996.

In June 1993, Repligen announced plans to launch a trial of its vaccine in already infected people. The company's sudden interest in the therapeutic potential of its vaccine—and the fact that the company had yet to test it in a single uninfected person—signaled trouble. A press release explained that the company had modified its agreement with Merck to let the biotech firm pursue the therapeutic vaccine on its own, but stressed that tight bonds between the two companies remained. "Repligen and Merck are continuing research on a prophylactic HIV vaccine," the release assured.

On July 19, 1994, Repligen announced that it had decided to "terminate" its entire HIV vaccine research program. The "restructuring" of the company, which included letting go of more than a third of its 300 employees and focusing on three of its "most promising" products, "was driven by both financial and ethical consideration," explained the release.

Repligen's demise reflected the faddishness of modern biomedical science. After labs around the world began studying the V3 loop, the mass of data began to suggest that it varied so much between different strains of HIV that it didn't make sense to base a vaccine on this moving target. After all, an antibody against one HIV's V3 loop might have no impact on another strain of the virus—and this mutable virus clearly had many strains. Merck also began to see the entire concept of an antibody-based vaccine as simplistic. And no one, interestingly, ever tested the V3 loop vaccine by itself in a chimpanzee challenge experiment.

Had Repligen's startling turnaround been an isolated event, you could write it off as a normal development in a long-distance race; after all, the early leaders in the New York City marathon seldom stay the course. But Repligen's troubles reflected the experience, on various levels, of most of the companies in this field. The market forces, regardless of what the analysts said, simply did not attract substantial, sustained private investments in AIDS vaccine R&D.

• • •

A FEW HOURS INTO A SEPTEMBER 10, 1990, meeting entitled "Toward an AIDS Vaccine: The Policy Supporting the Research," Roy Widdus bluntly explained to the 100 heavy-hitting participants who had gathered at the National Academy of Sciences in Washington, D.C., why he and his colleagues at the Institute of Medicine had organized this unusual "roundtable" discussion. "I think everyone has recognized that this meeting is held around a fear that the vaccine manufacturing capability of industry will not necessarily be brought to bear on the development of an AIDS vaccine, either because they don't see a potential product that will at least not lose money, or because they don't have the resources to take the development process forward," Widdus said. So much for the billion-dollar marketplace and the race.

Widdus's comment came on the heels of a provocative presentation by George Todaro that suggested these fears were well founded. A former leader in the National Cancer Institute's Virus Tumor Program, Todaro was president of Oncogen, the Seattle biotech firm that had a leading AIDS vaccine and, in 1986, had become a subsidiary of Bristol-Myers (which in 1989 merged with Squibb). Todaro's startling perspectives called into question both Bristol-Myers Squibb's commitment to the Oncogen AIDS vaccine project and the attitudes of big pharmaceuticals in general.

Todaro outlined a fictional AIDS vaccine efficacy trial with different outcomes, and the various reactions he proposed would occur in the medical community, a small biotech company, and a large pharmaceutical house. In his theoretical, 5,000-person trial, half the participants received the vaccine, while half served as a control group and received a placebo. If 50 people in the control group and 50 in the treated group became infected, said Todaro, everyone would obviously agree to abandon the vaccine. But what would happen if 50 in the control group became infected, but only 5 did in the treated group? "The medical community, the worldwide community, I think would consider that a major success," said Todaro. "I think no industry, no large pharmaceutical company or small biotech company, would possibly proceed under those circumstances." And if the vaccine protected all but one person in the treated group, Todaro suggested, the medical community and the small biotech firm would support moving the product into wide use—but the pharmaceutical company would not.

Todaro overstated the case when he suggested that an AIDS vaccine would have to prove 100% effective before a pharmaceutical company would sell it—no vaccine has 100% efficacy—and several participants called him on this. But even though he revealed more about the degree of cautiousness at Bristol-Myers Squibb than the industry as a whole, the rationale he offered for the misgivings applied widely. "Liability issues are key ones," said Todaro. "And even if one got down to 98% protection, perhaps a small biotech company would risk doing it, but basically you are betting the company in the absence of a resolution of all of the liability concerns."

Worries about liability surfaced repeatedly during the daylong meeting. Peter Kingham, an attorney at the megafirm Covington and Burling, offered a legal perspective. Kingham echoed Todaro's argument and explained why small biotech firms and large pharmaceutical companies viewed the issue through different lenses. "The willingness of a small-risk venture-capital type of firm to engage in early-stage research on a vaccine of this kind does not necessarily answer the question whether a large, established firm would be willing to undertake the risk associated with later stages of development," said Kingham. "By their very nature, small-risk capital firms exist to take risks. They have to take many risks. They live day to day, and, to put it quite bluntly, you can only die once."

Although Kingham stressed that the field had not existed long enough to judge whether liability fears truly had hindered AIDS vaccine development, he homed in on the reason why pharmaceutical companies have come to view vaccines in general as having a uniquely high liability risk. Even if a company properly makes a vaccine, it may well cause harm a small percentage of the time, as happens with the oral polio vaccine. And under a law known as "strict liability," manufacturers, who have done nothing wrong, can still lose multimillion-dollar lawsuits. "It is difficult to explain to a parent or a jury why a healthy child who was at no apparent risk of disease died or suffered a disabling injury as a result of a routine immunization, especially if it was an immunization against a disease that almost no one gets anymore and no one that they know has ever died of," he said, adding that companies have to weigh these risks against potential profits. "Traditionally, the vaccine business has not been as profitable as some other areas of the pharmaceutical business, and quite simply you are willing to take more risk if the return is greater."

Theodore Cooper, CEO of The Upjohn Company, provided an insider's perspective. Upjohn, one of the many major pharmaceutical companies that had decided not to pursue the making of an AIDS vaccine, had an aggressive R&D program for anti-HIV drugs. Cooper said Upjohn didn't have "the expertise" to make an AIDS vaccine—a dubious explanation because if Upjohn wanted the expertise, it could have purchased it—but, more importantly, he stressed the "risk capital" to making any pharmaceutical product. In addition to liability risks, he noted, a recent study had shown that companies on average were then spending more than $200 million to bring a drug to market.[8] Factor in that the worldwide vaccine market in 1990 for all vaccines, according to the authoritative Frost & Sullivan *Market Intelligence*, totaled a mere $1.52 billion; one best-selling drug could have brought in more revenue.[9] Before a company launches an R&D program for a potential new product, "there is a number that you use as a hurdle to get your investment back," summarized Cooper. "That is what the name of the game is."

Cooper, an M.D. who during the Gerald Ford administration had served as assistant secretary of health at what was then the U.S. Department of Health, Education and Welfare, had deep concerns that because of this equation, industry needed encouragement to get into the game. "If [AIDS] is a national emergency and an international emergency, then I think all of us in the industry have to be willing to forgo the usual reward hurdle." Cooper backed an idea that others at the meeting had been keen on: forming a consortium, such as was done in the U.S. computer industry, that used creative tax incentives and patent agreements to allow several companies and the government to work closely together to solve a national problem.

An even bolder proposal about how to reduce risks and increase incentives came from Jonathan Mann, who recently had left his job as head of WHO's Global Programme on AIDS and become a Harvard professor. Mann suggested that the ultimate discoverers of an AIDS vaccine donate their patents to an international organization, which then would manufacture and distribute the vaccine to the world. In exchange, the discoverers would receive the right to prolong the patent for an existing drug. The discoverers could use this patent extension for their own drug, or they could sell it to the highest bidder. "We are, from what I have been told, talking about hundreds of millions of dollars potentially at stake in the possibility of prolonging certain patents," said Mann. "The prize could be a considerable financial incentive."

While the consortium and patent-exchange ideas both theoretically could have alleviated concerns about liability, the topic loomed large, and speakers detailed a handful of specific solutions.

A half dozen of the attendees at the D.C. meeting had taken part in the Keystone AIDS Vaccine Liability Project, a "dialog" requested by Representative John Dingell, the hard-charging Michigan Democrat, that began in 1988 between scientists, attorneys, congressional staffers, insurance companies, consumer advocates, and AIDS vaccine makers. The Keystone participants wrestled mightily, as Kingham had, in their attempt to assess the true impact of liability concerns. As their final report, issued five months earlier in May 1990, stated:

> The group heard from several industry representatives in the group that liability was an important factor in decisions made by companies which have stopped or not begun vaccine development, but also learned that at least ten companies in the United States are engaging in AIDS vaccine research and development. There are currently no comprehensive data, and it is unlikely there will ever be data that systematically establishes [*sic*] the linkage between the liability system and vaccine development because liability concerns are only one of many issues which companies must evaluate when making decisions regarding work on a potential vaccine.[10]

Despite these uncertainties, the diverse group did reach a consensus that "a new legal framework" could promote the development of an AIDS vaccine. "The group felt it was worthwhile to address the issue of AIDS vaccine liability now, before a crisis occurs," they concluded. "Too often, our society does not anticipate problems and fails to plan ahead, often with poor results."

The Keystone project detailed comprehensive solutions for both clinical trials and marketed products, many of them based in principle on the National Childhood Vaccine Injury Compensation Act. Passed in Congress in 1986, the act came in the wake of liability disasters arising from vaccines for polio, diphtheria-tetanus-pertussis (DTP), and swine flu, which drove many companies out of the business. The act established the National Vaccine Injury Compensation Program, a body that rewards victims without punishing responsible manufacturers, thereby aiming to ensure a supply of reasonably priced vaccines. Congress set up the no-fault program to cover childhood vaccines—initially, DTP, measles, mumps, rubella, and polio—and it requires claimants to have suffered specific, well-characterized injuries within a certain period after being vaccinated.[11]

No one of course knew the specific risks of an AIDS vaccine, but the Keystone group used the childhood vaccine program as a template, aiming to compensate injured parties through a newly established tribunal that could speedily and fairly handle claims. The group explored several other schemes, too, including reforming tort law to protect manufacturers who made a vaccine according to "government standards," such as those required by the FDA. In this scheme, injured people could receive compensation for medical expenses and lost wages, but there would be caps on pain-and-suffering awards, and no company would have to pay punitive damages. Although some participants worried that diluting tort law would reduce the pressure on industry to maintain high standards, others countered that manufacturers could still be punished fully if they engaged in "unacceptable behavior."

Wendy Mariner, a professor of health law at Boston University who attended the Keystone meetings, presented yet another ambitious solution at the D.C. gathering: having the U.S. government provide health care to everyone. Yet Mariner was not sanguine that Congress or other powerful bodies would act on any of these solutions. "Ultimately," she said, "you need a crisis to solve the problem."

Jonas Salk, who bounded to the microphone repeatedly during the meeting, suggested that people think carefully about how they attempted to bring about reform. "I cannot help but draw attention to the role played by the

March of Dimes, and at that time, the attitude was war is too important to leave to the generals," said Salk. "It was too important a problem to be left to the government . . . and therefore, I think it is important to recognize that when the public interest is involved, it may be necessary to develop a whole new way of thinking, a whole new set of strategies, and whether it is done by consortium or one means or another, it does seem to me that the questions that are being raised here are of vital importance to going forward, as well as how are you going to apply the benefits of this to those parts of the world that are disadvantaged."

Salk, like others at that forward-looking meeting, left feeling as though the field had taken an important step forward. "I must say, this has been one of the best meetings I have attended," he told the crowd, "and I really think that this is an extremely fruitful way of discussing these matters." There was a lot of fruit to be harvested, with ideas ranging from a consortium to patent exchange to tort reform to compensation programs to universal health care. But if fruit is not picked, it drops from trees and rots.

• • •

IN DECEMBER 1991, management at Bristol-Myers Squibb asked Oncogen's Shiu-Lok Hu to fly from his lab in Seattle to corporate headquarters in Princeton, New Jersey, for a meeting with a senior vice president. For three years, Hu and his coworkers had been conducting human tests of their AIDS vaccine, a vaccinia virus that had HIV's gp160 gene stitched into it. At the international AIDS conference the previous summer, scientists presented data from human trials of the Oncogen vaccine, followed by booster shots of the gp160 vaccine made by MicroGeneSys. The "prime-boost" combination, which theoretically would stimulate both the antibody and cell-mediated arms of the immune system, registered such a whopping immune response that a plenary session moderator hailed it as "the most promising" approach yet tested in humans.[12] And now, Hu had monkey data that backed the same prime-boost strategy. As a paper about to be published in *Science* showed, priming with vaccinia-gp160 followed by a gp160 boost had protected four monkeys from a challenge with SIV. This experiment marked the first evidence in monkeys that a vaccine other than ones made from the whole, killed virus might work. But Bristol executives did not have a chilled bottle of champagne waiting for Hu in Princeton. "Basically, I was dressed down," said Hu. "It was completely ironic. I was told not to do these sorts of things."

As Todaro's September 1990 talk at the Institute of Medicine had foreshadowed, Bristol-Myers Squibb—the maker of the popular aspirin Excedrin and of several top anticancer drugs—had little interest in selling a vaccine of any sort, and by the end of 1991, the company wanted out, completely, of the AIDS vaccine project it had inherited. Hu said that three main factors influenced Bristol: liability, marketing, and patent issues. Hu said Todaro had explained the company's liability perspective in stark terms. "The example was given by him, look, you have 100 people with AIDS," recalled Hu. "You come up with a drug, and if in 50 of them, you can extend their lives for two years, you're a hero. With a vaccine, if one thing goes wrong for one person, you can sink not just the product but the whole company."

In the marketing arena, the prime-boost approach created a nightmare, as it required Bristol to collaborate with another vaccine maker. "For Bristol, and I assure you for any other company, the prospect of working on a product with another company is not very appealing at all," said Hu.

Finally, Bristol had become embroiled in a protracted feud with a New York biotech company about who owned the patent rights to an engineered vaccinia AIDS vaccine.[13]

By the spring of 1992, Bristol had killed the Oncogen vaccine project. "Bristol is short-sighted in terms of research," Hu told me at the time. "We're not even supposed to mention the dreaded 'V-word.' "[14]

No longer could anyone claim that the liability issue raised only theoretical concerns. Reducing the liability risk may not, of course, have kept Bristol in the running. And liability did not seem to be slowing the efforts at France's Pasteur Mérieux Connaught, a major pharmaceutical that had the most ambitious AIDS vaccine program; Emilio Emini at Merck, who headed their program with Repligen, also insisted that it had little impact there too. Still, the pharmaceutical houses that had not seriously committed themselves to HIV vaccine R&D and yet had research teams devoted to finding anti-HIV drugs comprised a long list: Glaxo, Burroughs Wellcome, Abbott, Hoffmann-La Roche, Boehringer Ingleheim, Upjohn, Schering-Plough, and Dupont. Add to that list the major pharmaceuticals that already had established themselves as major vaccine makers but had no serious anti-HIV vaccine R&D program: SmithKline Beecham, Lederle, Hoechst, and Medeva.[15] Something was amiss.

To Dan Hoth, head of the Division of AIDS at the National Institute of Allergy and Infectious Diseases, worries about liability had moved from the theoretical to the concrete. "At [the NIH's] AIDS Program Advisory Committee meetings everybody brings it up and wrings their hands," I quoted Hoth saying in an article about the AIDS vaccine liability issue that *Science* ran in April 1992.[16] "It's very hard to be specific, but there's certainly smoke and I have the perception that there is a problem."

In addition to Oncogen, my article detailed several other specific instances in which liability fears slowed AIDS vaccine progress. As first reported in the March 6, 1992, *Wall Street Journal*, the Immune Response Corporation during the preceding year had hoped to launch trials of its whole, killed HIV vaccine in uninfected people. But because of "logistical issues" involving liability, IRC could not move forward. Although the company had agreed to take the risk and had gone so far as to line up volunteers—including its cofounder, Jonas Salk—another outfit that actually made the preparation refused to provide it because of fears of lawsuits.

Liability concerns apparently had put another old-fashioned vaccine approach, weakening the live virus, on indefinite hold, according to Stanley Plotkin, who then had recently left the University of Pennsylvania to join Pasteur Mérieux Connaught. Plotkin, who had helped develop vaccines for six diseases, said he repeatedly failed to convince NIH grant reviewers to fund him to study an attenuated HIV vaccine. "One of the reasons for not funding the project was safety issues," Plotkin insisted. "Therefore, the likelihood is that liability was a factor. Frankly, the reviewers—in addition to their scientific criticisms—must have had liability in mind."

Though many AIDS researchers then believed that lab successes and failures accounted for Genentech's on-again-off-again AIDS vaccine program, liability fears had played a critical role there too. The pioneering biotechnology company, based in South San Francisco and one of the first out of the gate in the "race" to make an anti-HIV vaccine, badly wanted extra legal protection. In 1986, Genentech attorneys helped move a bill through the California legislature to protect HIV vaccine developers from litigation unless they blatantly failed to inform people of the risks or if their negligence or misconduct led to the injury.[17] Paul DeStefano, then Genentech's general counsel (and a member of the Keystone project), recalled being asked at an official hearing whether Genentech would drop out of the business unless such protection was mandated by law. "Yes, that's what I'm telling you," DeStefano said he replied.

The bill did pass, but DeStefano contended that the consumer-oriented California Trial Lawyers Association had significantly watered it down. "For that reason—though I'm not saying it's the sole reason—Genentech dropped its AIDS vaccine program," said DeStefano. Only two years later, the same combination—legal climate plus research results—turned Genentech around.

In 1988, the California Supreme Court handed down a decision, the so-called *Brown* ruling, that said manufacturers of "unavoidably dangerous" drugs (and presumably vaccines) can only be held responsible for the damages caused by a properly made preparation "if it was not accompanied by a warning of dangers that the manufacturer knew or should have known about." In effect, the *Brown* ruling created a "government standard" rule, and was so favorable to manufacturers that the California legislature repealed its AIDS vaccine indemnification law.[18] DeStefano said the *Brown* decision, coupled with success in the lab, led Genentech back into the business.

State liability laws had an impact across the country on the MicroGeneSys vaccine. In the late 1980s, liability fears blocked a trial of the company's gp160 in infected people at the University of Connecticut. In that case, it was the university, not the biotech company, that did not want to take the risk.[19] But the company learned a lesson that it would apply later when it wanted to test its vaccine in pregnant, HIV-infected women.

In the summer of 1991, researchers at Tennessee's Vanderbilt University had hoped to begin these trials to see whether the vaccine could both treat the women and simultaneously stimulate antibodies that might prevent transmission of HIV to their fetuses. Liability concerns, however, threw those plans off because Tennessee law didn't offer the company—which by then had begun to receive backing from Wyeth-Ayerst, a division of the pharmaceutical powerhouse American Home Products—much protection.[20] And MicroGeneSys had just succeeded in lobbying the Connecticut legislature to pass a law that specifically offered substantial legal protection to companies testing AIDS vaccines.[21] So the company decided to hold the trials at Yale in New Haven, Connecticut, instead, but at a cost—the delay of a trial.

Word of the delay made it all the way to the White House. "I'm told of an experimental vaccine that might reduce the incidence of HIV-positive babies born to mothers with AIDS," said Vice President Dan Quayle in a tort reform speech he gave to the American Bar Association. "This is a wonderful development; but for fear of lawsuits, companies have been reluctant to proceed with testing."[22]

Efforts to improve the legal climate for AIDS vaccine manufacturers moved forward both at the White House and in Congress. In the Bush administration, Dan Quayle and his Council on Competitiveness aggressively pushed for broad tort reform that would make "unreasonable lawsuits" a thing of the past by, among other things, establishing a government standard's defense. Not only was Quayle wholly out of touch with the AIDS community; the shamelessly pro-business council had attracted the ire of Congress because of its secretiveness, which led critics to charge, as one *Washington Post* story reported, that it was "a back-door conduit through which industry can influence the way laws are implemented."[23] The Senate made its sweeping attempt at tort reform, the Product Liability Fairness Act, which mirrored the changes pushed for by Quayle's council.[24] The bill did not pass. So in the end, neither the executive nor the legislative branch moved the issue closer to a resolution.

The California Supreme Court recognized in its *Brown* decision the illogic of the campaign to overhaul liability law across the board: Not all products are created equal. Some, such as vaccines, have a societal benefit if they are used—and thus it makes sense to reduce the liability risks for vaccine makers where it might not make sense to offer the same relief to the makers of ladders, baseballs, or, for that matter, Pepto-Bismol.

Representative Pete Stark (D-CA) on August 12, 1992, actually introduced legislation specific to AIDS vaccine liability issues.[25] Stark's bill would have created a National AIDS Vaccine Development and Compensation Program, similar to the one that existed for the childhood vaccines. "If clinical trials are halted on the most promising and advanced of vaccines, which may be the case at present, and even if these vaccines are not ultimately the chosen candidates, significant ramifications result," Stark said when he introduced the bill. He also cited an NIAID study that showed how a mediocre vaccine today would prevent more infections than a terrific vaccine introduced five years later.

As NIAID's Hoth had detailed in his plenary presentation at that year's international AIDS conference in Amsterdam, if a country of 1 million people had 100,000 HIV-infected people and 5% more were becoming infected each year, a 60% effective vaccine would prevent 145,000 infections over 10 years' time. But if the same country waited five more years and had a 90% effective vaccine, at the 10-year mark the vaccine would only have prevented 87,000 infections. "Whether the delay in administering a vaccine occurs because of the inherent difficulties of science or because of liability concerns, the results are the same," said Stark. "In this instance, where the scientific questions pose such a great challenge, it would be a tragedy if liability concerns were allowed to compound this difficulty."

Stark's bill never went anywhere either.

• • •

LOGICAL AS IT WAS TO BLAME the slow pace of AIDS vaccine R&D on the difficult scientific obstacles, by the early 1990s hard evidence existed that other forces had slowed the search too. An AIDS vaccine simply was not that attractive a business venture. Most every major pharmaceutical had stayed away from the field, biotech companies like Repligen and Oncogen that had jockeyed for frontrunner position in the supposed "race" had already fallen by the wayside, and liability issues had slowed progress in at least half a dozen specific cases. The leading lights in the field had recognized these problems, and solid, progressive ideas had come forward. But once again, no organization with a powerful constituency like the March of Dimes existed to turn promising ideas into effective action.

Maybe improving the marketplace and reducing liability risks would not have had much impact on the AIDS vaccine search. That is a reasonable enough thesis. But the opposite thesis is reasonable, too—that ideas like a patent exchange or a consortium or indemnification of vaccine makers could have made a difference. And what would have happened if Bristol-Myers Squibb, which perhaps had the most promising AIDS vaccine approach back then and certainly had a talented team of scientists, had thrown itself at the project with all of the company's might? And what if Bristol was in a hot race with Pasteur Mérieux Connaught, Merck, Hoffman–La Roche, Upjohn, Medeva, Glaxo Wellcome (the merger in 1995 of Glaxo and Burroughs Wellcome), and SmithKline Beecham? What if industry, in short, behaved the way it did when it came time to make anti-HIV drugs and aggressively supplemented the research efforts funded by governments? Maybe a vaccine would be developed a little more quickly. And as NIAID's scenario emphasizes, even a modestly effective vaccine could give humans a leg up in the one real race underway: the one against HIV.

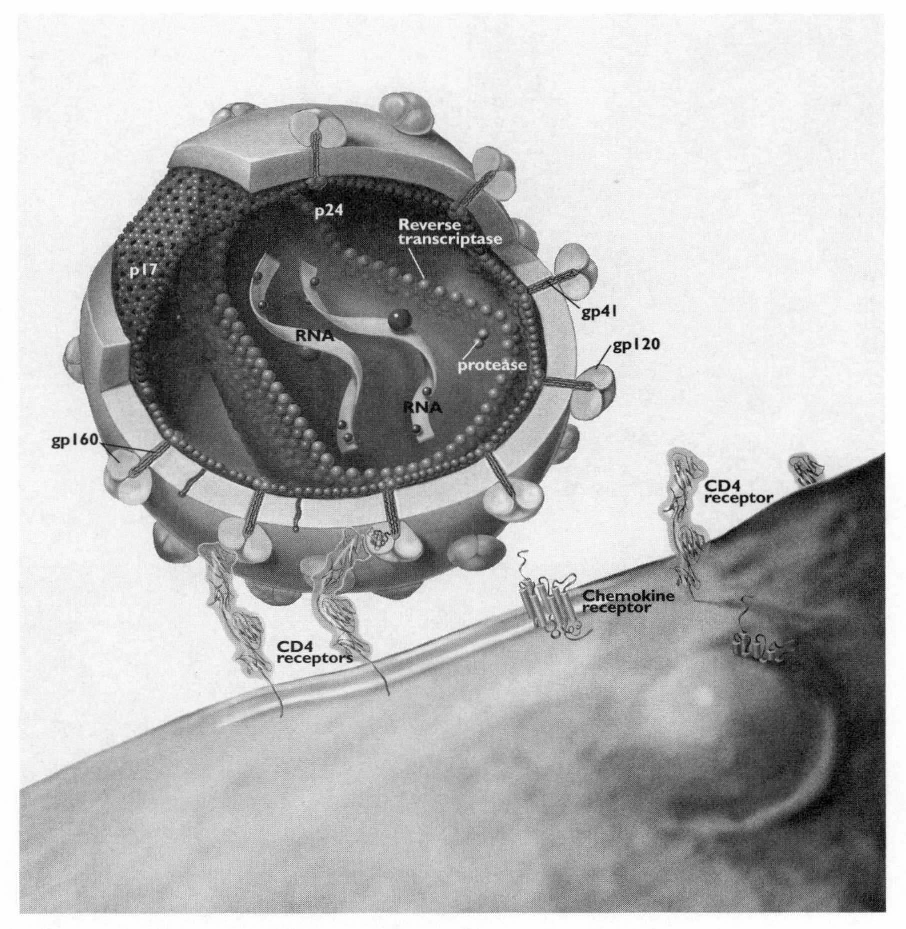

A cross-sectional illustration of HIV, shown here docking onto a white blood cell that has both CD4 and chemokine receptors jutting from its surface.
Credit: Terese Winslow

Dani Bolognesi, the longtime dean of AIDS vaccines, contemplating the variables at the fateful June 17, 1994, meeting of the AIDS Research Advisory Committee (ARAC).
Credit: Darrow Montgomery

Seth Berkley, head of the International AIDS Vaccine Initiative, in Chiang Mai, Thailand, September 19, 1995.
Credit: Darrow Montgomery

MicroGeneSys CEO Franklin Volvovitz speaking to the blue-ribbon panel assembled at the NIH on November 5, 1992, to discuss the fate of the $20 million congressional appropriation for a trial of his company's gp160 vaccine.
Credit: Darrow Montgomery

Robert Gallo and then Secretary of Health and Human Services Margaret Heckler at the April 23, 1984, press conference in Washington, D.C., where she announced that Gallo's lab had discovered "the probable cause of AIDS."
*Credit: Jim Marks/*Washington Blade

Anthony Fauci, head of the National Institute of Allergy and Infectious Diseases, speaking to the press at the June 17, 1994, meeting of the AIDS Research Advisory Committee.
Credit: Darrow Montgomery

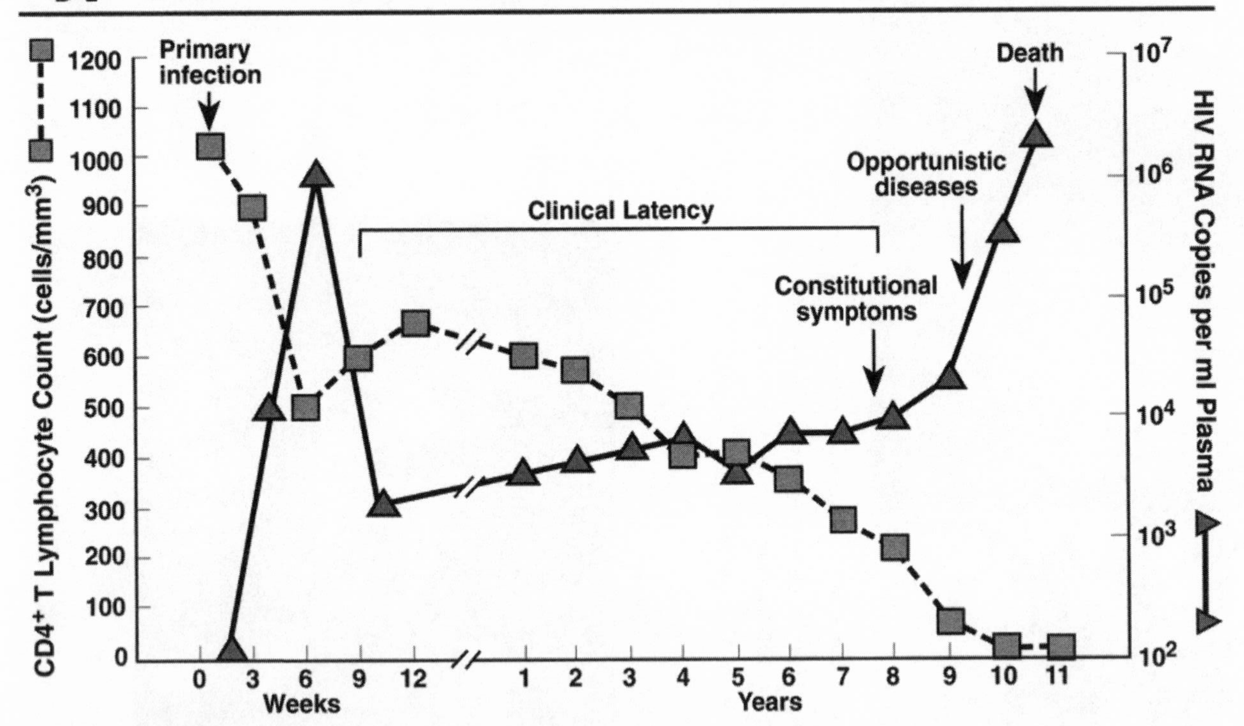

Shortly after HIV (triangles) infects a person, levels of the virus skyrocket and then, because of the immune response, plummet. CD4 cells (squares) steadily decline over time, with the average untreated person (in the United States or Europe) developing AIDS after about 10 years and dying about 12 years after the initial infection. *Credit: Anthony Fauci, NIAID*

Rhesus macaques are the most commonly used animal models in which researchers test AIDS vaccines.
Credit: Yerkes Regional Primate Research Center, Emory University

John Moore (center) at a September 1999 AIDS vaccine meeting at Oxford University, with Neal Nathanson (right) and Jack Nunberg (left).
Credit: John Moore

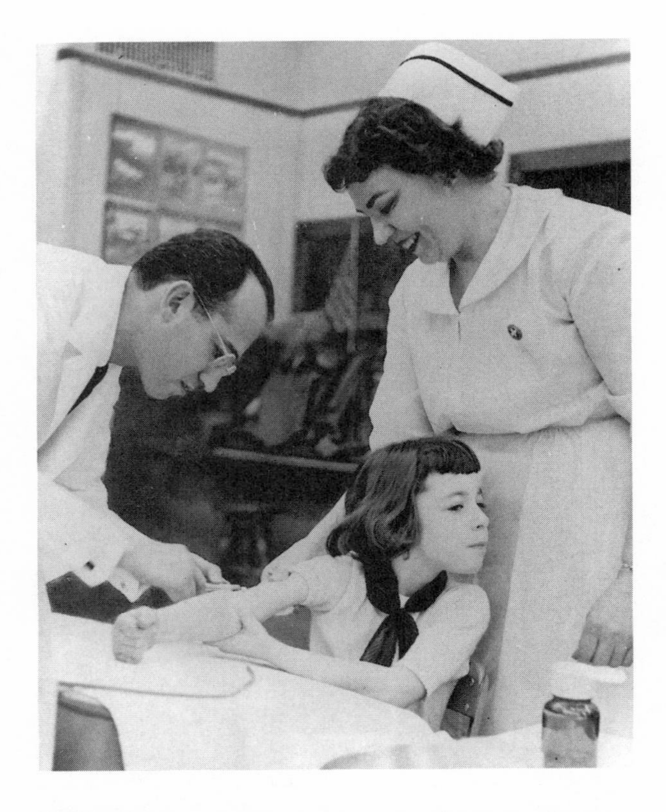

Jonas Salk injecting a child with his killed polio vaccine, circa 1954–1955.
Credit: National Library of Medicine, undated photo

Jonas Salk in 1992, standing outside of his office at the institute that bears his name.
Credit: Jim Cox, Salk Institute for Biological Studies

CHAPTER 7

Unwanted: Dead or Alive

James Stott, a good-natured hobbit of a man with a knowing smile often hiding behind his untrimmed beard, in 1991 shot to fame in the AIDS vaccine world for his little group's tremendous success in monkey studies. Stott headed the United Kingdom's nascent SIV vaccine program, and, unlike many of his colleagues, he had experience making vaccines. Before moving into AIDS research at the U.K.'s National Institute for Biological Standards and Control in 1988, Stott had spent 18 years at an agricultural research institute developing a whole, killed vaccine that powerfully protected cattle from pneumonia caused by respiratory syncytial virus. "I enjoyed basic science, but I was never one of the people who saw science purely as an intellectual exercise," said Stott. "Maybe I didn't have the intellect for it." The type of mind he did have, though, was practical, goal oriented, and to the point.

With only a fraction of the budget that the U.S. National Institutes of Health devoted to its network of U.S.-based primate research centers, the U.K.'s Medical Research Council set up a highly coordinated, standardized program between Stott's group outside London and another one near Salisbury run by Martin Cranage at the Centre for Applied Microbiology and Research. These two groups quickly tackled several of the Big Questions facing the field. As Stott explained at an NIH meeting held on May 13, 1991, to discuss whether animal and human trials were on the same track, he and his coworkers had successfully protected monkeys from the most rigorous challenges yet performed.[1]

Not only had their vaccines prevented infection from an intravenous challenge with a strain of SIV that matched the one in the vaccine; they also could protect against distantly related strains. This suggested that the great concern about the supposedly hypermutable AIDS virus forever dodging a vaccine-trained immune system might be overblown. The Brits additionally had shown that the immunity triggered by their vaccines lasted relatively long, which they demonstrated by incrementally increasing the time between the last vaccination and the challenge. So the issue of a vaccine's durability might not present as huge a hurdle as some imagined. More impressive still, the British group revealed for the first time that a vaccine could, in addition to preventing intravenous infection, stop SIV when the researchers placed the challenge virus into the rectum. This meant that the same vaccine potentially could thwart transmission of the AIDS virus whether it came through a needle puncturing a vein or a penis touching a mucosal surface.

All of these protections owed their success to the same strategy: a whole, killed SIV vaccine. Indeed, Stott pointedly noted at this NIH meeting that the various attempts by the British researchers to protect monkeys with genetically engineered viral proteins had all failed. And he had no respect whatsoever for the chimpanzee experiments that showed protection with gp120, gp160, or anything else. "We're all the same—we all as scientists develop our prejudices, and we hate to have them rearranged by anybody," Stott told me then. "We're said to be basing everything on scientific proof, but we don't. I mean look at the way those crazy chimpanzee experiments are treated. They're scientific crap. Really. The chimpanzee field, I mean it's just appalling."

Stott's presentation at that NIH gathering was bookended by talks from two other monkey researchers, Murray Gardner and Mickey Murphey-Corb, who had similar results. "There's no question this [whole, killed protection] holds up across the world," said Gardner of the University of California at Davis. Murphey-Corb of Louisiana's Delta Regional Primate Research Center ratcheted the enthusiasm for the whole, killed approach a notch higher. Monkey researchers were well on the way to proving that a whole, killed vaccine could address all of the Big Questions. True, no one had yet demonstrated that vaccinated monkeys could ward off challenge virus placed in their vaginas. And the challenge viruses used in the experiments to date all consisted of plain SIV, not SIV inside a cell; in the real world, transmission occurs both ways, and "cell-associated" HIV presumably would have an easier time establishing an infection because the immune system often cannot tell that a cell harbors the virus. Still, the researchers had begun these studies. What if they succeeded? "At that point, you would need to say that, yes, a good working vaccine for HIV is out there, is ready to go," said Murphey-Corb, noting that "many of you hesitate to go into something like that because of liability."

A roundtable discussion ended the daylong conference. Stott rhetorically posed a provocative question: If he knew that six months from now he would be exposed to HIV, what vaccine would he take today? "At the moment my choice would be a whole, inactivated vaccine," said Stott. Duke University's Dani Bolognesi, one of the most forceful advocates for the genetically engineered approach and the wonders of the V3 loop—and convinced that a whole, killed vaccine would remain too dangerous—countered with a message that seemed convenient and tired: Maybe the monkey model is not relevant. "It strikes me that it is so easy to protect in SIV and it's so difficult with HIV," said the avuncular Bolognesi, who had a reputation for cutting to the quick.

But Australian immunologist Gordon Ada, who chaired the roundtable, refused to let go of the idea. "Say you had five vaccines to try in phase III efficacy trials in developing countries," Ada said. "Is there a case for making a whole, inactivated HIV vaccine one of them?" Dennis Carlo, the chief scientist at the Immune Response Corporation—which then intended to start vaccinating uninfected people with its whole, killed HIV vaccine—took the floor. "It's almost unethical not to," said Carlo. "Based on the SIV data, how can you not?"

Leaving aside Carlo's self-interest in seeing the whole, killed approach move forward, his answer to Ada's question laid bare the most frustrating, disturbing reality then facing the field. An old-fashioned vaccine, which opinion makers like Bolognesi—a great synthesizer of data who had close ties with the NIH's two most influential AIDS researchers, Bob Gallo and Anthony Fauci—had deemed too risky, had outperformed every other approach in the monkey model. At best, this line of reasoning went, this vaccine could reveal the immune responses that a safer vaccine should aim to trigger.

On the other side of an intellectual ravine that would grow wider with time stood more empirically minded primate researchers like Murphey-Corb, Gardner, and Stott. Jonas Salk, high priest of the whole, killed approach, greatly amplified their views as his mystique—he had done it before, hadn't he?—spread the gospel far and wide. This group believed that industry could tackle the safety issues—just as the experts had told the NIH in 1990—and from their vantage, decoding the mechanism behind a vaccine's success represented something of an intellectual nicety. As the clock ticked and efficacy trials in humans moved ever closer, it appeared that a head-on collision would soon take place between the leading lights in the AIDS vaccine enterprise.

Still more momentum built for the whole, killed strategy the next month at the Seventh International Conference on AIDS held in Florence, Italy. A poster presented by the U.S. National Institute of Allergy and Infectious Diseases, which listed every monkey study then completed, said it all.[2] Genetically engineered versions of SIV's surface protein had protected 7 of 52

(13%) of the monkeys, the NIAID poster recounted, and all of these successes relied on intravenous, low doses of a challenge virus that matched the strain used in the vaccine. In contrast, whole, killed vaccines had protected 86 of 124 (69%) of the monkeys in experiments that often challenged the animals much more rigorously in an attempt to address the Big Questions.

Then James Stott did an experiment that, overnight, brought the field to a halt and explained why, as Bolognesi had noted, SIV had been easier to protect against than HIV. The experiment did not validate the opinions of the opponents to the whole, killed vaccine. The experiment also did not excuse away their biases toward high-tech, molecular biological approaches. Nor did the experiment forgive their steadfast refusal to accept the obvious conclusion that the whole, killed approach until then had proved the best available strategy. Rather, the experiment rescued them from having their prejudices rearranged. In short, James Stott bought time for colleagues whose deepest convictions about how best to proceed ran counter to his own.

• • •

DANI BOLOGNESI FIRST HEARD OF what became known as "the Stott experiment" on September 5, 1991, while sitting with some of the world's leading AIDS vaccine researchers atop picnic tables that flanked a golf course at Maryland's Potomac Valley Lodge. Each year, Robert Gallo hosted a marathon, weeklong AIDS conference that offered one light day in the countryside for colleagues to put away their slide trays and just brainstorm in tutorial sessions. Near the end of the tutorial on AIDS vaccines, a British researcher, Angus Dalgleish, briefly reported on what he called "a rather interesting observation" made by the Stott group.

"Whoa, whoa, whoa, whoa," said Bolognesi when Dalgleish described the experiment. "Say again?"

Dalgleish repeated the data.

"Could you say that again?" asked Harvard's Ron Desrosiers.

The group started laughing and then, as often happens when a truly new idea confronts a pack of scientists, many people started speaking at once, some in disbelief, others in awe.

Despite the widespread, convincing success with whole, killed SIV vaccines, scientists still had yet to find a satisfactory immunologic explanation for the protections. The two most obvious candidates were antibodies against SIV, which could neutralize the virus and prevent it from infecting a cell in the first place, and killer cells, the smart bombs that search and destroy already infected cells. But when researchers tested the blood of the protected monkeys, neither the levels of neutralizing antibodies nor killer cells seemed to correlate with the protections. Stott then decided to look at a potential correlate that everyone else had ignored.

Stott's insight, like many in science, mixed one part serendipity with one part smarts. In the field of veterinary vaccines that Stott came from, researchers routinely use preparations that would never be deemed clean enough for humans. Stott, for example, made a respiratory syncytial virus vaccine for cattle by just infecting a bunch of cells with that virus and then killing them. This was much simpler than making a pure batch of virus and killing it. That type of clean product required growing up huge amounts of the virus in cells and then painstakingly washing off those cells. It was the difference between scrambling a few eggs in a pan and separating whites from yolks to bake a soufflé. So when it came time to make a whole, killed SIV vaccine, Stott used the quick-and-dirty procedure, growing SIV in human white blood cells and then killing the entire concoction. "It was a very bizarre thing to do in many ways," said Stott. "Most people thought this was stupid, and in some ways it was pretty crude. But we got protection."

After the initial protection under idealized conditions, the British researchers began pushing the vaccine harder, testing it against such real-world situations as a rectal challenge and a mismatched strain. They also asked how well the vaccine would work if they only gave two doses before a challenge, rather than the standard four. They added another twist to this experiment: They made the control more rigorous. In each challenge, the Brits compared vaccinated monkeys to control monkeys that had not received the vaccine. But what would happen, they wondered, if the control monkeys received the human cells alone, with no SIV in them?

As it turned out, two doses of the vaccine protected three of four monkeys. But to the astonishment of the British group—and of the scientists who attended the tutorial at the Potomac Valley Lodge—two of the four control monkeys did not become infected. And a close analysis revealed an equally astonishing immunologic correlate of protection, which ultimately explained much of the protection seen with whole, killed SIV vaccines.

Two critical factors explain the results of the Stott experiment. First, the outer coats of SIV and HIV contain pieces of the cells they infect. When these viruses first infect a cell, they leave their coats at the cellular door, shooting in only the strands of nucleic acid that hold their genetic code. The nucleic acid weaves itself into the chromosomes of cells, and then directs the production of new viral proteins. Once these proteins properly assemble themselves into a new virus, it buds from the cell's surface, which forms the viral coat. The virus, then, has in its coat many cellular proteins.

The second key to Stott's results is that immune systems, whether they belong to monkeys or humans, develop the strongest reactions to things that are most foreign to them. So monkeys given human white blood cells develop a whopping antibody response to them.

The Stott experiment showed that those human-cell antibodies played a large role in protecting the monkeys from SIV. In hindsight, it made perfect sense. The vaccine, which contained human cells, triggered production of human-cell antibodies. The SIV used to challenge the monkeys contained pieces of the human cell in its coat, which the already present human-cell antibodies could glom onto, preventing the virus from infecting the animals.

Stott's group formally reported their results the next week at a meeting in Coventry, England. *New Scientist*, which broke the story, said the experiment had thrown the search for an AIDS vaccine into "disarray." A cartoon accompanying the piece showed two monkeys reading an article headlined "SIV Shock," with one asking the other, "Perhaps we should be doing the research."[3]

Had the Stott experiment only applied to the crude type of whole, killed vaccines made by his lab that contained entire human cells in the preparation, the field likely would have written it off as a curiosity. But as the researchers described at the Coventry workshop, they also made whole, killed SIV vaccines from purified virus. A reanalysis of blood from animals protected by those vaccines showed that they, too, had significantly higher levels of human-cell antibodies than vaccinated monkeys that had become infected upon challenge.

How could this be? The researchers had grown the purified virus in human white blood cells. Although they had separated these cells from the vaccine, the very process of growing the virus in human cells meant that the new SIVs had pieces of human cells in their coats. So monkeys vaccinated with these whole, killed vaccines that had purified SIV still could develop antibodies to the human cells. Unfortunately, every primate researcher around the world had made whole, killed vaccines by growing SIV in human cells. All of these vaccines, basically, had contaminants in them. Suddenly, a black cloud hung over the enthusiasm of the entire field. "It calls into question the basis of all the previous protection studies, even my own," Harvard's Desrosiers told *New Scientist*.

The baffling results became even more confusing when no less than John Maddox, editor of *Nature* magazine, wrote an editorial that week entitled "AIDS research turned upside down."[4] Maddox, demonstrating once again that ignorance about the AIDS vaccine field spread depressingly far and wide, attempted to link the results from the Stott experiment to the views of Peter Duesberg, a University of California at Berkeley professor who maintained that HIV does not cause AIDS.[5] Maddox argued that the Stott experiment supported the cofactor theory, which holds that HIV by itself does not cause AIDS unless the immune system already is compromised. The compromising factor, in this far-fetched scenario: the presence of "foreign" white blood cells that accompany the virus when a person becomes infected. Maddox (who, curiously, later became a staunch critic of Duesberg supporters) failed to note that molecularly cloned SIV and HIV—in other words, pure virus with no white blood cells or their components present—caused AIDS. The editorial also twisted the ideas of Duesberg, who did not ascribe to the cofactor theory: He called HIV "a pussycat" and claimed it has nothing to do with the cause of AIDS. Maddox's editorial, one of many freakish sideshows in the search for an AIDS vaccine, caused a predictable uproar.[6]

Nature published the British group's first formal report about the finding in its October 3, 1991, issue, putting Stott and his colleague Martin Cranage on the hot seat at the NIAID annual AIDS vaccine meeting, held later that month in Florida.[7] In one particularly fiery workshop on the topic, Murphey-Corb and Gardner circled the wagons and strongly doubted that the results would apply to their work. "I know what the Judge Thomas treatment is now," Stott told me, referring to the controversial nomination hearings for the Supreme Court justice that were then taking place. "It's important to not get it out of proportion. I do think it's a question of taking a step back to understand better what's going on in SIV."

But critics of the Stott experiment had trouble criticizing Jim Stott: he had solid credentials and shared their bias that the field should rely more on empiricism than reductionism. And before the meeting's end, data surfaced that supported the British researchers' observations. Dani Bolognesi's lab had analyzed antibodies in blood from 200 monkeys used in vaccine experiments and could easily show high levels of human-cell antibodies in those that had received whole, killed vaccines.[8]

The logical, scientific response to the Stott experiment would have been to grow SIV in monkey cells and use the resultant virus both to make whole, killed vaccines and challenge "stock." This monkey/monkey approach would most closely mirror the situation if a whole, killed vaccine ever went into humans: Scientists would make the vaccine from HIV grown in human cells, and the challenge virus, so to speak, would also be HIV grown in humans.

But a technical difficulty stood in the way of this rational plan. It requires huge amounts of SIV to manufacture enough whole, killed vaccine to carry out even one experiment with, say, four vaccinated monkeys. To grow that much virus, scientists ideally would have a biological machine called an "immortalized" cell—one that has been transformed, by various manipulations, into a cancer cell that copies itself faster than the SIV can kill it. Such cell lines exist for many types of human white blood cells. But in 1991, no lab had yet concocted a similar monkey cell line, which explains why everyone relied on the human material.

Several labs did grow enough SIV in freshly harvested monkey cells to make challenge stock. When they injected this virus into monkeys previously vaccinated with the whole, killed human-grown SIV, the animals routinely became infected.[9] These experiments rightly reinforced the idea that the earlier challenges had worked because of antibodies to the human cells. But the experiments effectively obscured the need for the monkey/monkey challenge, and wrongly contributed to the increasing sense that the whole, killed approach had gone bust.

Murphey-Corb independently attempted to conduct the monkey/monkey experiment without the help of an immortalized cell. After 10 months of saving the spleens of SIV-infected monkeys that died from AIDS, Murphey-Corb had enough virus to make a vaccine. "I spent a considerable fortune and time trying to make virus in primary spleen cells," she recalled. But the SIV she harvested had tons of cellular junk in it that she could not purify away. The resultant vaccine, she said, "was really crap." In particular, the vaccine contained so few viral proteins that the immunized monkeys did not even develop a detectable immune response to SIV. Still, she decided to challenge the monkeys anyway, and they all became infected.

That failure, and the roundly held opinion that the whole, killed vaccine was history, ultimately led Murphey-Corb—the biggest advocate for the approach—to shift gears. "I didn't want to be sacrificed," said Murphey-Corb, who moved to the University of Pittsburgh in 1997. "Everyone took Stott's work and said, 'OK, this vaccine is not doable.' You have to respect their opinion. I abandoned my whole, killed vaccine work. Tons of stuff never got published. I literally wiped off three years of my life. . . . I've always had a personal sense that I should have hung on longer."

By 2000, no one had yet done a clean monkey/monkey challenge with whole, killed SIV.

Had there been an organized, coherent AIDS vaccine search, the leadership could have declared it a top priority to make a monkey cell line that could grow SIV. This was precisely the type of scientific endeavor that cried out for a targeted project. For starters, had an investigator at a university submitted a grant proposal to make this cell line, it likely would have fared poorly in the NIH's peer-review process, which favors sexier, more basic research. A targeted project would have offered the money up front to complete the job. Targeted projects also work well when they have a clear and practical goal, as this one did, and the task is feasible, as this one was (scientists had had years of experience fashioning similar cell lines). Just such conditions existed in 1948, when the March of Dimes organized its successful poliovirus typing project. Clearly, that project addressed a more urgent question, but the point remains: The March of Dimes had a mindset that a vaccine would emerge most quickly if scientists could collectively identify and knock down some hurdles.

Underscoring the disorganization in the field, Desrosiers's lab in 1996 finally made such an immortalized monkey cell line for other purposes, yet this critical whole, killed challenge still did not occur. Why had the community failed to tackle this problem? "I don't know," said Desrosiers. "It's a reasonable question to ask."

• • •

ASIDE FROM PREMATURELY DECLARING the death of the whole, killed vaccine, the Stott experiment paved the way for a host of new ideas about how best to proceed. Two of the most talked about strategies appeared back to back in the December 18, 1992, issue of *Science*.[10]

Desrosiers's lab authored one of these two landmark papers. Even before the Stott experiment, Desrosiers had long been underwhelmed by the whole, killed approach—and not because of the safety issues raised by other critics. He simply did not believe that the vaccines could trigger the long-lasting, powerful immunity needed to truly protect humans from HIV. He had even stronger doubts about vaccines made from genetically engineered proteins of the virus, because, like others, he had tested several in monkeys that failed. And the only success with this approach, the prime-boost gp160 vaccine that Shiu-Lok Hu and his colleagues had reported in *Science* that January, relied on a relatively wimpy challenge virus. "Simply by adjusting the vaccine strain, you can get any result you want," Desrosiers told me in 1992. "I became convinced a number of years ago that it's really, really tough to see vaccine protection except under optimized laboratory conditions."

While studying the mechanism by which the AIDS virus decimates the immune system, Desrosiers stumbled upon another vaccine approach that proved more powerful than anything yet tried. Desrosiers at first wanted to see what would happen if he removed one of the genes from a highly virulent strain of SIV. After excising a little-understood gene called *nef*, which the virus does not need in order to copy itself, he injected this altered SIV into six monkeys. As a control, Desrosiers and his coworkers simultaneously gave another group of 12 monkeys a completely intact version of the same strain of SIV. More than two years later, all of the monkeys that received the *nef*-deleted SIV were healthy, with normal CD4 counts and low levels of the virus in their blood. In contrast, the controls were becoming ill or had died from AIDS. Maybe, thought Desrosiers, he had used high-tech to create, inadvertently, an old-fashioned attenuated vaccine that contained a live but harmless virus.

To test whether the *nef*-deleted SIV indeed behaved like an attenuated vaccine, the Desrosiers group challenged these animals with a small dose of the intact version of the virus. Within 36 weeks, the four controls were sick or died, and the four "vaccinated" animals remained healthy. Ultrasensitive tests of the vaccinated monkeys' blood could find no evidence of the strain of virus used in the challenge, indicating that they had managed to resist infection.[11]

Desrosiers upped the ante, rechallenging the vaccinated monkeys with a 100-times-higher dose of the intact SIV. Again, the animals showed no sign of the infection.

News of Desrosiers's findings bowled over AIDS researchers. "This is three orders of magnitude better than any protection we've seen," said Alan Schultz, then acting chief of the vaccine branch at NIAID's Division of AIDS. Duke's Bolognesi called the data "head and shoulders above everything else."

But, for safety reasons, neither NIAID nor Bolognesi had much faith that this approach would ever make it into humans.

If most researchers viewed the whole, killed AIDS vaccine as playing with fire, they saw the attenuated AIDS vaccine as playing with plutonium. Three main safety concerns—which Gallo, in particular, had been articulating for years—stood out. First, the AIDS virus in an attenuated vaccine could revert to a virulent state. Second, even if it didn't cause AIDS, the vaccine might cause cancer by the process of inserting its genes into the host's DNA; such "insertional mutagenesis" occurs with chicken and mouse retroviruses. Finally, a virus that initially looks safe might, decades later, turn out to cause AIDS.

Desrosiers took each of these concerns seriously, but tried to parse the real from the perceived risks. Desrosiers maintained that he could reduce the risk of the virus reverting to virulence to a practical impossibility. Reversion can occur by several mechanisms. The virus can mutate. It can incorporate pieces of the host's genetic material that coincidentally match the deleted parts of the virus. Or, if the virus meets one of its relatives—as theoretically would happen if a person who received an attenuated HIV vaccine then became exposed to HIV—they can swap genes. But the more genetic material that is missing, the harder it is for any of these mechanisms to resuscitate a weakened virus.

Desrosiers decided to treat the virus like a racing sloop, stripping off every unnecessary component. This meant removing more and more genetic elements, like *nef*, that lead the virus to cause disease. Yet Desrosiers could only remove so much: He had to find the highest number of deletions that still allowed the weakened virus to function as a vaccine. It does not make sense, after all, to leave a sloop's steering wheel on the dock to save weight.[12]

As for the vaccine causing cancer, Desrosiers stressed that no evidence suggested that lentiviruses—the retrovirus family that both HIV and SIV belong to—have such a capability. "There's a long record of lentiviruses in tens of thousands of animals and people and there's not a single case of directly caused cancer," he argued to me. To back this assertion, he noted that the African green monkey apparently suffered no ill effects from SIV infection. Gallo, who earlier had slammed the idea of an attenuated HIV vaccine as "seriously nuts," predictably had a less sanguine outlook, noting that even an attenuated HIV might trigger B-cell lymphomas or Kaposi's sarcoma—both AIDS-related malignancies. Desrosiers countered that aside from the murky understanding of what causes these cancers, they seem to depend on HIV copying itself, and his attenuated virus did not replicate much.

Desrosiers did acknowledge that the long-term safety issues were "legitimate," and he said before a vaccine went into humans, it would have to prove its worth in chimps. And even if the vaccine worked in chimps, predicted Desrosiers, it would still take up to 15 years of safety testing in humans before moving the vaccine into thousands of people—assuming that industry or the government would even make such a vaccine. All of which led Desrosiers to conclude that there was no time to waste. "Every year, every five years we delay will delay that," he said.

Desrosiers proposed starting safety tests in volunteers who had the highest risk of becoming infected. He offered a nonscientific, but compelling, scenario to make his point. Desrosiers reasoned that some "high-risk" populations, such as injecting drug users or young gay men in the United States, had new infection rates of up to 5% per year. Over 10 years time, then, if people in such populations did not change behaviors and the prevalence of the virus remained constant, they would have a 50% chance of becoming infected. "If you're unable or unwilling to change your lifestyle or you're trapped, let's do a guesstimate of risk/benefit versus an HIV vaccine with four deletions," suggested Desrosiers. "If it were myself, I'd take the vaccine. I'd take it tomorrow."

Yet Desrosiers did not want to appear overly zealous about the attenuated vaccine. "The more I push, the more resistance there is," he said. "I hope gp120 saves the world from AIDS. But it's probably not going to work. Then what are we going to do?"

Although Dani Bolognesi did not oppose small safety trials of the concept in humans, he saw the attenuated vaccine as a means to the holy grail: it might reveal the immune responses that correlated with protection. "So far, we've been flying without a compass," Bolognesi said. Desrosiers's *Science* paper offered no such correlate though. Neutralizing antibodies produced by the attenuated SIV did not explain the protection; indeed, other vaccines his lab had tested triggered production of much higher levels of these antibodies yet still failed to protect the monkeys. Desrosiers had yet to even hunt for killer cells, the other obvious possibility, as researchers had yet to develop an easy way to test for them in monkeys.

As much as Desrosiers, a molecular biologist, enjoyed teasing out mechanisms, at heart, he too had an empirical outlook when it came to moving the AIDS vaccine field forward as quickly as possible, and he had little faith that his experiments would uncover the holy grail. "I just kind of chuckle at it," Desrosiers said. "It's dreaming that science can do a lot more than it can. Surely it's a worthy concept, but dream on. I want to try myself, but don't hold your breath."

There was one holy grail type of experiment that Desrosiers and many other researchers were anxious to see conducted with his monkeys, but once again, a technical hurdle stood in the way, and no organized effort was made to topple it. If killer cells did protect the monkeys given the attenuated vaccines, then, theoretically, researchers should be able to bleed the animals, separate out those cells, and inject them into unvaccinated monkeys. If the monkeys that received the transplanted killer cells could resist infection after a challenge with SIV, that would provide a strong argument that they were *a*—if not *the*—correlate of protection.

The major hitch to this experiment: all cells have proteins on them that are unique to an individual. That is how the body distinguishes "self" from "nonself." So if you inject killer cells from one monkey into another, the second animal will mount an immune response against the nonself cells, eliminating them. Identical twins, which can freely share cells, suggest a way around this problem. Desrosiers scoured the world for identical twin monkeys in captivity, but found none. The next logical step would have been to call in help from reproductive endocrinologists, who during the past few decades have learned much about how to manipulate many species to create multiple births. No such concerted effort was made.

Desrosiers, for his part, concentrated on the more empirical exercise of engineering SIVs and HIVs with three, four, and even five deletions, and then evaluating how well they worked in monkey and chimp challenges. "Our backs are getting to the wall in a dramatic and dangerous situation," he said. "We need to be ready to accept some radical approaches. The sooner we can start, then the better off we are. This really does look like a vaccine that would work. How long will it take for the mindset to change and people to be won over? It's going to be tough, tough, tough."

During the next few years, it would become tougher still as data emerged from both monkey and human studies that emphasized how an attenuated vaccine deemed "safe" today can appear dangerous tomorrow.

• • •

THE SECOND INTRIGUING AIDS VACCINE PAPER published in that December 1992 issue of *Science* stemmed directly from the Stott experiment and suggested a way to turn the lemon into lemonade.

Larry Arthur, an easy-going, can-do researcher who grew up on a Louisiana chicken farm, headed an AIDS vaccine team at the National Cancer Institute that asked a fundamental question about SIV and HIV.[13] Researchers knew that each virus contained about 216 molecules of gp120 on its surface. But no one had yet asked how much of the viral surface contained proteins from the cells that in essence gave birth to the new virus. Arthur and coworkers separated out these cellular proteins and, strikingly, found between 375 and 600 molecules of the most common cellular proteins studding the viral surface. In other words, the surfaces of HIV and SIV held more cellular proteins than viral proteins. Arthur's group further showed that when they mixed antibodies against these cellular proteins with HIV or SIV, the viruses no longer could infect cells. No wonder the cellular proteins had played such a large role in the protections seen with whole, killed SIV vaccines: they were abundant, and antibodies against them hobbled the virus.

To Arthur, these findings reached far beyond providing a clearer explanation of the Stott experiment though. They raised the possibility that cellular proteins, rather than being contaminants, could serve as an integral part of an AIDS vaccine. Maybe they could even, by themselves, operate as a vaccine.

By the fall of 1993, Arthur had preliminary data that suggested the idea might work. Arthur injected two monkeys with nothing more than human cellular proteins, and then challenged them with SIV grown in human cells. The animals showed no sign of infection.[14]

Gene Shearer, a prominent AIDS immunologist at the NCI who pioneered the study of people who had not become infected despite repeated exposures to HIV, coauthored a detailed argument for just such a vaccine.[15] As Shearer and his coauthors explained in a letter to the editor that ran in the October 8, 1993, issue of *Science*, the vaccine could contain "a pool" of different proteins commonly found in human white blood cells. If vaccinated people then later became exposed to HIV, their anticell antibodies could stop the virus exactly as had happened in Stott's and Arthur's experiments.

Although the immune system remains constantly on a state of red alert for foreign proteins, this bizarre-sounding strategy revved up that mechanism. Yet there was a catch-22. Organ transplants and skin grafts frequently fail because people's immune systems reject the foreigners. So injecting people with this type of AIDS vaccine would critically limit their options if they suffered heart failure or a bad burn. "However, in several parts of the world, effective immunization against HIV infection might take precedence over the unlikely prospect of a future organ transplant," countered Shearer and his coauthors.

Adding to the validity of this idea, researchers in Kenya had uncovered an interesting phenomenon when studying the white blood cells of prostitutes there who resisted infection despite repeated HIV exposures.[16] The immune system relies on a system similar to social security numbers to identify self from nonself: each cell houses a collection of proteins on its surface, that, together, identify the person. But two people can share digits in their social security numbers and still have unique numbers. And the more digits two people share, in effect, the more their immune systems will tolerate each other's cells.

The scientists in Kenya found that these exposed, uninfected prostitutes had an extremely rare protein on their white blood cells. It was as though their social security numbers contained a fraction in one spot instead of a round number. Theoretically, because their immune systems had such an unusual makeup, they would have a stronger immune response than normal when confronted with the proteins on the average person's white blood cells.

Both animal and human data backed this innovative idea for an AIDS vaccine. One of the world's most prominent scientific journals publicized the idea. Respected researchers promoted the concept. It went nowhere.[17] "It's too simplistic," said a frustrated Shearer, years later.

Maybe the idea, like most bold scientific ideas, ultimately would have proven worthless. But that is a cynical outlook. A more pragmatic approach would have been for scientists to compare the protection offered by this type of vaccine with other strategies. Maybe they would have found that cellular proteins offer a more robust immune response than viral proteins. Maybe the combination of viral and cellular proteins would have proven more potent than the cellular proteins alone; this is what the initial Stott experiment suggested, which found that SIV plus cells protected more monkeys than the cells by themselves. Maybe the cellular proteins would have proven unnecessary or too dangerous. As it happened, no organization existed to push for such comparative studies, and this promising lead, like so many others, amounted to little more than a soon-to-be-forgotten pile of maybes stacked along the side of a road that was not heading anywhere in particular.

• • •

IN DECEMBER 1993, AT one of the routine meetings that NIAID director Anthony Fauci held with the members of his lab to talk about their work and the state of the field, it became clear that the Stott experiment had reached far beyond revealing contaminants in monkey challenge studies and influencing the design of potential vaccines. The experiment had ushered in a new point of view about the most profound question facing researchers: must an AIDS vaccine prevent infection, or is the real aim to delay and possibly prevent disease? "We spent a considerable amount of time on just this question," Fauci told me the week after the meeting. "Most people agreed we need to rethink this."

The rethinking of such a fundamental issue nine years into the search for a vaccine reflected a fading of the overconfidence and even cockiness that had driven the young field. Until the Stott experiment, researchers roundly agreed that an AIDS vaccine should completely block infection—so-called "sterilizing immunity." After all, the reasoning went, every case of AIDS begins with one HIV entering a person and nestling its genetic material into a cell. Logistical reasons also led to sterilizing immunity's emergence as *the* "endpoint" in vaccine trials. HIV infection, untreated, takes an average of 10 years to cause disease. So staging a trial with disease as a clinical endpoint would probably require several years—and tens of thousands of people—to arrive at a statistically meaningful answer. And that makes it tough, because researchers have trouble retaining people in a vaccine trial for more than a few years; running multiyear tests costs astronomical amounts too. On the other hand, a trial with infection as an endpoint might take as little as three years, according to NIAID estimates, and require fewer than 3,500 people. So researchers long deemed the zero-tolerance policy of sterilizing immunity the only way to go. And as more and more labs showed that whole, killed SIV vaccines could achieve this, the grand goal seemed doable.

But then came the long list of studies in which whole, killed vaccines failed to prevent infection from monkey-grown SIV. Despite the imperfect design of these experiments, they humbled researchers and made them ask whether an AIDS vaccine should actually shoot for the lower standard of delaying or preventing disease. The mounting failures in monkey studies with other vaccines added to the despair; even Shiu-Lok Hu's celebrated prime-boost strategy did not work when researchers challenged the monkeys with a more robust strain of SIV than he originally used.[18] NIAID's Alan Schultz had reached the conclusion that true sterilizing immunity was "patently absurd" because the immune system relies on some level of infection to mount a full-fledged response. "If sterilizing immunity is the way a vaccine is going to work, we should put all our money into condom distribution," said Schultz.

Hints also began to emerge from several labs that vaccinated monkeys who had "failed" challenges and become infected often fared better than the unvaccinated control animals. "There may be a lot of primate data we need to reanalyze and find out whether they get sick and whether they are protected from subsequent infections," concluded Fauci.

One hint came from NIAID's own Vanessa Hirsch, who in 1991 gave animals a whole, killed vaccine and then challenged them with monkey-grown SIV. The experiment resoundingly "failed" under the sterilizing immunity paradigm, which then prevailed: the five vaccinated animals and six unvaccinated controls all became infected. But 20 months later, three vaccinated animals remained healthy, while AIDS-related opportunistic infections had riddled all of the controls, killing five of them within the first year.[19] "We should redefine what we call protection," Hirsch told me. "Even from a fairly ineffective vaccine—that I don't think is anywhere near optimal—there's some evidence that it's protective."

Confirmation that "failed" vaccines could delay or prevent disease came from the Henry M. Jackson Foundation for the Advancement of Military Medicine, a laboratory in Rockville, Maryland, that worked hand in glove with the Walter Reed Army Institute of Research's AIDS vaccine program. Researchers there injected three monkeys with a vaccine made from small fragments of SIV's surface protein and then challenged the animals. All became infected. Four years later, one of the vaccinated monkeys had died of AIDS, but two showed signs of SIV in their lymph nodes only, not the blood—indicating that the animals had contained the infection. In contrast, two control animals died and a third had SIV in the blood. This milder infection in the vaccinated monkeys also could not spread SIV as readily, which the researchers showed by taking blood from the vaccinated animals and injecting it into uninfected ones.[20]

Gerald Eddy, lab director at Henry Jackson, acknowledged that the vaccine had serious shortcomings, as it didn't protect all the animals from disease. "But if AIDS gets you after 30 years rather than after 10, as a vaccinee, you're ahead of the game," he said. "And if you are much less likely to transmit the virus, everybody's ahead of the game."

Other studies supported the thesis that vaccinated monkeys could better contain their subsequent infections. Several experiments, for example, found that vaccinated monkeys which became infected by the challenge virus had lower "viral loads"—the total amounts of virus in their blood—than the controls. In tests with 18 infected monkeys given various vaccines, Edward Hoover, a pathologist at Colorado State University, found that they had 10 to 200 times lower viral loads than found in five control animals. "If one can heighten the immune response enough, it may be a realistic goal to try to achieve lifetime suppression of virus," said Hoover.[21] Similar data came out of SIV vaccine experiments done by Murphey-Corb, Stott and Cranage, Gardner's group at UC Davis, and Desrosiers.[22]

Unfortunately, most of these researchers had neither the lab space nor the funds needed to keep their infected monkeys alive to see whether they developed AIDS. "A lot of monkeys have been killed because it seemed like the experiment was over," said Patricia Fast of NIAID's Division of AIDS. "In retrospect, we wish we would have kept them alive." But longer trials would require more cages for the monkeys, which researchers must isolate after they become infected, as well as more personnel to care for the surviving animals—all of which added up to more money.

Many of the researchers involved told me that they thought the investment made imminent sense, as the resultant data would allow them to make distinctions that they might otherwise miss, ultimately helping them select the best vaccines to move forward most aggressively in humans. Cranage, for example, advocated considering protection on a "sliding scale" ranging from high viral loads to no detection of virus. And some judged the sterilizing immunity paradigm as a case of the perfect being the enemy of the good. "With HIV, we're attempting to set such strict criteria for efficacy that it's unrealistic," argued Eddy, "and a lot of money is possibly being wasted by attempting to reach that standard."

• • •

IMAGINE IF A MARCH OF DIMES TYPE OF ORGANIZATION had existed in 1993. A scientific advisory board could have designed a sophisticated master monkey experiment that pitted a myriad of vaccine strategies against each other: whole, killed; attenuated; recombinant envelope proteins like gp160 and gp120; core proteins like p27 and p17; mixtures of envelope and core proteins; various live "vectors" like vaccinia that expressed different SIV proteins; priming with such a live-vector vaccine and then boosting with a recombinant protein; cellular proteins by themselves; cellular proteins mixed with recombinant viral proteins—the list of possibilities stretched on and on. New vaccine approaches could enter the testing pipeline soon after they surfaced. This mythical scientific advisory board also could have dictated that all of the researchers use the same laboratory conditions. The same species of monkey. The same immunization schedules. The same route of challenge. The same doses of the same challenge viruses. The work could then have been assigned to primate researchers around the world. After the challenges, the researchers could have assessed protection on a sliding scale, as Cranage suggested. Had they kept the monkeys alive, scientists could have determined whether the various vaccines could delay or prevent disease. In the end, the AIDS vaccine field could rationally have compared the various strategies. Then, as

Hilleman had recommended in 1987, the most promising candidates could have moved into head-to-head comparisons in chimpanzee challenge studies. From this mass of data, scientists could, finally, select a few vaccines to push forward in human studies as quickly as possible.

Yes, it would have been a gamble: the monkey and chimp models might not have reflected the human situation. But the alternative was the status quo, in which scientists moved into humans with most anything that appeared safe, and then measured immune responses, hoping they would see something that would give them enough confidence to launch full-scale efficacy trials. Animal data, which often had no statistical significance, served as little more than campaign bumper stickers to argue for or against a given approach.

As it happened, primate researchers continued to conduct challenge experiments with little coordination, and thus comparing one lab's results to those of another became more and more difficult. Critical experiments, like the monkey/monkey whole, killed study and expanded studies of cellular proteins, fell through the gaps. No coordinated push was made to funnel more money to primate researchers for the express purpose of keeping monkeys alive after vaccine experiments.

A major reason the AIDS vaccine community failed to organize itself and conduct an ambitious, definitive experiment like this is that the NIH provided an illusion of organization and leadership. From the beginning of the AIDS epidemic, the NIH had devoted more research dollars to the disease than any other institution in the world: By 1993, its AIDS budget had swelled to $1.073 billion, about 10% of which went to vaccines. (In comparison, the 1997 budgets of the four largest spenders on AIDS research in Europe totaled a mere $90 million.[23]) But the NIH did not like to direct research. Rather, it believed that science was best moved forward by individual researchers submitting proposals for funding, which their peers would then evaluate. This system of investigator-initiated science would, theoretically, shuttle the best ideas to the front. That it often did. But the system had a significant limitation, which the AIDS vaccine search starkly highlighted: it favored the creative over the practical.

The NIH's high-profile leaders added to the sense that science had attacked the AIDS vaccine problem as vigorously as possible. In particular, Anthony Fauci, director of NIAID, which received the bulk of the NIH's AIDS money, exuded competence and confidence. Fauci additionally headed the NIH's Office of AIDS Research, which putatively coordinated research at nearly two dozen different institutes. But Fauci could not operate with the freedom or authority that Basil O'Connor or Harry Weaver had at the March of Dimes.

A Brooklyn native educated by Jesuits, Anthony Fauci had a distinguished record as one of the world's most devoted, diplomatic, and disciplined AIDS researchers. A favorite on Capitol Hill and at the White House (President Bush had singled him out as a personal "hero"[24]), a frequent guest on shows like *Nightline* and the *MacNeil/Lehrer NewsHour*, a must-quote in any big AIDS story that broke, a keynote speaker at every major AIDS conference, one of the most widely published and cited AIDS researchers, head of a large lab, a clinician, a voracious reader of the AIDS literature, an avid jogger, and father of three young girls, Fauci's days started early and ended late. But the bureaucratic constraints of working for a government institution, where he had a severely limited ability to move money at will or to express opinions, coupled with the need to please too many constituents, including the conflicting demands of academia and industry, made Fauci more of a figurehead for the AIDS vaccine field than a true leader. He could no more have launched this type of targeted, visionary, multimillion-dollar, sure-to-be-controversial monkey experiment than he could have solved the conflict in the Middle East. Simply put, it was not in his job description.

Evidence that the NIH and its leaders had a tenuous hold on the reins that steered the AIDS vaccine search dramatically came to the fore in 1992–1993 when a clamor rose in Congress, the activist community, and even academia to better organize the field. But the most telling indication that no one was in charge came about through a peculiar appropriation that Congress made to one AIDS vaccine company, MicroGeneSys, which was derided by Fauci, NIH director Bernadine Healy, FDA commissioner David Kessler, and prominent researchers from coast to coast. The objections of scientists of this stature, speaking on behalf of an enterprise devoted to finding something as critical as an AIDS vaccine, should have derailed this scandalous $20 million handout, which made a mockery of the peer-review process and undermined the NIH's authority. Yet the earmarking went forward, and ended up distracting the field for more than a year with endless blue-ribbon panel hearings, investigations, and protests that finally required intervention from no less than the White House.

CHAPTER 8

All MicroGeneSys's Men

In early September 1992, a staffer in the U.S. Senate who worked on the Appropriations Subcommittee on Defense phoned Army Colonel Donald Burke and asked whether he would like 20 million extra dollars to stage an efficacy trial of the MicroGeneSys therapeutic AIDS vaccine, which hoped to treat already infected people. "No, no, no," said Burke, who headed the AIDS research program for the U.S. military. "Line in the sand, no fucking way, no."[1]

Behind closed doors on September 16, 1992, U.S. senators and staff who attended a meeting of that subcommittee wrote a paragraph into a report that, in time, would lay bare how the AIDS vaccine field suffered from disorganization, fractiousness, sleazy politics, sloppy science, a shaky marketplace, greed, unbridled ambition, and leaders with shockingly limited powers. As part of the $253 billion Defense Department budget for the next year, recommended the subcommittee, $20 million should go to the Army "to begin a large-scale phase III clinical investigation of the drug GP-160."

The "drug" they were referring to was none other than the MicroGeneSys vaccine, dubbed VaxSyn, which researchers had begun testing as both a traditional preventive for the uninfected and as a therapeutic for already infected people. The Walter Reed Army Institute of Research spearheaded the studies of VaxSyn as a therapeutic, and then had a 600-person, phase II trial underway. The 10 senators who attended the subcommittee meeting apparently saw no reason to listen to Burke or to wait for data from the phase II trial to evaluate whether the vaccine merited a phase III trial. And although they did put what amounted to a safety check into the language—a stipulation that the secretary of defense, the director of the National Institutes of Health, and the commissioner of the Food and Drug Administration "must approve the investigation protocols, based on an evaluation of the preliminary evidence of safety and effectiveness of the drug"—they apparently had few qualms about steering around the peer-review mechanism and telling scientists which treatments appeared most deserving of public funds.

The curious subcommittee report went so far as to suggest a private group to operate the trial. "The Committee understands that the developer of GP-160 has proposed the Greater New York Hospital Foundation as the administrator of this test program," the report stated. This would end up explaining much about the genesis of the recommended appropriation.

The Greater New York Hospital Association, the parent of the foundation named in the legislation, represented more than 100 hospitals and nursing homes in New York City. It had nothing whatsoever to do with AIDS vaccine testing. But the association had hired two high-powered lobbyists who had a great interest in the field and, in particular, in MicroGeneSys.

One of the association's lobbyists was John O'Shaughnessy, who from 1983 to 1986 served in the Reagan administration as assistant secretary for management and budget in the Department of Health and Human Services, the NIH's overseer. O'Shaughnessy had moved on to become a vice president of the Greater New York Hospital Association, which he left in 1988. That same year, MicroGeneSys entertained the idea of going public and filed a prospectus with the Securities and Exchange Commission that listed O'Shaughnessy as a member of its advisory council.[2] Records at the U.S. House of Representatives show that in July 1992, O'Shaughnessy registered both himself and the firm he was then president of, Strategic Management Associates Inc., to officially lobby for MicroGeneSys. The declared aim: "Assist MicroGeneSys in understanding federal activities in the area of AIDS research and in arranging meetings with federal officials on the subject." On top of these links, a brochure from Strategic Management Associates listed O'Shaughnessy's brother, Michael, as an associate. Michael O'Shaughnessy had by then established himself as a heavy-hitting AIDS researcher in Vancouver—he was director of the British Columbia Centre for Excellence for HIV/AIDS—and had previously been involved with primate and human studies of VaxSyn.

The Greater New York Hospital Association's other high-powered lobbyist packed even more punch than O'Shaughnessy: Russell Long. A former U.S. senator from Louisiana who served from 1948 to 1986, Long was the son of the infamous Louisiana governor, U.S. senator, and presidential hopeful Huey Long, the "Kingfish," whose heavy-handed exploits Robert Penn Warren fictionalized in *All the King's Men*. Huey's son Russell, first elected to the Senate the day before his 30th birthday, was, in the words of journalist A. J. Liebling, "a toned-down, atypical King . . . a Samson with a store haircut."[3] But when he left the Senate, he had become, as the *Washington Post* put it, one of its "most influential veterans."[4] He soon took advantage of this influence, lobbying his former colleagues. In addition to the Greater New York Hospital Association, since April 1991, Long, through his law firm, was a registered lobbyist for MicroGeneSys.

If anyone had doubts about how the $20 million appropriation for MicroGeneSys came to be, Senators Sam Nunn and John Warner clarified the matter on September 18 when they jointly introduced the report language as an amendment to allocate the funds. " . . . Former Senator Russell Long has brought this matter to our attention," said Nunn, a Democrat from Georgia. " . . . Not only did our former colleague and friend bring it to our attention," added Virginia Republican Warner, "but he was present on the floor of the Senate today, as is his right as a former senator, not in the capacity of lobbying, but indeed his presence connoted the importance of this amendment. I wish to commend him personally."

Nunn, famous for sharply criticizing colleagues who earmark Defense Department funds, went to some lengths to justify this earmark. "According to Army medical experts, phase II has shown that the vaccine should go to phase III as soon as possible," said Nunn. "Mr. President, in the case of this kind of research, time literally means saving the lives of people now sick with this deadly disease. The sooner we get these tests completed, the better off so many people in our Nation will be. Mr. President, I urge adoption of this amendment."

The amendment passed without objection.

Four days later, Senator Daniel Inouye, the Hawaiian Democrat who chaired the subcommittee that made the appropriation, introduced an amendment to the original amendment. The modification, which Inouye said he introduced for Senator Bennett Johnston—like Long, a Democrat from Louisiana—correctly changed the language "gp160 drug" to "gp160 vaccine." But, more importantly, the new language relaxed the safety check provision that required taking the idea to the defense secretary, the NIH director, and the FDA commissioner and having them each bless it. Now the trial could only be stopped if one of the troika found out about it and, in writing and within six months, explained why it should not proceed.

This amendment passed without objection too.

But within the scientific and AIDS activist communities, the objections would soon build into a roar of outrage directed at everyone affiliated with MicroGeneSys and the $20 million boondoggle.

• • •

WORD OF THE APPROPRIATION first spread after an article appeared in the September 30, 1992, *BioWorld Today*, a newsletter for the biotechnology industry. The short item quoted two critics of the legislation, both representatives of biotech companies that themselves had AIDS vaccine R&D programs underway.[5] The next day, I phoned Anthony Fauci for his opinion. "I can't comment," he said.

Fauci's initial cautiousness illustrated how his job often required diplomacy rather than leadership. His strategy: better to have academics fire the salvos. And that they did. "This is the most ridiculous thing I've ever heard of," Ron Desrosiers told me. "There are a lot of products that deserve to be tested, and to sneak one through is ridiculous." John Moore, a sharp-tongued biochemist at the Aaron Diamond AIDS Research Center who would make attacking this appropriation a part-time job, had an even more caustic reaction. "I think the whole thing stinks," said Moore. "They're abusing the goodwill of individuals who are willing to volunteer for these trials." Moore's boss, David Ho, called the legislation "politically pretty screwy, morally pretty corrupt, and scientifically slippery." Barry Bloom, a preeminent vaccine researcher at the Albert Einstein College of Medicine, saw the $20 million as nothing short of corruption. "This kind of a rip-off going on in the defense budget is just outrageous," said Bloom. "The one credibility we should have in science is the process." AIDS researcher William Haseltine of the Dana-Farber Cancer Institute slammed the appropriation as a bad precedent as "it opens the door for a tremendous amount of lobbying abuses in the system."[6]

But Fauci's tried-and-true strategy of letting outsiders fight the good fight failed him this time, because none other than Bernadine Healy, the NIH director, railed against the appropriation. "It's really scary," Healy told me, when I first spoke with her about the appropriation on October 2. "If we're going to have legislators determining what drugs we test in people, I think that, as physicians and scientists, we're potentially facing as large a moral dilemma as we have ever faced in medical science." Although many scientists had accused Healy, a cardiologist, of championing causes to aggrandize her own shaky status within the basic research-oriented NIH community, her outspokenness on this issue juxtaposed against Fauci's silence undermined his stature as chief of the AIDS vaccine search. (I phoned him again after speaking with Healy. "Go with the no comment," he said.)

The widespread anger over the appropriation created many strange bedfellows. One of Healy's arch enemies was Representative John Dingell, the Michigan Democrat who chaired a powerful subcommittee that oversaw the NIH and other government science agencies (and conducted a highly charged investigation of the Gallo lab's discovery of HIV). Yet when the $20 million MicroGeneSys appropriation first surfaced, Dingell joined Healy as one of its loudest critics. "I believe this Senate amendment is deeply flawed," Dingell wrote in a letter to Representative John Murtha. At that point, the House and Senate were about to hold a conference committee to work out the differences in their defense appropriations bills. As part of that process, they would discuss the $20 million, which then only appeared in the Senate language. Murtha, a Pennsylvania Democrat, chaired the House side of the conference committee. "I am suggesting that these determinations should be left to scientific peer review and not made by the Congress in this matter," Dingell implored. " . . . None of us is equipped to make this decision."[7]

Even AIDS activists, who constantly berated the government for not spending enough on AIDS, wrote to Murtha urging him to delete the language. "Judgement [*sic*] as to the best therapeutic vaccine(s) for expanded study must be left to scientists at Walter Reed, NIH and elsewhere based on ever-changing latest scientific literature," wrote Mark Harrington of the Treatment Action Group. "Earmarking this sum for one manufacturer's product in advance of the results of ongoing phase II studies might result in government funds being spent on suboptimal research amidst this terrible crisis."

But as the din grew, the bill tilted to make it even more likely that the trial would happen. When the final bill emerged on October 5, it stated that the trial would go forward unless the *entire* troika agreed that it should be blocked and explained why in writing. What's more, if they blocked it, the money would roll into DoD's $50 million AIDS research budget, boosting it by 40%.

Louisiana's Senator Johnston gave a lengthy speech on the Senate floor that day touting VaxSyn and the great promise it held. Following treatment with VaxSyn, said Johnston, "CD4 counts, which are a measure of the functioning of the immune system, have been shown to be stabilized rather than declining, and the amount of virus in recipients has been stabilized rather than increasing." The senator, making claims that researchers had yet to validate and indeed would soon prove overblown, sounded like a MicroGeneSys lobbyist himself. "In dealing with a lethal disease which is claiming lives in epidemic numbers, delay, even delay motivated out of the desire for greater scientific certainty, means that more lives may be lost," said Johnston. "If the gp160 vaccine is not tested now and later proves to be effective or even partially effective, another group of HIV infected patients will have been sacrificed. It may be a gamble, but it is a reasonable gamble we cannot refuse to take." Johnston completed his heart-stirring speech with a request to enter into the record "a summary of the results of major tests of this gp160 vaccine." This "summary" is a verbatim copy of a press release that MicroGeneSys issued at the previous summer's international AIDS conference in Amsterdam.[8]

On October 6, the legislation passed both the House and Senate without a whisper of dissent from the floor.

MicroGeneSys quickly issued a press release in which CEO Franklin Volvovitz said he was "delighted," intoning that "AIDS is a devastating disease affecting the lives of hundreds of thousands of Americans and millions of others around the world." The $20 million, concluded Volvovitz, "will enable the Phase III research to proceed with the urgency that is warranted."

FDA Commissioner David Kessler publicly lambasted the appropriation. "I have grave reservations about Congress mandating the testing of specific products," Kessler told me. And Kessler said he had discussed the issue with Healy in detail. "We will speak with one voice on this." They decided to convene a blue-ribbon panel to shine a light on the appropriation and, if they could, move the money into a more worthy research project.

• • •

THE MICROGENESYS LOBBYING BLITZ actually began in the spring of 1991 to promote efficacy trials of VaxSyn as a preventive vaccine. Deconstructing how the company's campaign evolved into what became the $20 million appropriation to stage efficacy trials of the vaccine as a therapeutic offers an unusually detailed look at how the lobbying machine works. More importantly, a close look at the process reveals many somber realities about the AIDS vaccine marketplace, the strained marriage between science and politics, and the disconnect between the research programs run by the U.S. military and the U.S. National Institutes of Health.

Russell Long's efforts first came into public view on March 14, 1991, at a hearing of the appropriations subcommittee that funds the NIH. At the time, MicroGeneSys badly wanted to launch an efficacy trial of VaxSyn as a preventive vaccine. In a confusing exchange that led nowhere and dramatically underscored the lunacy of legislators micromanaging science, the subcommittee's chair, Tom Harkin, directly questioned Fauci about this. "I was contacted recently about the AIDS vaccine which has been developed by MicroGeneSys," said Harkin. "I understand the drug is ready to move into phase II clinical trials to determine whether or not the drug is effective."

Rather than correcting Harkin's two obvious misunderstandings—a preventive is not a drug, and an efficacy test is a phase III trial—Fauci smoothly replied that a group of experts recently had advised the NIH about whether to stage efficacy trials. "And the overwhelming recommendation of the group was to establish efficacy after a challenge in an animal model before we went ahead right now and did a large-scale efficacy trial on any vaccine," said Fauci. He did not mention that VaxSyn had yet to prove its worth in a monkey or chimp challenge study. Nor did he mention that a recently published study of the vaccine in humans showed that it triggered decidedly underwhelming immune responses.[9] Instead, he offered the politically palatable answer that the vaccine had proven safe in human studies and "certainly would be one of the candidates that we would look very favorably upon when we get to the stage of looking at an efficacy trial."

Harkin pressed the point, asking whether the vaccine would have to pass a challenge study before entering efficacy trials in humans, to which Fauci answered yes. Then Harkin showed his hand, asking why, if the vaccine had proven safe in humans, they did not simply move it into an efficacy trial.

"We know it's safe, fine, but there is no indication given the complexity of the immune response to HIV to think that it would be effective at this point," answered Fauci. He then emphasized that if you staged efficacy trials in the United States with every candidate AIDS vaccine, you might run out of qualified volunteers for the tests. To determine whether a vaccine works, a trial must recruit people who are at a high risk of becoming infected. In the United States, contended Fauci, there was a limited number of such high-risk populations. "I know that MicroGeneSys is now negotiating with doing an efficacy trial in foreign countries in which the rate of new infection is extraordinary," Fauci said.[10]

Fauci, constrained by the fact that he was addressing his funders, had dodged the real issue, demonstrating again that the putative leader of the AIDS vaccine search did not have the freedom to truly lead. Rather than publicly taking on Harkin and warning him about the dangers of the subcommittee's pushing a specific vaccine into efficacy trials, Fauci offered a technical argument—and, as would soon become apparent, it was an argument he did not even believe.

At any rate, Fauci's maneuvering failed, as Harkin was clearly baffled. "I don't know that I understand that completely," Harkin said. "I am told—and again, this is not of my knowledge—that almost all of the vaccines currently in use were not developed using animal models and that some vaccines which looked extremely promising in animals did not achieve significant results in humans. Now you are saying that the animal tests should come before human efficacy tests even when the vaccine is proven to be safe."

Once again, Fauci attempted to explain, but Harkin interrupted. "I guess I just don't understand," he said. Fauci tried yet again.

"I may have to get some more information on this," said Harkin. "I think I understand it. It's just my limited scientific knowledge is what is preventing me from fully comprehending it."

Daniel Inouye, a member of Harkin's subcommittee and chair of the defense appropriations subcommittee that later would introduce the $20 million appropriation, quickly sent Fauci a letter about efficacy trials of AIDS vaccines. "As we are both aware, AIDS is our nation's number one public health hazard and affects the lives of many innocent people," wrote Inouye in his April 16, 1991, letter. "Accordingly, your every consideration of progressing as rapidly as possible in pursuing the development of a vaccine [*sic*], would be deeply appreciated." Fauci replied with much the same information he gave Harkin, adding the significant detail that staging efficacy trials is "extremely costly."

Inouye followed up on May 20 with 10 specific questions that, to Fauci at least, seemed geared toward promoting efficacy trials of the MicroGeneSys vaccine. The tortuously worded questions boiled down to five main queries. First, was there truly a shortage of high-risk populations in the United States that would be suitable for efficacy trials? Second, rather than waiting for the best potential candidate to stage an efficacy trial in the United States, should the government instead finance these tests in foreign countries with vaccines that had already proven safe and able to stimulate an immune response? Third, how much would an efficacy trial cost? Fourth, might challenge studies in chimpanzees and monkeys have little meaning for humans and therefore be misleading? Finally, when would efficacy trials be launched and completed?

That same day, Fauci complained about the MicroGeneSys lobbying campaign at a meeting of the National Advisory Allergy and Infectious

Diseases Council, a group of outside investigators who direct NIAID. (Hereafter, for the sake of simplicity, this group will be referred to as NIAID's advisory council.) It was classic Fauci diplomacy, displaying both the strength and the weakness of his leadership: no naming names, no offering of his opinion. "It's no secret because it's public knowledge, we are coming under extraordinary pressure from the Congress and others about going into efficacy trials with a given vaccine preparation," said Fauci. "I was asked a specific question at the Senate appropriations hearings as to why we are not going into phase II, phase III trials on candidate X. And since then, I have received at least two and probably more letters from members of the Congress, Senate, House, or what have you, asking me for the reasons why we're not doing this particular trial on this particular candidate.

"So we're under some very difficult pressure. Obviously we want to fulfill and will, the mandates that we just spoke of, of making sure that we're founded in the right science. But perhaps I think it's important for you to know the kinds of pressure we're under. And maybe you could even make some suggestions of how we can circumvent that pressure."

Illustrating the field's lack of resolve and direction, advisory council member William Haseltine, the Dana-Farber researcher who later would blast the MicroGeneSys lobbying effort, responded that given the urgency of the epidemic, they should stage efficacy trials as soon as possible. "I agree with the people who are applying the pressure," said Haseltine. "Because the seriousness of the problem is so great . . . we have to go ahead with something."

Fauci recounted that experts had advised NIAID not to stage efficacy trials until a candidate had worked in an animal model, and that the group had serious concerns about using up high-risk populations in the United States.[11] "I don't put much stock in that, because I don't think we're going to run out of those populations," Fauci said.[12] Still, Fauci stressed that he wasn't here to offer his point of view. " . . . There are a lot of opinions going on and a lot of pressures both from scientific colleagues on the one end of the spectrum, who are very puristic and say you shouldn't do anything until you've proven efficacy in the animals, all the way to perhaps a public, semipolitical urging and pressure that you should definitely go into a phase II–III with this particular product," said Fauci.

The next day, May 21, Fauci received a letter from Senator Mark Hatfield, a Republican from Oregon who also sat on the appropriations subcommittee that oversaw the NIH budget. "It is my understanding that there are several pharmaceutical companies currently developing AIDS vaccines," wrote Hatfield. "However, no clinical trials to determine efficacy in humans have been conducted to this date. In an effort to facilitate development, suggestions have been made to the Subcommittee that federal funding be made available for the commencement of such trials immediately."

The spring 1991 lobbying effort had long tentacles. The Food and Drug Administration on May 6, 1991, received a letter from Democratic Senator Quentin Burdick of North Dakota, yet another member of the appropriations committee chaired by Harkin. Questions in the letter, addressed to the head of the FDA division that regulates vaccines, closely resembled—and, in places, were identical to—the questions Inouye had sent Fauci.[13] The Executive Branch registered its interest in the subject, too, with a letter from Jack Chow of the White House Science Office to NIAID's Division of AIDS.[14]

Fauci, who first spoke candidly with me about the MicroGeneSys lobbying campaign one week after Congress passed the $20 million appropriation, fought the onslaught immediately through back channels, calling Senate staffers and even meeting with Hatfield. "[Fauci] said, 'if you want to be helpful, don't tell us which vaccine to promote,' " recalled Bill Calder, Hatfield's press aide.

"I was against this from day one," Fauci said. "I couldn't believe it." He chose not to attack the company publicly, though, as it might reflect poorly on the Walter Reed Army Institute of Research, which had a stellar history of vaccine development. "I didn't want to get into one agency criticizing another."

Fauci said it was "openly known" that Russell Long triggered the questions from the senators. "They never specifically mentioned gp160, but it was so patently obvious what was going on."

Hatfield heard Fauci's initial objections, and that summer the senator offered an amendment to add $6 million to the NIAID's AIDS research budget which, in his words on the Senate floor, would "assist the institute in developing the capacity to test candidate vaccines in clinical trials in the United States or abroad."[15] He made no mention, notably, of a specific vaccine.

The MicroGeneSys lobbying campaign, it seemed, had won the company little ground. But the prestigious *New England Journal of Medicine* gave MicroGeneSys a reward that money cannot buy: scientific credibility. In the June 13, 1991, issue, the *NEJM*, which publishes only original research reports that survive their peer-review process, ran an eight-page article by Army researchers about their tests of VaxSyn as a therapeutic vaccine. In 19 of 30 HIV-infected people who received the vaccine, wrote Walter Reed's Robert Redfield Jr. and his coauthors, the vaccine "augmented" virus-specific antibody and cell-mediated immune responses. Although the researchers stressed that this phase I trial was not designed to assess efficacy, they noted that these 19 "responders" had stable CD4 counts during the 10 months of the study, while the CD4s in the 11 "nonresponders" decreased by 7.3%. Since most researchers then considered CD4 counts the best indicator of an HIV-infected person's health, the Army investigators wrote that these preliminary results were "encouraging." In an accompanying editorial, none other than Anthony Fauci declared the finding to be "interesting and potentially important."

For the first time, the establishment had accepted that the idea promoted by Jonas Salk in 1987 to vaccinate already infected people might actually do some good. *ABC Nightly News*, the *Washington Post*, and other prominent media played up the story. VaxSyn suddenly had momentum. The news also did not hurt the Immune Response Corporation, whose stock price, a measly $2\frac{7}{8}$ at the start of the year, climbed steadily all summer and broke $40 a share by September.

MicroGeneSys, which still had not gone public and anxiously wanted a product to move closer to market, on October 15, 1991, asked the FDA to reconsider the endpoints of the ongoing trial being run by the Army to assess VaxSyn's worth as a treatment. The Army had planned for the trial, which had enrolled 130 patients and hoped to recruit 470 more, to last three to five years. Now the company wanted to know whether the FDA would grant them a license to market the vaccine if, rather than proving that the vaccine reduced the incidence of disease and death, studies showed that it could slow the decline of CD4 cells—information that they could gather as early as nine months into the trial. This very "surrogate marker" of disease progression had been the basis of the FDA's decision on October 9 to grant Bristol-Myers Squibb a marketing license for the anti-HIV drug ddI (didanosine).

The danger of having the FDA approving drugs based on surrogate markers like CD4 counts is that the marker might be misleading: treatments can improve markers without delaying disease or improving survival rates. And the FDA had black eyes from the anti-arrhythmia drugs flecainide and ecainide, both of which suppressed irregular heartbeats and thus won approval to prevent heart disease. A subsequent placebo-controlled study that assessed survival found that people on these drugs died at nearly three times the rate of people taking the placebo.[16]

On November 12, 1991, the FDA held an unusual joint meeting of two of its advisory committees, one that specializes in vaccines and the other in antiviral drugs, to wrestle with the question of whether changes in CD4 counts could lead to approval of a therapeutic HIV vaccine. In addition to presentations from MicroGeneSys, the committees heard from scientists at Immune Response, whose stock by then had skyrocketed to $61\frac{3}{4}$, and Genentech, which had just started a trial of its gp120 vaccine as a therapeutic. "This is a very important meeting," Volvovitz, the fourth representative of his company to present at the meeting, told the committees, winding up for a trademark, inflated speech. "What you recommend here today has the potential, if FDA agrees, to influence the future course of AIDS research and therapy. We all understand the enormous impact that AIDS is having on our society. We all have heard the cry for help. Whether VaxSyn proves useful against AIDS is not the issue; the issue is whether there is some valid measure for assessing the efficacy of products like VaxSyn without a study with a very large and unmanageable sample size or that will take many, many years to complete."

The committee members had deep misgivings about granting approval based on CD4 counts alone. The ddI approval, stressed several members, had the advantage of a "biologically plausible" mechanism: the drug shuts down production of HIV, which allows the immune system to recover. With therapeutic vaccines, the mechanism of action—boosting the immune system—was fuzzier. The ddI approval also relied heavily on the fact that a similar drug, AZT, already had shown that it could impact disease progression *and* CD4 counts, offering some assurance that this surrogate marker with anti-HIV drugs had validity. Although the committee did like IRC's idea of looking at both CD4 changes and the impact that a therapeutic AIDS vaccine had on a person's viral load, they unanimously agreed that they would not rely on CD4 counts alone to grant licensure. As the committee's chair put it, "there is no good evidence to show that if you prop up or stabilize CD4 counts, especially by immunologic means, that it will be beneficial to patients."[17]

Stock analysts sitting in the meeting room whispered the news into their cell phones. Wall Street was rocked, with biotech stocks plummeting right and left.[18] A financial show on CNBC called it "a bloodbath." IRC, which had hoped to ask the FDA as early as 1993 for licensure based on surrogate marker data, saw its stock plunge $15\frac{3}{4}$ points the next day. By the end of the week, the company's stock had lost 44% of its value and was trading at $34\frac{3}{4}$. That same day, November 15, the stock market took its biggest plunge in two years, which several analysts traced back to this seemingly parochial FDA meeting.[19]

• • •

THE FDA ADVISORY MEETING dashed whatever hopes Franklin Volvovitz had for speeding his therapeutic vaccine's journey to market—which, in turn, would potentially make MicroGeneSys as attractive to Wall Street as IRC had been that year. "Volvovitz, he just went crazy," recalled Ed Tramont, who, as the former head of the U.S. military's AIDS research program, had helped set up the collaboration between the Army and MicroGeneSys. "He said, 'This isn't fair. Why are we being subjected to different rules than for ddI?' "

Tramont, who then was developing the Medical Biotechnology Center for the University of Maryland and was running his own therapeutic trial of VaxSyn, had much sympathy for Volvovitz's point of view. Although the NIH had funded the initial trials of VaxSyn as a preventive, the institution—and most academics—clearly favored the gp120 vaccines under development at Chiron and Genentech. Researchers based this preference on the scientific rationale that MicroGeneSys engineered its envelope protein in insect cells, and the resultant HIV protein compared poorly to the "native" one made in mammalian cells, which Genentech used from the start and Chiron had adopted. On top of that bias, NIH researchers and the network of academic institutions they supported to test experimental treatments had done only limited clinical studies of VaxSyn as a therapeutic.[20] "Frank felt strongly that he wasn't being treated fairly by the process."

In the winter of 1992, Volvovitz called Tramont and asked that he join him and Russell Long for a meeting with Senator Orrin Hatch, the Utah Republican and ranking minority member of the committee that oversaw the NIH. Tramont, who had much experience on Capitol Hill, at the outset gladly showed his support for the vaccine studies. After Hatch and Long chatted a bit, Volvovitz made an impassioned pitch. No longer was Volvovitz seeking congressional help to stage efficacy trials of VaxSyn as a preventive. Now he wanted aid to launch a large-scale trial of the vaccine as a therapeutic. With no hope of winning FDA approval based on CD4 counts alone, Volvovitz recognized that he had to stage the large, long—and expensive—trial that he had decried at the FDA meeting. But, as he explained to Hatch, his company could not afford it.

When they left the meeting, Tramont gave Volvovitz some advice. "I told him, 'Frank, this is not the way to do it,' " recalled Tramont. " 'Let everything stand on its scientific merits. That's the only ground that's level. Let it be debated in a scientific way.' He said he had tried that and never felt that he was given a fair shake or that a real debate was allowed to take place." Tramont did not join Volvovitz on lobbying visits again.

During the next several months, MicroGeneSys shifted its lobbying into high gear. Long handled Capitol Hill. In addition to Hatfield, Hatch, Nunn, and Warner, Long met with the staffs of Senators Edward Kennedy, Christopher Dodd, Joseph Lieberman, and Bennett Johnston.[21]

Long also made a trip that winter to the NIH to lobby Bernadine Healy about VaxSyn. "I thought it was a peculiar meeting," said Healy, who did not even know that Long had left the Senate and was confused to have him pressing her about an AIDS vaccine. Soon afterward, a senator whose name Healy could not recall invited her and Fauci to give a confidential briefing on AIDS vaccine research. Then she learned that the senator had invited Long too. "I was outraged," said Healy. "I made a fuss and said I didn't want to go. It was totally inappropriate." She skipped the meeting.

(Russell Long refused my repeated attempts to speak with him. "Long Law Firm does not discuss its clients' representations with the public," I was finally told by a partner in his office.[22])

Russell Long was not the only one talking up VaxSyn in Washington. On February 24, 1992, the therapeutic promise of VaxSyn received conspicuous attention at a House hearing held by the Subcommittee on Health and the Environment on "AIDS research opportunities." The people here advocating more aggressive testing of the vaccine were not registered lobbyists, though. They were Donald Burke, the colonel who took over from Tramont as head of the U.S. military's AIDS research program; Lieutenant Colonel Robert Redfield, the principal investigator of the therapeutic VaxSyn studies for the military; and W. Shepherd Smith Jr., president of Americans for a Sound AIDS/HIV Policy, a nonprofit that strongly supported the military's AIDS research program and had Redfield as its chairman and Burke on its executive committee.

Burke, who had the look of a senator himself with a prominent profile and a neatly cropped head of silver hair, first overviewed data from their studies and stressed that vaccine therapy, although greeted skeptically by many AIDS scientists, had been a popular research avenue for more than a century.[23] Burke carefully stressed that their studies had yet to answer whether VaxSyn worked.

Subcommittee chair Henry Waxman, a California Democrat, then asked Burke whether the military had the money to stage an efficacy trial. Burke explained that the military had a trial in several hundred people under way that they hoped would yield efficacy data within "a year or two," but that they did not have money for needed follow-up studies, which he suggested would cost as much as $20 million.

Redfield, a charming doctor with a soothing voice whose habit of fluttering his eyelids when he spoke gave him an affected, professorial air, next fielded a few questions. Representative Gerry Studds, the Democrat from Massachusetts, noted that the Army AIDS researchers sounded "very optimistic" about VaxSyn. "[T]his is, to put it mildly, a very exciting thing that you may develop," said Studds. "I don't want to get too excited too quickly, but I believe you are within a year of establishing whether or not a therapeutic vaccine can slow down or stop disease in the early stage, correct?"

"We think we are within 12 to 19 months, yes," said Redfield.

Shepherd Smith spoke next. Like his friend Bob Redfield, Smith enjoyed close ties to officials in the Reagan and Bush administrations and publicly espoused Christian values—he and his wife coauthored the 1990 book *Christians in the Age of AIDS*, to which Redfield wrote the forward.[24] But Smith, whose father was an M.D. in the military, was neither a scientist nor a doctor. Rather, as he wrote in his book, he had gone from wondering whether AIDS was "God's judgment" of gays to seeing it as "an issue crying for the involvement of Christians," who could promote responsible behavior for the uninfected and compassion for the affected.

Redfield led the cheerleading section for vaccine therapy, and Smith headed the marching band, encouraging the trumpeters to blare and the drummers to pound louder and louder. "Undoubtedly, any military witnesses today will understate the significant advancements being made in regard to vaccine therapy at Walter Reed Army Institute of Research," Smith told the subcommittee. He then lambasted the decision made by the FDA advisory committees the preceding November. "Such misunderstandings must yield to the recognition that we are on the brink of a significant breakthrough in respect to HIV therapy," said Smith.

The military's collaboration with MicroGeneSys led many, mistakenly, to call VaxSyn the "Army's vaccine." Redfield furthered this perception when, with Surgeon General Antonia Novello and Assistant Secretary of Health James Mason, he had a series of meetings that summer with representatives of key agencies, including the FDA, the NIH, and the CDC, to advance the idea of a clinical trial of an AIDS vaccine in pregnant, HIV-infected women. That proposed trial aimed to both treat the women and prevent transmission of HIV to their fetuses. Since the NIAID had such trials already in the pipeline, the urgency of this campaign baffled some of the researchers dragooned into the meetings. Redfield insisted that although he believed VaxSyn was the only HIV vaccine that had been studied thoroughly enough to be given to pregnant women, he was not shilling for the company. "What I've really tried to do in those meetings, from the beginning, is force the debate," said Redfield shortly after Congress made the $20 million appropriation. "That's my position and still is. I don't have any additional preference for one product or another."

With the Army association and Long's clout in Congress, MicroGeneSys had assembled all the elements needed for a winning strategy. Only one was lacking: access to the executive branch. That's where John O'Shaughnessy, president of Strategic Management Associates, fit in. The former assistant secretary at HHS during the Reagan administration not only knew his way around the executive branch; O'Shaughnessy also had help from Donald Clarey, a vice president at his company who himself had served as a special assistant for cabinet affairs to President Reagan. "You let someone of Senator's Long ability and experience handle Congress, which he's the expert at," explained Clarey. Strategic Management focused on the budget office at the Department of Defense. "Basically, the Defense Department in a case like this—$25 million is about their cut-off for doing things," said Clarey. "But the budget people had been fully briefed as to what was going on. Their attitude was, if the money was allocated, they'd entertain the possibility of spending it."

Canadian AIDS researcher Michael O'Shaughnessy, John's brother and a listed member of Strategic Management Associates, said he had no idea about the firm's lobbying campaign for MicroGeneSys until I phoned him. "I've never done anything with my brother's company with respect to MicroGeneSys," said O'Shaughnessy, who stressed that he, too, was "bothered" by the appropriation. "I've never received a nickel from that company."

Despite the widespread disdain that scientists had for lobbyists influencing the testing of experimental medicines in humans, the MicroGeneSys plan worked beautifully. Clarey took pride in his firm's accomplishment, and he asserted that the researchers who griped the loudest about the appropriation did not truly care whether Congress had subverted the peer-review process. "You know why it rankled 'em?" Clarey suggested. "Because they didn't get the $20 million, if you want to know the truth. I used to work in the White House and used to referee that kind of stuff all the time."

Volvovitz also staunchly defended the lobbying campaign. "We actually have been providing information about that product for quite a long time," he told me. "I think what we're seeing here is an extension of that activity, to provide on a broad basis available information about the product to educate as wide an audience as possible in terms of the clinical results that have been achieved with gp160."

But the scrutiny directed at Congress, the lobbyists, the military, and MicroGeneSys would turn the $20 million appropriation into a Pyrrhic victory.

• • •

WHEN SENATOR NUNN FIRST introduced the $20 million legislation, he stated that according to "Army medical experts," data from the phase II trials of VaxSyn had shown that it should move into an efficacy trial "as soon as possible." In response to my written questions, the Department of Defense sent me a fax on October 2 that starkly contradicted Nunn's assertion. "Army feels that it is premature to initiate a large-scale trial until current studies of gp160 are completed and the data analyzed," wrote the DoD. "The Army was not aware of this proposed appropriation until the amendment was introduced and passed."[25]

Many AIDS researchers presumed that Walter Reed's Redfield must have supported the legislation, but he categorically denied the supposition. "Neither I nor any of the Army scientists associated with the HIV program had anything to do with it," Redfield told me. "I'm not a proponent of product-specific legislation." Redfield further explained that he first heard of the push for the efficacy trial from Enrique Mendez Jr., the assistant secretary of defense for health affairs, who called him on the weekend after the legislation had been introduced. "I didn't know anything about it," said Redfield. "I didn't see that it was in the Defense Department's best interest. And I didn't believe that it would actually go through, because I saw no reason for the Defense Department to basically take this money."

The DoD would not let me speak with Mendez but, in another faxed exchange, did acknowledge that a budget officer in Mendez's department had met with Volvovitz, Clarey, O'Shaughnessy, and Russell Long in late spring 1992. The officer, wrote the DoD, "explained that the President's FY93 budget had been submitted and that the Department would not support any increase in the President's request for DoD medical research."

I repeatedly asked Senator Nunn to name the Army medical experts he had mentioned. His staffers gave me dubious leads, all of which led to dead ends.

On October 23, *Science* ran a four-page story I wrote that detailed how the $20 million appropriation came to be. The same day, Nunn issued a statement. "Prior to sponsoring our amendment we contacted medical experts within the Department of Defense who advised us that they supported the amendment because it would continue the Department's AIDS research program at the level of previous years," wrote Nunn. Not only had Nunn avoided naming names; the assertion about the budget was false: funding for DoD AIDS research was $44.4 million in 1992 and $50 million in 1993 (without the extra $20 million). The statement had a bigger whopper in it, though. "I carefully avoided earmarking these funds for any specific AIDS research program," Nunn claimed.

The media now hauled out their flamethrowers. On October 26, the *New York Times* ran a page 1, above-the-fold story, "Scientists Assail Congress on Bill for Money to Test an AIDS Drug: One Vaccine Favored After Big Lobbying Effort." Long provided the paper with a written statement. "I discussed the GP-160 situation with a number of senators and representatives who were convinced of the necessity for continued testing of this drug in an expedited manner," he wrote. "My discussion in this regard were [*sic*] conducted in an entirely ethical manner."

No one questioned Long's ethics. But editorial writers soon began assailing the senators and representatives who listened to him. "The matter is too important to be slipped into a heavily weighted defense bill as if it were a new fire station for the Klamath Falls, Ore., airport," chided an October 28 *Washington Post* editorial. The next day's *New York Times* editorial page weighed in with "Medical Madness on Capitol Hill," which drubbed Congress for the "foolish" appropriation. "Decisions as to which vaccines are worth testing are best left to medical experts, not to a Congress whose judgment is so demonstrably suspect."

Senator Warner on October 30 sent Secretary of Defense Richard Cheney a six-page letter about the appropriation. "With respect to those who question the apparent circumvention of the scientific peer review process in determining the allocation of limited research funds, I would like to clarify misleading information which has been reported in the media," wrote Warner.

Warner claimed that the original amendment he cosponsored with Nunn "carefully avoided earmarking these funds" as it "left to the discretion of the Secretary of Defense the wisest allocation of these funds to ongoing AIDS research programs within DOD." In fact, as Nunn had explicitly stated when he introduced the amendment to boost the defense budget by $20 million, the money "would provide for the third phase of testing for a new vaccine which has shown great promise as a way of delaying the onset of the deadly implications of the AIDS virus." Warner had the audacity to suggest that it was Inouye's subcommittee that actually "elected to earmark" the funds. He further asserted that "DoD medical experts" had "advised us that they supported the amendment," but he, too, did not name names. "Those who question this Congressional action have apparently chosen to ignore the enormous potential benefit of this additional $20 million directed at accelerating completion of testing of the gp160 vaccine," he concluded.

On November 1, the *Washington Post* ran an op-ed that I wrote in which I detailed the myriad financial ties between MicroGeneSys, the Long Law Firm, the Greater New York Hospital Association, and Strategic Management Associates. According to lobbying reports, the Long Law Firm had then received $14,100 from MicroGeneSys. Records showed that the Greater New York Hospital Association had paid the Long Law Firm between $3,000 and $13,000 a month for services between October 1989 and March 1992. Strategic Management had earned at least $96,996 from the association during that period. O'Shaughnessy later reported that MicroGeneSys paid him $32,000 in 1992 and $45,000 in 1993. The Long Law Firm reported earning another $5,125 from the company during that period.[26]

Clarey insisted that the initial inclusion of the Greater New York Hospital Foundation was "innocuous," a simple suggestion made by John O'Shaughnessy to Senate staffers. "And some staff guy writes that down and puts that in, and it's there," said Clarey.

FDA Commissioner Kessler, who once ran a hospital in New York, bristled at the inclusion of the foundation in the legislation. "It's ridiculous," said Kessler. "I know these guys. What do they know about research?"

Kenneth Raske, head of the Greater New York Hospital Association, acknowledged that if the association's foundation ran such a trial, it would benefit by receiving operating costs, but he said it was "only trying to do a good deed." Citing concerns in the research community, the foundation later decided not to pursue this good deed.

The next day, the *Washington Post* ran yet another op-ed, "Pork-Barrel Research," by Jessica Mathews, a molecular biologist with the World Resources Institute who had served as both a House staffer and a National Security Council member. "[W]e now know the sorry story of how ill-informed legislators, too well-connected lobbyists and a huge pot of money combined to set a precedent that subverts scientific due process and is bad for sick people," Mathews wrote. She went on to chastise Nunn, Warner, Inouye, and Long about what she called "the Gp160 fiasco."

The $20 million appropriation was starting to burn people. And the flames were spreading.

• • •

THE U.S. ARMY on October 30 added another dimension to the controversy when it faxed an "information paper" to key members of Congress. "SUBJECT: Allegations of Scientific Misrepresentation of GP160 Vaccine Effectiveness," the fax read.

As the fax cryptically explained, the Army had initiated a "fact finding investigation" to determine whether Robert Redfield had "overstated" the effectiveness of VaxSyn as a therapeutic. The Army took this action because "fellow researchers" questioned Redfield's "statistical analysis and interpretation of data."

The dispute dated back to Redfield's presentation about VaxSyn at the previous summer's international AIDS conference in Amsterdam, in which he purported to show that an AIDS vaccine could impact the amount of HIV in a person's body. In his July 21 talk, Redfield presented data from what he said were the first 15 patients who had received the vaccine during the phase I trial that he had described in the *New England Journal of Medicine* article the year before. Using a new, ultrasensitive version of the polymerase chain reaction test—which can fish small pieces of genetic material from blood samples—Redfield's lab quantified the levels of viral RNA in the blood of those patients. Because that study did not have a control group, they compared the viral loads from these 15 people with "natural history" data they had gathered from untreated patients.

Of the 15 patients, Redfield said, 93% showed decreases or no change in their HIV genetic material. By contrast, 47% of the natural history group had *increases* in viral genetic material. Redfield described the results as statistically "significant"; in this case, the statistical analysis said there was only a 2% chance that the difference between the groups was due to happenstance. "We think this data is exciting, it's provocative, and for the first time demonstrates I think in a good scientific way that these vaccine patients, nonselected, have a reduction in [viral load]," Redfield said at a press conference following his talk. "I personally would challenge the drug development people to show me data where they see a reduction that's maintained for two years as opposed to the natural history."

In an interview on *CBS Evening News*, Redfield stated the findings more plainly still. "The virus goes down," said Redfield. "These are quite striking, significant, real, and reproducible observations."

When Redfield returned from Amsterdam, several of his coworkers confronted him about his presentation. In particular, the scientist who ran the viral load tests, Maryanne Vahey, had deep misgivings that the data were not statistically significant and wondered why Redfield had only presented 15 of the 26 patients she had analyzed before he left. She circulated the raw results to the staff, and on July 31, Redfield met with his group to discuss the data. During the spirited talk, several critical questions surfaced. How had he selected the natural history group? If they had suffered more immunologic damage than the people who received the vaccine, how could you fairly compare the two groups? Was the experimental assay used to assess viral load reliable? Did the data truly show a statistically significant difference between the groups?

Redfield decided to ask Bill McCarthy, the chief statistician at the Henry M. Jackson Foundation for the Advancement of Military Medicine—the private group that did contract work for the Army's AIDS program—to analyze the data from all 26 patients. On August 20, McCarthy wrote a memo that concluded "there is no significance [*sic*] reduction in viral load in the intervention cohort."

Donald Burke, Redfield's boss, asked McCarthy to analyze viral loads only from the first 15 patients. The next day, McCarthy reported that the results still did not reach statistical significance.

On August 24, Shepherd Smith of the Americans for a Sound AIDS/HIV Policy (ASAP) phoned Maryanne Vahey to discuss the data. Vahey, upset by what she viewed as an intrusion from an outsider, fired off a memo to Burke. "He asked if I had any awareness of the millions of dollars at stake in the vaccine and drug company arena," wrote Vahey. "He asked if I was aware of the pressures generated by the 'poor' studies conducted by the NIH (Dr. Fauci) and the problems that have arisen from the lack of attention paid to the ARMY studies by the NIH. . . . He asked if I was aware of the presentations required to be made to Congress and of political pressures in general."

Although Vahey did not believe that Smith was "motivated by maliciousness," she wrote that the call took her "a good deal off guard," especially since Smith knew the data so well. "While I am aware, albeit to a very naive extent, of scientific peer pressure and financial, political and social issues, I do not feel that, in my role as a scientist, such issues must ever be allowed to influence my findings."

Three days later, Burke reviewed Redfield's original calculations used for the Amsterdam presentation and found "no evidence of falsification or intentional misrepresentation."[27] The main reason for the discrepancy was technical. Redfield had used an arbitrary threshold for determining whether a person's viral load had increased or decreased. McCarthy used standard statistical criteria. Redfield also agreed that he had done the analysis hastily before the Amsterdam meeting. They agreed to set the record straight at NIAID's upcoming annual AIDS vaccine meeting.

It seemed that the matter would blow over as an internal squabble. Redfield and Vahey presented their data at the NIAID meeting on September 1 and 2, reporting that the full analysis of the 26 patients did not reveal significant changes in viral load. At the same meeting, Aaron Diamond's John Moore gave a presentation in which he assailed the use of VaxSyn as a preventive. Because of its nonnative structure, said Moore, the gp160 made by MicroGeneSys "could not be a worse choice" for the large-scale efficacy trials in uninfected people then being considered.

Later that month, Smith, under the aegis of ASAP, organized an investor seminar for MicroGeneSys at the luxurious, ocean-view home of a Los Angeles physician. Richard Gere, the model Cindy Crawford (Gere's then wife), and other Hollywood stars attended. After hearing presentations by Volvovitz, Redfield, and Gary Blick, another physician testing the vaccine, the guests were asked to invest. They exploded. "It got very heated," Blick told the *Hartford Courant*, which ran an in-depth, three-part series about Volvovitz that said he had hoped to raise $20 million from investors. "People were pointing at Frank Volvovitz and saying, 'You're the problem.'" Volvovitz told the *Courant* that the seminar was "a total misread."[28]

Congress then appropriated the $20 million, and the critics again went after Redfield and Smith's actions. This time, however, the issue transcended an in-house matter. An Air Force major who ran the HIV program for that branch of the military and collaborated on the VaxSyn trial wrote Burke a memo, cosigned by an Air Force colonel, that said Redfield's Amsterdam presentation was potentially a case of "scientific misconduct." They criticized Smith's interactions with Vahey, which had, their October 21 memo charged, "on the surface, the appearance of a very gross impropriety." The scientific credibility of the military's AIDS research program was at risk, they said. "Severe, painful steps must be taken less [*sic*] we dishonor the honest labors of so many colleagues and patients within our research consortium," they concluded. "We cannot continue to deceive."[29] Three days later, *New Scientist* threw more fuel on the fire when it ran a story based on McCarthy's August 20th memo that concluded the viral load data were not significant.[30]

The allegations swirling around Redfield and Smith left both men aghast. "I stand by all the data that we did," Redfield told me. "We are extremely honest individuals. We are trying to share our data, probably much more openly than maybe some of our colleagues." Smith said he made the call to Vahey strictly out of "curiosity" about the data and that he had no intention of pressuring her. From his perspective, the feud amounted to little more than "professional jealousies and inter-services rivalries." The ordeal, Smith wrote Burke—who two weeks before the Army announced its investigation of Redfield had resigned from ASAP—was "ludicrous and regrettable."[31]

The Army investigation pledged to determine whether the charges were ludicrous or Redfield had committed serious wrongdoing, in which case his punishment could be as severe as a court-martial. But there was an obviously ludicrous aspect to the level of attention that the matter received, which had everything to do with the unchecked zeal of a small biotech firm from Connecticut to make money from an AIDS vaccine—even if it meant harming its allies and forcing the entire field to drop everything and examine the needs of the company.

• • •

WITH AIDS ACTIVISTS WAVING PLACARDS that read "YOU DIDDLE WE DIE," TV cameras rolling, and the rapt attention of a blue-ribbon panel that included NIH director Healy, FDA Commissioner Kessler, Assistant Secretary of Defense Mendez, NIAID Director Fauci, NCI Director Sam Broder, Upjohn CEO Ted Cooper, Aaron Diamond's David Ho, and Nobel laureate Fred Robbins of polio vaccine fame, Frank Volvovitz came unglued. "Please try to provide the data without bias so people can see everything that's there," Volvovitz said, at the end of a lengthy soliloquy that Fauci repeatedly attempted to cut short. "Our election-year politics have certainly established that whatever you say up front sticks."

It was November 5, 1992, and Healy and Kessler, seeking advice about the $20 million efficacy trial of VaxSyn legislated by Congress, had assembled this 32-member blue-ribbon panel, which sat around an enormous oval table in a conference room on the top floor of an NIH building that featured panoramic views of the Bethesda countryside. Healy opened the meeting by taking a few shots at the legislators who passed the appropriation. "From the most cynical point of view, it would appear that the Congress has signed an uninformed-consent form for patients with AIDS," she chided, adding "it's a case of the tail wagging the dog."

Following Healy's introduction, researchers who had tested VaxSyn in humans gave presentations. Larry Corey of the University of Washington gave a particularly damning overview of the vaccine's performance in uninfected people, noting that the antibody responses were "significantly less" than those seen with the Genentech and Chiron vaccines. And it was Corey's presentation that led to Volvovitz's outburst.

Volvovitz's speech particularly incensed Jesse Dobson, a panel member who worked with the San Francisco–based AIDS activist group Project Inform. "I think if anyone is going to talk about presenting all of the data, you have to come under question as well," snapped Dobson, a once muscular man whose face and limbs had withered from AIDS. "I just want to point out that there may be perhaps some bias in the president of the company interpreting his own data."

"There's going to be bias by whoever looks at the data, whether it's a company president or an investigator," Volvovitz replied.

"So why are you wasting our time with this crap?" Dobson shot back.

The most novel information presented to the panel was a statistician's description of what it would take to compare several therapeutic vaccines in a head-to-head trial that had disease and death as endpoints: 30,000 patients to answer the question in two years or 14,000 to answer it in five. Either way, the presentation underscored that answering this question was an ambitious—and expensive—undertaking.

Twelve AIDS activist groups submitted a consensus statement to the panel that emphatically backed staging such a trial. "There are waiting lists for different vaccine trials, hundreds or thousands of people long around the country," said panel member Mark Harrington of Treatment Action Group. "Whether or not their expectations have been raised cruelly by a pharmaceutical industry that wasn't really presenting both sides of the story, or whether or not this is a valid therapeutic approach that's the most promising thing since sliced bread, we should do something about the fact that there's nothing else on the horizon. This is a precious opportunity . . . to really answer the question once and for all with clinical endpoints whether this approach is going to pay off."

From Fauci's perspective, no scientific issue even existed. Normally, scientists would not consider doing a large-scale trial unless researchers had completed a phase II study, such as the one being conducted by the military, and presented positive results. More galling still, the NIH was then organizing a 120-person trial that would compare the therapeutic effects of the vaccines made by MicroGeneSys, Chiron, and Genentech. But now that Congress had anted up $20 million, HIV-infected people saw an opportunity, and scientists had to respond. "It's the need to give people hope," Fauci told me. "You're asking a committee to give more than scientific advice. You're asking a committee to give sociological advice. The scientific question is really very clear."

At the rambling meeting's end, the only conclusion was that the group had to meet again. And next time, they would hear from a conspicuously absent scientist, Robert Redfield, whose travails were detailed in that morning's *New York Times*.

• • •

REDFIELD, WHO TYPICALLY WORE A FEW EXTRA POUNDS around the middle, looked gaunt at the second blue-ribbon panel meeting, held at the NIH on November 23, 1992. His friend Shepherd Smith was worried. "This has been a scurrilous episode," he told me. "It's damaged someone who is honest, straightforward, and well meaning."

Redfield, the featured speaker at the meeting, gave a thorough review of military studies with VaxSyn, noting that they had not yet found a statistically significant impact on viral load. He also repeatedly stressed the value of skepticism in science, and said he had had a change of heart since the FDA meeting in November 1991 that gave a thumbs-down to using surrogate markers as endpoints for trials of therapeutic vaccines. "I believed very strongly that the surrogate endpoints would eventually be accepted," Redfield said. "I think personally, in 1992, I'm not as convinced that we were right. . . . This is a new form of therapy. It has not been evaluated. And the bottom line is whether it makes a clinical difference in prolonging survival and clinical disease in these patients." His own ongoing, 600-person trial would not be able to answer that question until 1997, he said.

Redfield stopped far short of endorsing the trial legislated by Congress, but he made his point clearly: the $20 million could help answer the question more quickly.

Other principal investigators of VaxSyn trials, including researchers who had come in from Canada and Sweden, reviewed their data. All told, hints had come from these studies that the vaccine might have positive effects on the immune system's population of CD4 cells and even reduce or stabilize the amount of HIV in a person. Yet the researchers followed almost every hint of that kind with caveats, noting that the data had no statistical significance, the trial had no placebo control, or the experiment included too much "noise" to arrive at a meaningful conclusion.

At Fauci's urging, the panel went on record that the data did not merit a large-scale trial of VaxSyn, as Congress had legislated. Instead, they unanimously agreed the $20 million would do more good if it went toward a trial that compared several therapeutic vaccines.

Such a trial likely would cost more than $20 million, and Mark Harrington suggested that the industry might chip in some money, an idea that won wide support. Representatives from Chiron, Genentech, and Immuno who were present each said their companies would consider a public-private partnership and would provide vaccine for free. IRC later sent in a letter stating the same thing. Volvovitz remained silent.

Before the meeting's end, the panel decided to appoint a subcommittee to design the trial and work out cost estimates. Healy concluded the proceedings by suggesting that the NIH might even open up its own coffers to help pay for it.

MicroGeneSys had recently hired a new public relations firm, Powell Tate, the collaboration of Jody Powell, the former press secretary for President Jimmy Carter, and Sheila Tate, the former press secretary for Nancy Reagan and campaign secretary for George Bush. That evening Powell Tate attempted some spin control. In a company press release, Volvovitz said he was "delighted" with the panel's decision. The press release contended that the MicroGeneSys vaccine is the "principal drug to be tested" in the trial. "Volvovitz," the release said, "emphasized the importance of caution in the design of clinical protocols for inclusion of other vaccines in the trials." The MicroGeneSys product is "farthest along the development path," the release argued and "unless additional funding is available to enlarge the study endorsed by the NIH advisory panel, inclusion of other vaccines will dilute the power of the study to arrive at a definitive answer." The release concluded with a pat on the back to the NIH. "The scientific process worked," Volvovitz added, "and NIH should be proud of its decision."

The release offended both Healy and Fauci, who made the unusual decision to publicly criticize the company. Healy called the claim that

MicroGeneSys was further along than the others "preposterous." She complimented Redfield's presentation, but said "his data were not very compelling," and added that "no one product seems to deserve preference over another." As for the claim about diluting the power of the study, Fauci said, "one could say that the danger of picking out the wrong product over the others is a greater danger than diluting the study."

Yet both believed the meeting had been an overwhelming success and had come to a sound conclusion. Now, maybe, they could get back to their real jobs. "I personally did nothing but this for the past two and a half weeks," Fauci told me, exasperated.

A *Washington Post* editorial furthered the sense that the scandal had come to a close. The panel's decision, the editorial concluded, "should remind Congress and drug lobbyists that scientists, not legislators, have the primary responsibility for decisions about medical research and public health."[32] On December 2, an advisory committee to the NIH director—which included Donna Shalala, who was about to be named the new head of HHS—gave their stamp of approval, too. Now the NIH director and the FDA commissioner could write Congress, well before the six-month deadline stipulated by the law, and suggest that the $20 million not be spent on a trial of VaxSyn alone. If the secretary of defense concurred, Congress's ill-conceived generosity would end up funding research that the scientific community had chosen.

But the military would soon breathe new life into the controversy over the $20 million, changing what could have been an unfortunate incident into an absurd saga that, as Fauci originally feared, pitted one U.S. government agency against another.

• • •

ALTHOUGH MILITARY BRASS had attended the blue-ribbon panel meetings at the NIH, they kept mum about whether they would endorse its recommendations. On January 24, at the opening of a two-day meeting of the Department of Defense's HIV Vaccine Therapy Advisory Group held across town at the aging Walter Reed Army Medical Center, Army Major General Richard Travis ended the mystery with a shot across the blue-ribbon panel's bow. "There are those who would say this meeting is unnecessary because [the issue] has already been decided by NIH and FDA," Travis told the 20 civilian and military scientists on the panel. "I would say it's been decided in their minds, but it has not been decided by this group."

The military clearly felt uneasy about the role Congress had thrust upon it. As Don Burke explained, the military's $44 million AIDS research program emphasized prevention rather than treatment of HIV infection. "A $20 million appropriation is a big chunk of money for us," noted Burke, stressing that a large-scale vaccine therapy trial could take the military off course.

As for the details of how researchers should conduct a comparative trial, the military advisory group stressed that the blue ribbon panel had not considered which vaccines were ready to be tested for efficacy. Redfield, in particular, wanted to know which companies had enough product available and what data they had accumulated on safety and the ability to stimulate immune responses. This line of argument highlighted that other vaccine manufacturers were, by these standards, behind MicroGeneSys, and thus perhaps did not deserve to be included in an efficacy trial.

After the panel heard presentations from several researchers testing MicroGeneSys's gp160 and from four other vaccine manufacturers, it closed the meeting for a vote on how to proceed with spending the $20 million. As I pieced together from interviews at the time and later confirmed from an official memo, the group voted 10 to 9 that if a trial occurred, it should not be limited to the MicroGeneSys vaccine.[33] But because the vote was so close—and because the military did not want to offend Congress by flat-out rejecting its suggestion—the panelists voted 15 to 4 in favor of a trial of VaxSyn together with any other vaccines that met the minimum requirements, stipulating that "full implementation of all aspects of the trial will depend on adequate resources."

Three days later, back at the NIH, the blue-ribbon panel's trial design subcommittee met. After subcommittee members outlined different possible protocols for a multivaccine trial that would cost between $34.2 million and $53.4 million, Healy asked Burke to recount the results of the DoD's meeting. When it became apparent that he would not reveal the group's conclusions, the fireworks started to fly.

Healy asked how many of the vaccines that the military reviewed met their criteria for inclusion in the trial.

"At least one," Burke quipped, and the room broke into laughter.

Healy was not amused. "That's not proprietary," she said. "More than one?"

"Ideally, we'd like to proceed with several products, if possible, but as pointed out here, the budgetary constraints of proceeding with multiple products is difficult," said Burke. "We too did the cost consideration and were appalled at the notion of trying to do a $50 million study with a $20 million budget."

"Are you saying that you're committed to multiple vaccines?" Healy asked.

"We'd like to have a minimum capability to set up a trial design, and then if other parties wanted to contribute, say the manufacturers, then we don't shut anybody out," said Burke. "We'd hate to shut out the right product."

Healy, bearing down like a prosecutor, asked whether he was suggesting that they *had* to test gp160, and if other companies wanted to contribute, they would consider including them. Burke dodged the question.

According to Healy, "many sources" had told the NIH that the MicroGeneSys lobbying campaign included the goal of forcing the DoD to buy vaccine needed for the trial. Would the military pay the company for its experimental product?

"We haven't addressed that issue," said Burke, noting that they had never paid for other vaccines used in clinical trials.

Panel member Diane Wara of the University of California at San Francisco joined in the attack on MicroGeneSys and its odd relationship with the military. "I'm really appalled frankly . . . that as much as 50% of this money could go to purchase vaccine," said Wara.

"MicroGeneSys has refused to commit, like every other company has, to pay for the vaccine," added Healy. "This is really amazing behavior in clinical trials. . . . How would you set the prices? I think we're going through an exercise in futility if we're worried about $20 million, $30 million, $50 million, and it turns out the Department of Defense comes along and says, 'We're going to recommend a single vaccine and we're going to use half the money to buy the MicroGeneSys, and any other company that wants to join, fine, tell them we'll be charitable, to take a look at this protocol, and they can give their vaccine free.' "

Healy now was in high dudgeon, and she went for Burke's throat. "I really don't think, considering the openness of the challenges that we face, that we should have secretiveness on the part of the Department of Defense," Healy said.

"There's no secretiveness," said Burke. "It's just that it has not gone through our Secretary of Defense, and until it does, it would be inappropriate for me to talk about it."

"He's busy with other things right now," Healy sneered.

"That's exactly right," said Burke.

This, then, is how the U.S. government officials asked by Congress to steer the search for a therapeutic AIDS vaccine coordinated their actions in January 1993. And the leadership would devolve further still.

The next month, MicroGeneSys's reputation took another hit when the company decided on March 2 to pull out of the trial being organized by the NIH to compare in 120 HIV-infected people the effects of VaxSyn versus the vaccines made by Genentech and Chiron. "I'm not surprised," said Robert Schooley, the University of Colorado researcher organizing the trial. "They've managed to get what they want through approaches that don't involve the scientific process." David Ho, whose lab planned to do analyses for the study, decided to end his consulting relationship with Wyeth-Ayerst, MicroGeneSys's corporate partner. "In good conscience, I couldn't consult with Wyeth if they're involved with a company like MicroGeneSys," Ho told me. Ho's acerbic sidekick, John Moore, who had recently published a stinging critique of VaxSyn in *Nature*, saw the company as running scared.[34] "Obviously, they fear a comparison trial," said Moore. "They probably prefer to play with their friends than with people of objectivity."

Volvovitz said the charge that MicroGeneSys feared a comparison was "preposterous."[35]

• • •

IN THE FINAL LEGISLATION THAT CONGRESS PASSED, the NIH director, FDA commissioner, and secretary of defense could call off the gp160 efficacy trial if each one submitted a written objection to the legislators within six months. As a practical matter, then, the defense secretary could unilaterally allow the trial to proceed simply by not doing anything and letting the April 6th deadline pass. On March 29, the DoD's directorate for defense information, in a response to my questioning, gave the first official indication that the defense secretary was marching to his own drummer. "The Department of Defense is not aware of any submission to Congress from either NIH or the FDA that such a trial should not proceed," the directorate disingenuously explained. "Likewise, the Department of Defense does not plan to submit such a statement."

On March 31, Healy wrote her letter to Congress, stating the blue-ribbon panel's conclusion that a multiproduct efficacy trial should proceed. Kessler submitted his letter the next day.[36] Donna Shalala, the new HHS secretary, sent copies of the letters to Defense Secretary Les Aspin on April 2 and urged him to review their recommendations.

By then, as an item I had in that day's *Science* revealed, the DoD had decided that the $20 million would only cover the costs of testing VaxSyn.[37] An Associated Press story on April 3 quoted a Pentagon spokeswoman as saying the Army had in fact made that decision on February 27—three days before MicroGeneSys backed out of the NIH-sponsored comparative trial. The AP quoted Volvovitz as saying "science has prevailed." Healy was apoplectic. As she told the *Hartford Courant*, the Army's move "makes a mockery of all the work that was done over the past few months by so many people."[38]

On April 5, Dani Bolognesi, Larry Corey, Jesse Dobson, Mark Harrington, and David Ho sent Aspin a biting letter that lambasted the DoD for choosing to "utterly ignore" the NIH and FDA recommendation. "Defeating the AIDS epidemic demands that governmental agencies work together, not in isolation," they wrote. "This unilateral decision of the US Army bodes ill for the prospects of future cooperation between Federal agencies involved in HIV research." They also slammed MicroGeneSys for simultaneously rejecting federal funds for a comparative trial run by the NIH, and seeking federal funding for a noncomparative trial. "We can only assume that the reason behind these decisions is that the companies have a strong desire to profit from their product, but not to have it rigorously evaluated," they complained. "Individuals infected with HIV deserve more respect."

April 6 came and went, and the Secretary of Defense did nothing.

Now the White House became involved.[39] The next day, Healy, Kessler, and Admiral Edward Martin, the DoD's acting assistant secretary for health affairs, met at the Pentagon. Near the meeting's end, Martin said to Healy, "Why don't you take the money and do the trial?" "Sold," said Healy. When told of the meeting's results by a *Washington Post* reporter, Volvovitz said the decision to test several vaccines would mean "many years of delay before a drug that has shown great promise can be approved" and would cause "a lot of human suffering that could be avoided."[40]

The next week, another meeting was held at the Pentagon with Fauci, Martin, Volvovitz, Russell Long, Jody Powell, several lawyers, a White House representative, and staffers from the offices of Senators Dodd and Johnston. Fauci made the case for a multicandidate trial, and said the companies would have to supply their vaccines for free. Volvovitz maintained that MicroGeneSys could not afford to donate vaccine.

The saga appeared once again to come to a close on April 14 when the DoD announced that it would transfer the money to HHS, which would then stage a multiproduct trial, with vaccine donated by manufacturers. "We can only view it as an effort to drive us out of the test," MicroGeneSys spokesman Jody Powell told the *Washington Post*.[41]

But MicroGeneSys once again brought the wounded beast, which now had stakes puncturing its heart, back to life. Because the company refused to contribute its vaccine, HHS attorneys decided that the NIH could not run the multiproduct trial as planned, and notified the Army on June 21 that it did not need to transfer the $20 million. "[I]nstead, the Army could proceed with its original plan regarding a gp160 trial," wrote HHS general counsel Harriet Rabb to her counterpart at the DoD.

As the Army and MicroGeneSys haggled over whether the company would provide vaccine for free, the Army announced on August 10 the results of its Robert Redfield investigation. The Army concluded that the evidence "does not support the allegations of scientific misconduct" and that there was "no requirement for adverse action." The report did slap Redfield on the wrist for having a "close relationship" with Shepherd Smith's ASAP, which the investigation found had received scientific information from the Army researchers "to a degree that is inappropriate."

Several of Redfield's collaborators believed the investigation was a whitewash, and charged that his version of events was accepted over that of his critics.[42] But the grievances took a back seat to news from MicroGeneSys that it had decided to donate its vaccine to the Army trial. According to the company's press release, Wyeth-Ayerst had agreed to reimburse MicroGeneSys for the vaccine. The release did not mention that the company only agreed to this donation for the first year of the three-year trial, which MicroGeneSys said would begin before the year's end.[43] Volvovitz stressed that a comparative trial really was just a "smokescreen" anyway, an attempt by the NIH and its academic allies to let Genentech catch up with MicroGeneSys. "There's an incredible amount of incestuousness in this system," Volvovitz told me.

John Moore distributed a barbed letter to several AIDS researchers about the planned trial. "Does the US Senate now appreciate that it was conned into voting funds to test a product with dubious properties by Army scientists of dubious competence?" he asked. "There are some excellent scientists in the military program. However, to quote Lloyd George on the World War I British Army, they are 'Lions led by Donkeys.' I for one would sleep happier in my bed if the US Army stuck to its traditional role of killing people and left the task of saving lives to the NIH, who are competent to do so." AIDS activists from Treatment Action Group also lashed out in a letter sent to the blue-ribbon panel members that floated the idea of pressuring the White House and Congress to redirect the $20 million into traditional, peer-reviewed grants.

Staffers for Representatives Waxman and Dingell on September 17 gathered DoD and NIH representatives as well as the White House's recently appointed AIDS "czarina," Kristine Gebbie, to discuss the $20 million. Afterward, Gebbie wrote Waxman that a "single-vaccine trial is not a useful investment at the present time" and suggested it "would be far wiser to allow an appropriate review process to identify more basic research questions." Ten members of the blue-ribbon panel wrote Senator Inouye on September 27 recommending the same thing.

On September 30, Waxman introduced an amendment that would extend to 18 months the original six-month deadline to call off the trial. The DoD, NIH, and FDA "have concluded that the vaccine trial in last year's Defense bill is not appropriate and they have told my staff and Congressman Dingell's staff that they are prepared to certify that the trial should not proceed," Waxman said on the floor of the House.[44]

That same morning, 14 members of ACT UP, many wearing T-shirts that read MICROGENESYS AIDS EXTORTIONISTS, dropped in at company headquarters in Meriden, Connecticut. When employees refused them entrance, the activists attached a bicycle lock to the front door. "Basically, we wanted to bring the face of AIDS before their eyes and give them a jolt," said one of the protesters, Rick Loftus. "Frank Volvovitz is preying on our despair for a cure, and he's subverting science."

Failing entry, the activists took down the American flag flying in front of the company and hoisted ACT UP's "SILENCE = DEATH" banner. Three activists bicycle-locked their necks to the company's gates. The police soon arrived and, after breaking two pairs of wire cutters in their attempt to remove the bicycle locks from the activists' necks, finally chose to cut the gate apart. The police then arrested the activists and took them to jail. "The cops privately told us they supported us," said Loftus.

Adding to the groundswell of discontent, Representatives Waxman and Dingell sent Defense Secretary Aspin a letter on October 14 that all but called for a full-on investigation of the gp160 legislation and the DoD's role in its genesis. The letter asked Aspin for "all materials" related to gp160 research, including published and unpublished articles, abstracts, correspondence, budget plans, memoranda, notes from meetings, and allegations of misrepresentations of research data."

At the same time, Russell Long started making the rounds again on Capitol Hill, calling on prominent senators like Kennedy. But Kennedy, who had received an unusual letter from Fauci asking him to submit an amendment similar to Waxman's, was staunchly against the $20 million, single-product trial. Kennedy did not want to confront his colleagues on the floor, however, so during the conference committee in November for the appropriations bill, he persuaded colleagues to incorporate this House amendment into the final version. The bill passed.

• • •

THE MICROGENESYS SOAP OPERA ON January 4, 1994, entered what looked yet again like its final episode. That day, the leaders of the NIH, FDA, and DoD wrote Congress that the gp160 trial should not proceed. The DoD's Martin, writing for the defense secretary, instead said it would set up a peer-review panel and solicit proposals from researchers working on AIDS vaccine therapy.

MicroGeneSys was silent about the decision. But Wyeth-Ayerst was not. On January 14, the pharmaceutical maker issued a press release announcing that it had "terminated its involvement in the development of MicroGeneSys's VaxSyn."

In a move that broadcast the DoD's commitment to the integrity of the process, the Army hired an outside firm to assemble a peer-review panel, which, in turn, named John Moore, the fiercest critic of the appropriation, as its head.[45]

As the peer-review panel evaluated the 32 proposals sent in, the saga took another bizarre twist. Public Citizen's Health Research Group, an offshoot of Ralph Nader's consumer advocacy coalition, wrote Representative Waxman on June 7 requesting that his subcommittee hold a hearing to investigate "charges of grave impropriety" committed by Redfield. The Army's Redfield investigation, charged Public Citizen's Sidney Wolfe and Peter Lurie, "lends new meaning to the term 'whitewash.' " They were particularly outraged that the Army had, in response to a Freedom of Information Act request, blanked out hundreds of pages from the investigatory report, citing issues of privacy and internal communication.[46] "The events described here illustrate the increased potential for scientific misconduct when fame, financial reward and even a Nobel Prize await the discoverer of an effective HIV vaccine and suggests the need for special monitoring of research in this area," they concluded. "These incentives appear to have produced a campaign to promote GP160 that bears more resemblance to market research than it does to objective scientific research."

Despite the late hit over an issue that finally was nearing resolution, Public Citizen, which had mastered the art of influencing Congress and the media, received the attention it sought. CNN, the *Washington Post*, the *San Francisco Examiner*, and the Associated Press all ran the story.[47] Waxman issued a strongly worded press release, too. "These are serious charges," Waxman said. "If this vaccine has been misrepresented, we'll investigate that and correct the record."

Public Citizen's campaign had an absurd quality to it, however, because it was based on "internal memoranda, not previously made public" that documented *allegations* of wrongdoing. Since these allegations by that point had been explored at length by the Army, AIDS researchers, and the media, Public Citizen's action appeared to have more to do with boosting its stature than helping HIV-infected people, the group's purported aim. What HIV-infected people then needed most was for AIDS researchers and their leaders to bury this dead horse and return to work.

The Army then handed Public Citizen more ammo. That October, the Army announced that it had awarded five academic researchers working on AIDS vaccines $9.56 million of the $20 million. The balance would go to the Army's civilian partner, the Henry M. Jackson Foundation. Henry Jackson would spend $5 million of that money to complete the phase II trial, which fell short of funds after Wyeth-Ayerst pulled its support.[48]

Moore publicly vented his rage. "What the DoD does not seem to have understood is that the fuss over the original $20 million was largely about an abuse of the peer-review system," said Moore. "Now it looks like we're facing another one." Public Citizen wrote Waxman and the defense secretary, urging investigations.[49]

Neither Waxman nor the defense secretary ever instigated further investigations, but by December, the Army had had a change of heart, and decided that instead of funneling $10 million to Henry Jackson, it would take back $7.8 million and fund four more grant applicants who fared well in the peer-review process. In the end, the Army spent $17.7 million of the $20 million on 10 peer-reviewed grants, which proposed to investigate in test tubes, monkeys, and chimps a diverse collection of novel vaccine therapy strategies. The balance went to Henry Jackson.[50]

On December 14, MicroGeneSys announced that its board of directors had fired Franklin Volvovitz as its chairman, president, and CEO.

• • •

ON APRIL 17, 1996, BY which time the MicroGeneSys lobbying blitz and the fate of the $20 million appropriation had become a distant memory for most AIDS researchers and activists, the Army announced that it had completed its phase II trial of VaxSyn in 608 HIV-infected people. Results from the study, stated the DoD press release, "show no clinical improvement that could be attributed to the vaccine used as an adjunct therapy for HIV infection."

MicroGeneSys made a limp attempt to spin the story. "Although there were no statistically significant differences in clinical events, there were certain modest and time-limited effects favoring VaxSyn," their press release the next day claimed. "We certainly have not given up on VaxSyn." According to the Army's Burke, however, the independent board that monitored the trial said there were "not even any trends" to suggest VaxSyn benefited anyone.

Nine months later, MicroGeneSys issued a press release announcing that it had changed its name to Protein Sciences Corporation, "the finishing touch to a complete makeover of the company that pioneered development of a therapeutic AIDS vaccine." The January 15, 1997, release said the company's new CEO had a new business strategy and had "shelved high-risk, long-development time products, such as the AIDS vaccine."

Franklin Volvovitz's bold gambit had, in the end, badly backfired. The image of AIDS vaccine research at the Army, which ironically had one of the world's most directed and aggressive programs, suffered, as did the affiliated scientists. The leaders of the NIH and FDA spent an untold number of hours attempting to unravel the mess. The peer-review process triumphed, but the victory obscured the fact that the NIH did overemphasize the value of peer review. The researchers who eventually won the money did little to advance the field of AIDS vaccine therapy, which, save for the Immune Response Corporation, saw every early entrant abandon it. The VaxSyn trial was a bust. MicroGeneSys nearly sank. And Volvovitz himself lost a job.

But there was one positive outcome to this seemingly endless debacle: it came to a close when the White House became involved, demonstrating that if central leadership existed, it could better organize the field and speed the search for an AIDS vaccine—if only by policing the enterprise and thwarting the many forces that, inevitably, would attempt to hijack the agenda for their own ends.

CHAPTER 9

A Manhattan Project for AIDS

Two dozen AIDS activists greeted presidential candidate Bill Clinton when he entered his suite in the Sheraton Manhattan Hotel late on the Saturday evening of April 4, 1992. The activists, organized under the umbrella of United for AIDS Action, represented a diverse coalition that included gay men, mothers, blacks, American Indians, Latinos, adolescents, and the mentally ill. David Barr of Gay Men's Health Crisis, the coordinator of the activists, spoke first, welcoming Clinton and handing him two condoms.

"What if the press saw that I had these?" laughed Clinton, whose campaign was still reeling from charges that he had had an affair with Gennifer Flowers.

"It would have been worse if they thought that you didn't have these," said Barr.

The smile left Clinton's face. Dressed in a suit and tie, Clinton had large bags under his eyes, and at first seemed distracted as he sipped a Diet Coke and listened to Barr and the other activists. This was not the meeting he had envisioned. Clinton, whose campaign might die if he lost the New York primary election to be held that Tuesday, badly wanted the city's large gay and lesbian vote, a bloc that had much interest in another Democratic candidate, Jerry Brown. Clinton also nine days earlier, with the TV cameras rolling, had been jeered by a member of ACT UP, who charged that the Arkansas governor was dying of ambition while people were dying of AIDS.[1] So Clinton, whose staff had invited the openly gay, Democratic congressman from Massachusetts, Barney Frank, to the meeting, wanted to court the gay vote—but many of these activists were not even gay.

Still, as the activists spelled out their demands—increase funding for

AIDS research and treatment, involve HIV-infected people in policy decisions, support distribution of clean needles to injecting drug users and condoms to high school students—Clinton became intensely interested in the conversation. One question tossed at Clinton during the meeting was: If elected president, what would your priorities for HIV research be? "He said he'd support a 'Manhattan Project' for AIDS," recalled Barr, stressing that this was not one of the carefully chosen items on their agenda.

When the meeting broke up near midnight, Clinton had spent more than an hour with the activists, and had deeply impressed them. "A lot of us went in very, very skeptical of Clinton," remembered Barr. "We left with a very different impression. His knowledge was much, much better than we thought." Yet no one had a clear idea of what Clinton's Manhattan Project would have in common with the World War II crash program centered at Los Alamos, New Mexico, that developed the atomic bomb.

Speaking to gay and lesbian activists who attended a fund-raiser held the next month at Hollywood's legendary Palace nightclub, Clinton made his first public call for a Manhattan Project, outlining his vision. "One person should be in charge," Clinton told the 500 activists, who had paid between $100 and $250 each to hear his May 18 speech. "One person who can cut across all the departments and agencies, who has the President's ear and the President's arm." Aside from appointing such an AIDS czar, though, Clinton still did not really explain what he meant by a Manhattan Project.

He ended with a stirring, emotional pitch. "If I could wave my arm for those of you that are HIV positive and make it go away tomorrow, I would do it, so help me God I would," said Clinton, his voice breaking. "If I gave up my race for the White House and everything else, I would do that. Let us never forget, there are things we can and cannot do. But the beginning of wisdom is pulling together and learning from one another and being determined to do better, and this country is being killed by people who try to break us down and tear us up and make us be little when we have to be big. This is a big time. Let us rise to the challenge."

Clinton's call for a Manhattan Project, a stab at doing something "big," unfortunately was born in confusion and had small aims itself. Rather than reorganizing the AIDS research enterprise, Clinton's "Manhattan Project," as a campaign sheet distributed at that summer's international AIDS conference in Amsterdam explained, consisted mainly of a czar who would coordinate AIDS policy, cut through "bureaucratic red tape," and move forward on the recommendations made by the congressionally mandated National Commission on AIDS—a largely ineffectual group set up in 1988 supposedly to coordinate AIDS policy and cut through bureaucratic red tape. As an indicator of Clinton's lack of resolve about a Manhattan Project, by the time of the November election, the words "Manhattan Project" had been deleted from the same campaign sheet he distributed in Amsterdam.[2]

Despite the politically expedient nature of Clinton's call for a Manhattan Project for AIDS, the campaign promise caught the imagination of many activists and a few big-name scientists, who soon began advocating a wide array of targeted research programs that had specific structures, budgets, and even leaders. Frustrated by the failure of science to come up with a vaccine or an effective therapy in the decade since scientists first isolated HIV, these activists and researchers wanted to strip away several layers of bureaucracy at the National Institutes of Health, with some going so far as to call for bypassing its peer-review system altogether. Another factor explained why a bevy of creative proposals poured out shortly after Clinton took office in January 1993: the Senate that month introduced legislation, at the behest of activists and leading academic scientists, that would radically overhaul the way the NIH spent its AIDS money. Change, in short, was in the air.

In the end, researchers scotched the bolder, riskier ideas about how to speed the search for treatments and vaccines. The research community had deep fears that targeted programs would foster shoddy science and promote cronyism, just as had happened during the War on Cancer in the 1970s. This, in turn, would deplete funds for the independent investigators who pursued more creative, basic questions about immunology and virology. Many prominent researchers even resisted the congressional push to revamp the way the NIH ran its AIDS research program. This change did, however, take place, and it forced the institution to reevaluate how it spent its AIDS research budget, which had grown to more than $1 billion a year. Much good would come from this soul-searching exercise—but the NIH did not fundamentally alter the way it led the vaccine search.

So despite all of the high-minded talk about reorganization, cooperation, and delegation, the AIDS vaccine enterprise remained saddled with the same problems that had beleaguered it for the first decade. Promising leads languished. Research gaps remained unfilled. Coordination, especially among primate researchers, rarely occurred. Big pharmaceutical companies kept their distance from the field, and the involved biotech firms continued to drop out of the "race." Vaccine trials in humans stumbled forward, with little clear sense of what deserved to move into large-scale efficacy tests. And the campaign for a truly targeted effort to solve the problem steadily lost steam until it altogether disappeared.

• • •

VANNEVAR BUSH, A MATHEMATICIAN AND ENGINEER who oversaw all U.S. government science during World War II, helped run the real Manhattan Project. Near the war's end, President Roosevelt wrote Bush, who headed the Office of Scientific Research and Development, and asked him how they could apply the lessons they had learned from their "unique experiment of team-work and cooperation in coordinating scientific research." FDR wanted Bush to lay out a blueprint for the peacetime relationship between public and private scientific research to "create a fuller and more fruitful employment and a fuller and more fruitful life." And FDR had a particular interest in "the war of science against disease." As he wrote Bush in his November 17, 1944, letter, "The fact that the annual deaths in this country from one or two diseases alone are far in excess of the total number of lives lost by us in battle during this war should make us conscious of the duty we owe future generations."

In July 1945, Bush presented then President Harry Truman with his reply, a book-length report entitled *Science, the Endless Frontier*. Largely written by various committees set up by Bush, the report would form the foundation for the scientific culture in the United States—and it was distinctly *anti*–Manhattan Project.

The Committee on Science and the Public Welfare went to some lengths to separate applied from basic (which it called "pure") research. With applied research, the committee wrote, "the objective can often be definitely mapped out beforehand" and the "work lends itself to an organized effort." The group saw applied research mainly as the purview of industry, not government—although the committee did acknowledge that "the presence of a profit motive does not ensure the existence of adequate research and development." Basic research, on the other hand, did not aim to develop something useful, but rather to increase understanding.[3] It would suffer badly without government help:

> The distinction between applied and pure research is not a hard and fast one, and industrial scientists may tackle specific problems from broad fun-

damental viewpoints. But it is important to emphasize that there is a perverse law governing research: Under the pressure for immediate results, and unless deliberate policies are set up to guard against this, *applied research invariably drives out pure.*

The moral is clear: It is pure research which deserves and requires special protection and specially assured support.[4]

Endless Frontier established a paradigm for federal funding of scientific research that had a profound influence. The same year the report came out, Congress established a program that allowed the National Institute of Health—it was singular then, as it only had the cancer institute—to fund researchers in universities, who would submit grant proposals for peer-review groups to evaluate. Unfettered, investigator-initiated basic research flourished, and, as Congress added one new institute after another, NIH's budget steadily ballooned from $2.8 million in 1945 to $67 million a decade later and $773 million by 1965.

In 1966, President Lyndon Johnson challenged the Bush paradigm. "Presidents . . . need to show more interest in what the specific results of research are—in their lifetime, and in their administration," said Johnson. "A great deal of basic research has been done . . . But I think the time has come to zero in on the targets—by trying to get our knowledge fully applied. . . . We must make sure that no lifesaving discovery is locked up in the laboratory."[5]

Just as politicians would later react to AIDS activists, Johnson's oratory followed the lobbying of philanthropist Mary Lasker, whose husband had died of cancer and made boosting federal support of research into that disease her life's cause. "This conflict reached a high point in 1966, when President Johnson, after much prodding by Mary Lasker, called the directors of all of the NIH institutes to his office on short notice and demanded to hear 'what plans, if any, they have for reducing deaths and disabilities and for extending research in that direction,' " wrote *Science* reporter Robert Bazell in 1971.[6]

The relentless campaigning of Lasker and her allies in 1970 led the U.S. House of Representatives to pass a resolution declaring a national crusade that would lead to the conquest of cancer by 1976. More substantively, the "Laskerites" convinced both Congress and the White House that year to give the National Cancer Institute budgetary independence from the NIH and allow the President to directly control the institute. The move, explained Senator Ralph Yarborough, a Democrat from Texas, "would guarantee that the conquest of cancer becomes a highly visible national goal." After the formal declaration of the War on Cancer in the 1971 National Cancer Act, the NCI's budget skyrocketed from $233 million that year to $378 million in 1972 and $492 million in 1973.[7]

With the NCI's budget more than doubling in two years came increased scrutiny of how the institute spent the money. A targeted research program to find viruses that caused human cancer became one of the most heavily criticized expenses. In 1972 alone, the Special Virus Cancer Program spent $42 million—more than 10% of the NCI's budget—on research that the institution initiated.

Such a heavy investment in contract research ran directly counter to the *Endless Frontier* vision. And in March 1974, an ad hoc advisory group headed by Norton Zinder of Rockefeller University wrote a scathing review of the program. "There did not, nor does there exist, sufficient knowledge to mount such a narrowly targeted program," wrote the Zinder committee.

> The underlying assumption was, and in large part still is, that once human cancer viruses are found, they will be controllable by anti-viral vaccines. The analogy with the control of poliomyelitis is the guiding principle; that is, that knowledge of the virology and natural history of the infection is sufficient to achieve control, and that detailed understanding of the disease mechanism is unnecessary for the successful outcome of the program. . . . In short, we agree with the goal of identifying human cancer viruses and evaluating their role in the etiology of cancer, but reject the concept that this holds sufficient promise for a short term solution to the cancer problem to justify compromising basic research.[8]

The blistering critique accused the leaders of the virus cancer program of being "full of conflicts of interests," favoring their friends, and rubber-stamping contracts. "Success in science is an irregular and unpredictable phenomenon," the committee wrote, repeating a mantra that NIH leaders routinely chant to Congress to justify their heavy investment in investigator-initiated, basic research. "When and where important discoveries are and will be made is almost impossible to determine in advance. The goal should be to maximize the opportunities."

Like *Endless Frontier*, the Zinder committee report emphasized that the government needed to protect basic research from targeted projects—and the virus cancer program provided hard data that this was no mere philosophical conceit. "Simultaneously with the growth of targeted or non-investigator initiated programs, the support base for fundamental science is being eroded," the Zinder committee warned in its conclusion. "It is more than possible that none of our current approaches to the problem are on the right track simply because we lack the fundamental knowledge to see even indirectly our way to the clear water."

Concern about applied research driving out pure led *Science* in 1976 to publish a creative analysis of progress made in the fields of heart and lung disease. Julius Comroe Jr., who specialized in those fields, conducted the multiyear study with anesthesiologist Robert Dripps, who died nearly three years before the landmark paper appeared. After identifying 2,500 scientific reports that contributed to 137 "essential bodies of knowledge" in those fields, Comroe and Dripps settled on 529 "key" articles. Of these, Comroe and Dripps found that 61.7% focused on "basic" research:

> Our data compel us to conclude (i) that a generous portion of the nation's biomedical research dollars should be used to identify and then to provide long-term support for creative scientists whose main goal is to learn how living organisms function, without regard to the immediate relation of their research to specific human diseases and (ii) that basic research, as we have defined it, pays off in terms of key discoveries almost twice as handsomely as other types of research and development combined.[9]

Together, *Endless Frontier*, the Zinder committee report, and the Comroe and Dripps paper formed a sturdy tripod that elevated the status of investigator-initiated, basic research high above that of targeted projects. This belief became so integrated into the culture of science, at least in the United States, that it became gospel, and to question it was to reveal one's ignorance.

• • •

IN *PASTEUR'S QUADRANT*, political scientist Donald E. Stokes closely examined the Vannevar Bush model and found it seriously lacking. The dichotomous view of basic versus applied research "obscures as well as it reveals," wrote Stokes.[10] "The belief that the goals of understanding and use are inherently in conflict, and that the categories of basic and applied research are necessarily separate, is itself in tension with the actual experience of science."

Stokes divided scientific research into four quadrants based on whether the work was inspired by "considerations of use," a "quest for fundamental understanding," both of these aims, or neither. Neils Bohr's search for atomic structure, he argued, was aimed at understanding and nothing more. Thomas Edison's drive to commercialize electric lighting, in which he discouraged the researchers working for him from investigating the more fundamental scientific questions that arose, was all about use. Agricultural researchers, he noted, sometimes conduct experiments simply to enhance their skills, which has little to do with either use or understanding. Then there is Louis Pasteur, who often investigated basic research questions while pursuing a project that had clear application. Pasteur's quadrant, Stokes contended, was "[w]holly outside the conceptual framework of the Bush report" and, ironically, best described the Manhattan Project itself.

Stokes backed his argument by carefully parsing the compelling Comroe and Dripps study. Although Comroe and Dripps concluded that basic research accounted for 61.7% of the major advances in heart and lung disease that they analyzed, Stokes noted that the authors arrived at that by combining basic research that was related and unrelated to the solution of a clinical problem. In other words, one of their parameters was what Stokes called "use-inspired basic science"—Pasteur's Quadrant—and it accounted for fully 40% of the advances Comroe and Dripps tied to basic research.

In the 1980s, the National Cancer Institute commissioned an outside contractor to conduct a study that further challenged the ethos of investigator-initiated, basic research *über Alles*. Initiated by Vincent DeVita Jr., the NCI director from 1980 to 1988, the little-noticed study assessed which funding mechanisms had led to recognized scientific advances in cancer research.[11]

The report, completed in 1987, examined 13 "major advances" made during the past 15 years.[12] Chosen by a panel of experts, these advances ranged from mammography as a screening tool to the discovery of cancer-causing oncogenes in humans. The conclusion of the study: "no major mechanism of support utilized by NCI appears to be superior, overall."[13]

DeVita concluded from this study that targeted research projects had been wrongly tarred and that investigator-initiated research—known within the NIH as R01 grants—had received too much credit. "One of the fears I have now with the current NIH leadership is they're trying to turn every program into an R01," DeVita told me in 1998, by which time he was director of the Yale Cancer Center. "It's the biggest danger we face in biomedical research: go back in your lab and shut the door."

Richard Wurtman, a neuropharmacology researcher and clinician based at the Massachusetts Institute of Technology, attacked the Bush paradigm from yet another angle. An analysis he and a student conducted of new treatments discovered between 1965 and 1995 revealed that the majority were not even supported by the NIH or other public funders. Rather, Wurtman and his coauthor contended in *Nature Medicine* that the most widely used new drugs were slight variations of existing treatments, came directly from industry, or were discovered by doctors who noticed that a drug given to treat one disease happened to work against another. "Treatment discovery turns out to be very much a directed or mission-oriented enterprise, requiring the participation of investigators committed to that task," they concluded.

The Bush paradigm of basic research flowing into applied, they maintained, better described advances in physics than those in biology. "[T]heoretical biology alone could no more be expected to generate treatments for human diseases than theoretical architecture alone could design particular bridges," they sniped. "If decades of experience affirm the inadequacy of the Bush paradigm, then is it not appropriate for the government to adopt a new formulation of how it can best promote treatment discovery?"[14]

• • •

BY THE 1990S, frustrated researchers at the front of the AIDS vaccine search publicly began to challenge the gospel. Maurice Hilleman, the curmudgeonly vaccine veteran at Merck & Co., minced no words. "In the United States, investigator-initiated research is holy, you see," Hilleman told me. "You can't say anything against it. But when you try and apply it, it is all these little pieces. It's all R and no D."[15]

Wayne Koff, head of the AIDS vaccine division at NIAID, shared Hilleman's frustration. In 1989, Koff quietly began pushing the idea for a directed program similar to the one being organized by the NIH to map the human genome. In a draft version of an AIDS vaccine development agenda he wrote that December, Koff specified that the directed program would complement, rather than supplant, investigator-initiated research and would attempt to better coordinate the effort. The NIH AIDS Advisory Committee shot down the idea in June 1990, and instead encouraged Koff and his coworkers to "proactively coordinate" the field.[16]

In December 1990, Koff tried yet another tack to organize a directed program. This time, he called the program the Applied Research Strategy for AIDS Vaccine Development, hoping it would be more politically palatable. A letter to two dozen leading researchers, who worked both inside and outside of the AIDS vaccine field, described the program, stating that it aimed "to determine within the quickest time-frame which vaccine candidates should progress into large-scale clinical trials."[17] In essence, the program promised to streamline the search by better translating results of animal experiments into human studies, standardizing assays, and increasing funding (then estimated at $150 million a year, including all private and public efforts). At the end of the letter, Koff polled the scientists to see whether they agreed or disagreed with this "goal-oriented applied research approach." Save for one abstention, the scientists—who included Nobel laureates Renato Dulbecco and Joshua Lederberg, Hilleman, Jonas Salk, Duke's Dani Bolognesi, the U.S. Army's Don Burke, Lars Kallings of the World Health Organization, and SIV researchers Ron Desrosiers, Murray Gardner, and Reinhard Kurth—all agreed with the idea. Yet it, too, soon disappeared.[18] "Bottom line: the NIH wasn't swayed," Koff lamented.

With his discouragement steadily deepening, Koff made a dramatic plea to his colleagues on October 19, 1991, at the close of the NIAID's international AIDS vaccine meeting held in Marco Island, Florida. "[HIV] is winning," said the slightly built Koff, his voice shaking. "It's beating the crap out of us. The time has come where we have to look at ways of speeding up the effort."

Koff outlined the successful searches for other vaccines, including the one led by the National Foundation for Infantile Paralysis/March of Dimes. "You begin to see common themes," said Koff. "It doesn't make a difference if it's a Manhattan Project or a Brooklyn Project, if it's a mixture of the companies or a combination of organizations. . . . There has to be a process in which a group of individuals has access to all of the data—all of the data from all of the companies, all of the data from all of the academic institutions, all of the data from all of the researchers worldwide—and then look at this data and review the data and prioritize what has to be done."

Koff's speech was long on passion and short on specifics. He did not propose a way to fund this program, other than to note that the NIH system of individual investigators submitting grants wasted too much of their time. He also did not address the difficult question of patents, a large reason why scientists do not freely share data, nor did he have proposed fixes for the liability fears that he noted had scared industry away from the field. And Koff made no mention of the project, spearheaded by Jonas Salk, that hoped to confront these obstacles head on.

Salk had enlisted Koff to help create the World Foundation for the Advancement and Application of Immunization Research, the World FAAIR.[19] In a proposal sent to the John T. and Catherine D. MacArthur Foundation two weeks before Koff's Florida talk, Salk explained the vision for World FAAIR in detail. Salk noted that the government primarily funded basic research and that industry, dissuaded by liability and marketplace issues, did not have a "compelling profit incentive" to become involved. "The independent sector now needs to be mobilized for the purpose of catalyzing the organization of an applied research program toward HIV vaccine development," wrote Salk, who wanted the MacArthur Foundation to provide World FAAIR with $1 million in seed money. "The objective is not to take over the full responsibility but rather to see that those who are so motivated have an opportunity to act: equivalent in purpose to the March of Dimes for polio in the '40s and '50s."

World FAAIR, which Salk imagined several foundations would eventually support and might be run by Koff, hoped to sponsor forums that would speed the AIDS vaccine search. Specifically, these forums would evaluate the performance of different vaccines in monkey and chimp challenge experiments, assess the data from human studies, and design efficacy trials. World FAAIR also would explore liability issues and help find money for the efficacy trials. Salk listed potential members of the group's scientific advisory board, a diverse collection of researchers and policy makers that included Salk; Harvard's Ron Desrosiers; Australia's Gordon Ada; William Foege, former head of the CDC; Jean-Paul Lévy, director of France's AIDS research program; Lars Kallings of the WHO in Geneva; and Merck's Maurice Hilleman.

The MacArthur Foundation, whose board of directors included Jonas Salk, turned down the proposal.

• • •

WORLD FAAIR, WHICH NEVER RECEIVED ANY PUBLICITY, quietly disappeared. But Koff's talk at the Florida vaccine meeting—a daring *cri de coeur* from the top AIDS vaccine official in the world's best-funded AIDS research program—did not. In January, Maurice Hilleman praised Koff's speech during a talk he gave at an AIDS meeting in Hawaii. In Hilleman's presentation, a version of which later appeared in the British journal *Vaccine*, he called Koff's talk "perceptive and prophetic" and picked up where the younger scientist left off.[20] "A substantial part of the problem with AIDS vaccine development may lie with the organization of the research and development establishment, which is supported in the main by the grants programme," wrote Hilleman, who argued that people should look at the way that the U.S. Army and the March of Dimes funded vaccine research.[21] "Support by contract, rather than by grant, eliminates needless paperwork and frustration and makes possible coordinated, directed, and uninterrupted efforts in the vaccine endeavour. Such a directed programme seems urgently needed in the face of the expanding AIDS pandemic."

Hilleman acknowledged that HIV presented special challenges, and explained why vaccine developers had a much easier time training immune systems to outwit the viruses that cause measles and hepatitis B, both of which people sponteously recover from, have few mutants, and are susceptible to antibodies. Still, wrote Hilleman, AIDS vaccine researchers would benefit greatly by following the general principles that have guided these other successful vaccine searches. These include having a central authority guiding research, creating a strategic plan, defining gaps in knowledge, and steering redundancy. "The complexity, redundancy, and overinformation in modern molecular biology and immunology has created an immense problem for the vaccinologist to hear the tune amid the static," wrote Hilleman. "Elemental truths are simple and complexity can be deafening."

In April 1992, a disillusioned Wayne Koff quit the NIH and went to work for United Biomedical Inc., a New York biotech company that wanted to try its hand at making an AIDS vaccine. The next month, Hilleman made an even stronger pitch for a targeted AIDS vaccine program at NIAID's advisory council meeting.[22] Hilleman sang the praises of investigator-initiated research, which he said was superb at producing knowledge and scientific publications. But he stressed that making a vaccine was "a whole different breed of cat" that required "collective, coordinated work" and a "benign dictator."

Ideally, said Hilleman, AIDS vaccine researchers would be organized into a Manhattan Project, working in a big warehouse under one boss. But since that did not seem realistic to him, he reiterated his call for following the lead of the March of Dimes and the U.S. Army. He recounted how the National Foundation had a scientific committee that would meet every few months, review what had been accomplished, decide what needed to be done, and then split up the work. "It was terrifically wonderful, because they never got along on anything," said Hilleman, but he stressed that they had a contract officer sitting at the table with them. "When they decided that something should be done, they'd give somebody some money. . . . So they weren't sitting there, spending six months out of the year writing these damned grant requests and having them considered by peer review. That is absolutely sure death."

Spurred by questions from the advisory council members, a few of whom suggested that his ideas would not work in the modern world, Hilleman fleshed out this theoretical program. Pick some of the field's best researchers, he said, organize them into groups, and have them work on different projects. "Hopefully you would make them competitive so they'd try to cut each other's throats, because competition and winning is one of the motivations that people respond to," said Hilleman. Industry, he suggested, does not need to be part of the program: when you come to the point where you need a vaccine made, contract with a company to produce it.

Larry Corey, who ran clinical trials of AIDS vaccines at the University of Washington, said the idea of more coordination "sounds nice," but given the scientific uncertainties facing AIDS vaccine researchers, he remained skeptical that it would help. "I'm not sure coordination is necessarily ultimately the best approach, because I think it leads to conservatism," said Corey. "I think that's the problem at the moment: How do you take a risk and yet not cut off what really might be the best approach?"

The contention that scientists did not know enough about HIV and the immune system to stage a targeted vaccine development program was *the* central criticism of the proposition. Hilleman sharpened his tongue and explained his concept once again. "The polio foundation was put together because Roosevelt had a couple of paralyzed legs—they didn't know much more about polio than that," he said. "What I worry about is we come in and say, 'This question unanswered, this question is unanswered, gee, that's a good question, unanswered.' And it goes on year after year, 10, 15 unanswered questions. By George, somebody ought to be putting that together and seeing to it that it's done."

• • •

BERNADINE HEALY ON JANUARY 22, 1993, summoned NIAID chief Tony Fauci and leaders of the NIH's other 20 institutes, centers, and divisions to an emergency meeting. As the NIH director, Healy was caught between the scientists who ran her institution and the wishes of the incoming Clinton administration, which appeared to back a bill introduced the day before that aimed to better coordinate federally sponsored AIDS research. "Though some aspects of the legislation were unappealing to many and odious to some, it was inappropriate for NIH to take a do-or-die position at this time," Healy told me a few days later. Instead, the NIH brass decided to write her a letter that stated their concerns. She, in turn, forwarded the complaint to Donna Shalala, the newly appointed secretary of Health and Human Services.

The NIH directors chose this baroque form of communication because they wanted to downplay their unusual and bold revolt. At the heart of the battle sat the NIH's little-known Office of AIDS Research, which Fauci ran in addition to NIAID. The OAR putatively had the power to direct the way the various institutes spent the $1.1 billion AIDS research budget. But at that summer's international conference in Amsterdam, AIDS activists Mark Harrington and Gregg Gonsalves of the Treatment Action Group (TAG) distributed a 199-page "critical review" that they wrote of the NIH's AIDS research program, which portrayed the OAR as next to worthless.[23] "The NIH is actually a collection of fiercely autonomous fiefdoms," wrote Harrington and Gonsalves, who in 1991 had splintered off from ACT UP to form TAG. "OAR Director Fauci, who is supposed to coordinate AIDS research . . . has little real say in the half [of the NIH AIDS budget] he does not directly control as NIAID Director. Thus, there is no truly centralized planning and execution of AIDS research, and no adequate oversight from either Congress or the Administration."

After much lobbying by TAG, the report persuaded Senator Edward Kennedy to introduce a bill that conferred new powers to the OAR. According to Harrington, the veteran Massachusetts Democrat's staff even let TAG suggest the actual language. "We were punk kids from New York writing legislation," said Harrington, a San Francisco-bred Harvard graduate who was more comfortable in Doc Martin combat boots and faded Levi's than the sport coat and tie he would sometimes don while making the rounds on Capitol Hill. Much to TAG's delight, Kennedy's bill to refurbish the OAR caught the scientific community unawares, a move that prevented the expected opposition from killing the proposed legislation before its birth.

The most controversial part of Kennedy's bill had to do with money. Rather than Congress sending each NIH institute its slice of the AIDS budget, the bill would shunt the entire amount to the OAR, which then would distribute it. This budget authority would, theoretically, allow the OAR to persuade institutes to follow its commands. The new OAR also would have a discretionary fund to bankroll emergency AIDS research and fill in gaps. And, with the help of outside advisers, the OAR would have to develop each year a comprehensive, long-range strategic plan that set priorities and balanced basic and applied research.

Fauci—who suspected that such legislation would force him to relinquish the job as OAR director—and the other institute leaders strongly felt that giving the OAR budget authority would throw sludge in their gears. As they complained in the letter that Healy forwarded to Shalala, the change would add an "additional bureaucratic layer," thereby "impeding the progress of AIDS research." The relationship between the NIH director, the institute directors, and Congress, they predicted, also would be "severely disrupted." They congratulated themselves for the accomplishments of the NIH-directed AIDS research enterprise, which had been, they wrote, "without question, unprecedented in the history of biomedical research." And they closed with this warning: "It would be unfortunate if, despite the enormous successes, the lack of a cure or a vaccine at this point in time is interpreted that the system is defective. Therefore, changing the system with all good intentions may, in fact, work to the detriment of the execution of AIDS research."

In keeping with tradition, Fauci and his counterparts put out the call to their academic friends to write Congress and oppose the legislation. Letters poured in to the Senate and the House.

Nobel laureate Howard Temin of the University of Wisconsin contended that "reorganization, after an enormous amount of bureaucratic fighting and administrative costs, would have little or no effect on progress towards a cure and a vaccine and would significantly impact in a negative fashion on NIH and progress in preventing and curing other diseases." Temin's letter, which had a surprisingly overheated tone for this philosophical man, said the "most disturbing" provision was giving the OAR a discretionary fund, an arrangement that reminded him of the NCI's Special Virus Cancer Program and was "prone to great abuses." Nobel laureate Harold Varmus, who a few months later would leave the University of California at San Francisco to replace Healy as head of the NIH, asked colleagues to cosign a letter to Kennedy that criticized the bill for giving the OAR "tremendous authority with virtually no checks and balances." Duke's Dani Bolognesi warned that it "would be a tragedy indeed if the reorganization called for would have the end results of fractiousness, and needless duplication." NIAID's advisory council also went on record as opposing the bill, as did the Joint Council of Allergy and Immunology, the American Society of Tropical Medicine and Hygiene, the Association of American Medical Colleges, the Infectious Diseases Society of America, and the American Society for Microbiology.

TAG anticipated this response and staged a preemptive strike, convincing three dozen leading AIDS researchers—many of whom well understood that they would be crossing Fauci—to write in supporting the legislation. TAG's lobbying also led HHS Secretary Donna Shalala on January 27 to write Vice President Al Gore, president of the Senate. "We believe the bill provides a framework for directing AIDS research in an effective manner," wrote Shalala.

That same day, Kennedy's Committee on Labor and Human Resources endorsed the bill, which, as Fauci feared, stipulated that the OAR director's job required "the full-time attention of a dedicated individual." The committee stressed that these requirements "in no way reflect on the performance or capabilities of the incumbent director." But anyway about it, Anthony Fauci was put on notice that the Senate did not want him heading both NIAID and the OAR. The era of Fauci, leader of the largest AIDS research program in the world, was coming to an end.

• • •

PERPETUALLY PISSED-OFF LARRY KRAMER, the celebrated writer who founded AIDS activism first by starting the Gay Men's Health Crisis and then ACT UP, wrote the forward to the influential TAG report that spawned the OAR legislation. "Is it, at this late stage of this plague, intemperate or rude to suggest that this present state of AIDS affairs leads us even more to the fact that the only out that has not been tried, that still must be tried, is a Manhattan- or Apollo-type project, wherein the leading experts in all areas are granted emergency powers and sent off into the seclusion necessary to produce the cure that must be there if our civilization is to survive?" asked Kramer, a 1970 Academy Award nominee for his screenplay of the D. H. Lawrence novel *Women in Love*.

Kramer had no inkling that the TAG report would lead to a congressional bill to refurbish the OAR, and he hated it. In Kramer's mind, the Clinton administration marked the first time that the people running the White House truly cared about AIDS, and his disciples at TAG had set their sights too low. "I can't believe that I gave birth to them, because they've become such young bureaucrats," Kramer groused to me. "If you ask for the moon, you might get the mountaintop. They're not even asking for the mountaintop."

As Kramer argued in a *New York Times* op-ed piece the preceding

November and in letters to Shalala and other key players in the nascent Clinton administration, the NIH was ill equipped to lead the search for a cure and a vaccine.[24] Instead, he believed that Clinton should establish a Joint Chiefs of AIDS. Kramer had many provocative ideas, but to hear him discuss this concept was not unlike listening to the Wizard of Oz after he steps in front of the curtain and begins making pronouncements in the name of *E Pluribus Unum*: Appoint Nobel laureate David Baltimore—who recently had been disgraced by a misguided scientific misconduct investigation led by Congressman John Dingell—chief of AIDS research, and grant him emergency powers to hire and fire staff, set his own agenda, and authorize spending without peer review. Name former Surgeon General C. Everett Koop or former Defense Secretary Robert McNamara the AIDS czar. Kramer tapped Merck's Roy Vagelos as chief of drug and vaccine development, the FDA's David Kessler as chief of drug approvals, and Admiral James Watson as chief whip.

Rather than corralling researchers in a Los Alamos–like setting, Kramer advocated that people stay in their labs and be detailed to the project. When it came to funding, he said, take the money from the NIH budget.

Not surprisingly, Kramer's grand ideas made little impact in the Clinton administration. His reaction to the OAR reform legislation also did little to endear him to the people in power, as exemplified by his February 4, 1993, letter to Secretary Shalala: "From everything I have been able to find out it appears as if you are in the process of 1) creating an AIDS program as bad, if not worse, than the one you have been sent in there to improve; 2) listening to no one who has a brain, half a brain, or any intelligent idea of how to end this plague; 3) seeking suggestions and counsel mainly from those who fucked everything up in the first place."

Martin Delaney, cofounder of the most influential activist group on the West Coast, San Francisco's Project Inform, backed the Manhattan Project idea, but offered the Clinton administration a much more elaborate plan than Kramer's and sold it with a much softer pitch. As Delaney explained in a memo to Bob Hattoy, a prominent Clinton campaign aide with AIDS who spoke at the 1992 Democratic Convention, he agreed with Kramer on "big concepts" but they differed in many ways. "Larry doesn't like to be troubled with details or political realities," Delaney wrote.

A Chicago native educated in Jesuit-run schools, Delaney studied for the priesthood until dropping out to teach elementary school. From there, he became a management consultant, writing training programs for the likes of Bank of America. In October 1984, he learned that his lover had AIDS and soon began driving to Tijuana, Mexico, and smuggling experimental treatments back to the States. A year later, Delaney cofounded Project Inform to run tests of these smuggled drugs. When the FDA began confronting the group, the stature of Project Inform and Delaney rose—despite the fact that the drugs they were promoting soon proved worthless and even dangerous—and he became particularly chummy with the NCI's Robert Gallo.[25]

Delaney first publicly called for what amounted to an AIDS Manhattan Project in an after-dinner talk at Gallo's 1992 "lab meeting," the invitation-only, weeklong AIDS conference for 500 of Gallo's closest friends. Gallo liked to provide his guests with provocative speakers at the conference's banquet night, and this year he delivered a hat trick: military strategist Edward Luttwak, filmmaker Francis Ford Coppola (who attended much of the meeting as he then was planning to make an AIDS movie called *The Cure*), and, finally, Delaney. "In many ways, your world looks to me less like the military battlefield envisioned by the view of Dr. Luttwak and more like the Hollywood world of our other speaker, Francis Coppola," Delaney told the scientists. "I see a lot of stars and shining celebrities, concerned at their latest roles and reviews, but I don't see any generals, any colonels, or even any lieutenants commanding a program."

Delaney recognized that scientists had made many advances during the preceding year, but the field appeared in disarray to him. "Who is connecting the dots?" he asked. "Who's reading the whole book and not just their own chapters? Who's taking responsibility for the big picture and making sure that all these new leads and insights will be followed up? Who is taking steps to avoid duplication of efforts in the sea of research?" He reminded the scientists that most of them received government money and thus were public servants. "Ten years into this epidemic, it is no longer enough to demonstrate that you can trigger antibody production or find and define a V3 loop—without delivering on the 'So What?' " If they didn't organize themselves, he warned, Project Inform, TAG, or ACT UP would do it for them.

Delaney made good on his threat the next winter, energetically shopping his ideas—which he pointedly did not call a Manhattan Project because he thought the phrase was misleading—to the incoming Clinton administration. As a management consultant, Delaney had perfected the art of shepherding concepts through bureaucracies. In a lengthy executive memo to the "transition team" that paved the way for Clinton to take over the executive branch, Delaney counseled the administration to appoint a blue-ribbon panel of scientists, activists, and industry representatives to conduct a thorough evaluation of the existing AIDS research program. "I cannot stress strongly enough the dangers of proceeding with any kind of a new plan without first undertaking an assessment process," wrote Delaney, specifying how many people should be on the panel (16) and how long it would take for the committee to survey the field and hold hearings (14 weeks).

Although Delaney contended that the committee must reach consensus about how to proceed, he had his own vision. The NIH could not run the program, he argued, because training it to run a directed research program would be like "teaching an elephant to tap dance." Instead, Delaney advocated a setup similar to the March of Dimes, where a chief scientist working with a scientific advisory board would make research decisions. The program would be a "distributed institute" that had a central facility, which would link via video conferences to several laboratories around the world. Pet ideas for the headquarters included vacant buildings at the Presidio Army Base in San Francisco—Coppola offered to help with this, as he sat on a committee charged with determining the fate of the decommissioned base—or the NCI's Frederick Cancer Research and Development Center in Maryland. Commercial collaborators would receive special tax and patent benefits. To reduce red tape, the team would have a "one-stop" review process for clinical trials and a single team from the Food and Drug Administration assigned to monitor and assist the work. The price tag for this project: $50 million to $100 million for five years.

Gallo, in particular, strongly backed Delaney's proposal. "My bottom line on why something like a Manhattan Project is needed: it's getting enthusiasm and optimism at a time when there is pessimism and sadness in the field," he told me. "And even if it doesn't work, it will upset the apple cart enough that we'll all rethink, retry, recompete, and maybe do better." Gallo also badly wanted to simplify the regulations governing his ability to travel, hire assistants, share reagents, and conduct clinical research—regulations that Congressman Dingell and others had accused him of violating. As Gallo liked to tell the scientists in his lab, if a man has a heart attack while standing on a lawn that has a sign saying KEEP OFF THE GRASS, are you going to follow the rule and let him die? "Of course you jump on the grass and fuck the sign," said Gallo. "Forget the rule. Do what is right. But that is going to get you in trouble these days. . . . A Manhattan Project would solve all that shit."

Yet a third proposal for a grand, targeted research program began making the rounds at the White House and on Capitol Hill in early January 1993. The brainchild of the New York branch of ACT UP, this proposal amounted to a radical blend of Kramer and Delaney. Initially called the Barbara McClintock Project to Cure AIDS—named after the Nobel Prize–winning corn geneticist whom colleagues dismissed for much of her career—this scheme called for all scientists to work at one central location for "instant cross-fertilization of ideas." Scientists representing "divergent approaches" and diverse populations would govern the project.

Most radical of all, these scientists would have "extraordinary powers." They could have any government-funded facilities or staff they wanted. They could obtain data from public and private research outfits as they saw fit. They could exercise the right of eminent domain to test experimental medicines owned by private companies that refused to abide by the project's time lines, and they also could force a company to manufacture a promising medicine. The NIH would run clinical trials of medicines deemed promising, but it would not run the program. In a Q&A sheet the ACT UPers distributed with their proposal, they asked: "Why bury ourselves in some agency's ass, several layers down the bureaucratic chain, when we are promised a Presidential project by Clinton?"

• • •

THE OAR LEGISLATION sailed through the House and Senate, and on June 10, 1993, President Clinton signed it into law. For the activists and scientists who supported the OAR overhaul, the good news quickly became overshadowed by the deluge of disappointing results being presented that week in Berlin, Germany, at the Ninth International Conference on AIDS, where one study after another showed that AZT and its relatives had only a marginal impact on delaying disease and death. The doom-and-gloom mood filled the vaccine sessions too, where researchers plainly were groping to find their way.

Delaney by then realized that the Clinton administration did not intend to follow his advice, so he decided to stage his own blue-ribbon panel meetings in the form of a think tank called Future Directions in AIDS Research. The invitation-only meeting, held July 30 and 31 in Madison, Wisconsin, attracted respected scientists like Fauci, Gallo, Bolognesi, Hilleman, and NIAID's Bill Paul, a leading immunologist who was on the search committee for a new OAR director. The activists promoting all three Manhattan Project proposals attended, as did the ones who had pushed through the OAR reform. On the policy side, FDA commissioner David Kessler attended, as did Kristine Gebbie, Clinton's newly appointed AIDS Policy Coordinator (commonly known as the czarina), and a staffer from the office of Democratic Representative Nancy Pelosi of California. At the meeting's end, the participants agreed on nine consensus statements. "Targeted special programs should be considered for emerging, gap-filling, or risky innovative research areas," read one of the statements. "Mechanisms for prioritization are possible and should be utilized."

Delaney barred the media from the meeting, but attendees of what they called the Madison Project held a press conference at the end, which I and others participated in via telephone.[26] Delaney allowed that the group had had "heated debates" about the balance of targeted versus investigator-initiated research, but he said this largely amounted to semantics. "As we were able to calm down from some of the positions we all hold over those buzzwords, we were able to see that we weren't as far apart as we thought on those issues," said Delaney. Even the original angry activist, Larry Kramer, veritably gushed about the get-together. "This weekend is one of the most exciting couple of days I have spent in over 10 years of going to these kinds of meetings," Kramer said.

The Madison Project established four working groups; Delaney chaired one that focused on "alternative management models" for scientific research. At the second Madison Project gathering held in Boston that November, this working group unveiled what became known as the Accelerated AIDS Research Initiative. Rather than refashioning the entire AIDS research enterprise, the Accelerated AIDS Research Initiative proposed a pilot project to assess whether a small team of scientists could work together on defined, limited goals. The OAR and the head of the NIH would help the team scientists, who would spend 50% of their time on the project, choose these goals. An outside advisory group would then steer the project. Funding would come primarily from the federal government, though industry and the public would contribute too. The proposal was classic Delaney: reached by consensus, measured in tone, steeped in management-speak, realistic in goals.

Delaney on December 14 shopped the Accelerated AIDS Research Initiative to the White House, the Office of Management and Budget (OMB), and the NIH. "The proposed project is an effort to take what was best from such celebrated programs as the Manhattan Project and the Apollo Project, while capitalizing on the hard lessons learned in the War on Cancer," Delaney wrote in his executive summary.

Expecting strong opposition from NIH Director Harold Varmus, Delaney the next day sent him a personal letter. Delaney urged Varmus to see the limits of researchers competing to win NIH grants and to publish their findings. "They often seem more concerned with proving each other wrong than finding out what is right," wrote Delaney. "Teamwork is the heart of what I'm talking about, and teamwork is sadly lacking in current AIDS research."

Delaney assured Varmus that the Accelerated AIDS Research Initiative would not favor applied over basic research, nor would it dictate directions to participating scientists. "I raise these issues with you because it is important to me that you recognize that we aren't pursuing some pipe dream vision of a 'Manhattan Project,' " Delaney concluded. And just as he had done in his talk at the Gallo lab meeting, he closed with a threat. "My preference is to raise these issues first within the bureaucracy, where there must be at least a few people who understand what we're talking about. I hope that you are one of them. If I fail in this venue, then I will have no choice but to raise them in public, the Congress, and at the highest levels of the Administration."

Varmus saw no need for the proposal. "It's not a carefully worked-out model," he told me. "What kind of research? What's going to be done? Who's going to do the work?" asked Varmus, a basic researcher's basic researcher, who in 1989 won the Nobel Prize for codiscovering that humans carry genetic material—the so-called oncogenes—that can cause cancer. Varmus also emphasized that a search committee had yet to select a director for the newly refangled OAR, and he did not want to "preordain" how that person should spend the discretionary fund. And Varmus emphatically disagreed with what he saw as Delaney's negative portrayal of competition in science. Competition, he said, tended to improve the quality of research proposals.

Members of TAG similarly disliked "Marty's vague proposal" and they further complained that he distributed it without showing it to the people who supposedly helped draft it.[27] Academic researchers objected loudly, too. John Moore of the Aaron Diamond AIDS Research Center wrote an editorial entitled "Manhattan Project: It's a Bomb" for the *Advocate*, a popular gay magazine, that lambasted the whole idea of centralizing AIDS research.[28] "This approach is contrary to the way in which all great scientific discoveries have been made—free-spirited questioning of the unknown in investigator-initiated research projects," wrote Moore. "To apply office-management techniques to the utterly different discipline of the laboratory displays a profound ignorance of scientific research."

In face of the mounting opposition, Delaney, as promised, turned up the heat in Congress and the Clinton administration. Representative Nancy Pelosi on January 7, 1994, wrote Leon Panetta, a veteran California representative who left Congress to head the OMB, about the proposal. Pelosi suggested that Panetta add another $50 million to the OAR's $10 million discretionary fund. At the White House, AIDS czarina Gebbie fought a similar battle, arguing to boost the discretionary fund by a more modest $10 million to fund the initiative.

Both plans sank. Larry Kramer blamed TAG for "destroying" the Accelerated AIDS Research Initiative. "The thing that infuriates me is that TAG purports to speak for the whole gay community," Kramer told me. "These idiots, these children, have turned into a monster."

As much influence as TAG had, the strong resistance from academic scientists and Varmus surely contributed to the initiative's demise. Delaney's tight relationship with Gallo, who had many powerful scientific and political enemies, did not help, either. And, perhaps most importantly, when it came to closing the deal, Delaney's timing was off. Not only was the OAR leaderless, the attempts by Pelosi and Gebbie to increase the discretionary fund occurred late in the legislative session.

Representative Jerrold Nadler, a Democrat from New York, made a halfhearted stab in October at promoting a bill that would have turned ACT UP's Barbara McClintock Project to Cure AIDS into law. The bill never made it out of subcommittee. The next May, Nadler gave it one more go, renaming the bill the "AIDS Cure Project" (ACT UP deemed the Barbara McClintock reference too confusing), but changing little else. It, too, went nowhere fast.[29]

A Manhattan Project for AIDS had been promised by no less than the successful candidate for President of the United States. High-powered meetings with leading scientists, policy makers, and activists had been held to package it. The name had been changed repeatedly to make it more palatable. A respected congresswoman and the White House's own AIDS czarina tried to sell the idea to the OMB. The *New York Times* ran an op-ed by a famous AIDS activist that pleaded for it. Two congressional bills tried to legislate it. But a Manhattan Project for AIDS—no matter what it was called, no matter who sponsored it, no matter how it was packaged—was not to be.

The OAR reform did, in time, improve the field of AIDS research. When it came to the search for an AIDS vaccine, the OAR even proposed imaginative, targeted programs. But those, too, never came to life. The scientific culture in the United States simply hated the idea of directed research, and seeing as how no organized revolution could gain momentum—even though the revolutionaries had persuasive arguments—the culture easily held sway.

• • •

THE COMMITTEE LEADING THE SEARCH FOR the OAR directorship selected Harvard University's Bernard Fields, author of the classic textbook *Virology*, for the job, but Fields, who had pancreatic cancer, declined. In an unusual twist, search committee member Bill Paul, the eminent NIAID immunologist, was tapped by his fellow committee members for the job.

Although Fields could not in good conscience take the post, he badly wanted to help steer the field and put it on a surer course for effective treatments and vaccines. So the Harvard virologist wrote an impassioned commentary for *Nature*, published in the May 12, 1994, issue, that aimed to bring his wayward colleagues back to the church of basic research.

In "AIDS: Time to Turn to Basic Science," Fields contended that it no longer made sense to put as much emphasis on rushing experimental medicines into human trials. "In our zeal to control AIDS, we have invested enormous resources in the search for drugs and vaccines," wrote Fields. "This may have been reasonable 10 years ago, but is no longer."

Fields's main argument put an *Alice in Wonderland* spin on the polio vaccine story, which he used to illustrate the wonders of basic research. The National Foundation, after all, funded the Nobel Prize–winning work of Enders and his coworkers that discovered a way to culture viruses in test tubes, a basic research breakthrough that allowed Salk to make mass quantities of his vaccine. He closed by preaching the gospel. "Paradoxically, by targeting too narrowly, we may slow down progress in combatting AIDS," he wrote. "We must not compromise research in other areas of basic science at the expense of these directed programmes."[30]

Bernard Fields, who died from his pancreatic cancer on February 1, 1995, had a sharp and critical scientific mind that colleagues around the world admired. But the more profound paradox than the one he raised is how researchers of his acumen routinely framed basic versus targeted research in the *Endless Frontier* paradigm, without offering any hard data to back their claims or examining the shortcomings of the paradigm itself.

• • •

ORGANIZATION DOES NOT GUARANTEE ANSWERS to scientific questions. But the AIDS vaccine field could move more quickly at ruling possibilities in or out and finding probabilities. And that alone, when facing a virus that is racing through populations, is a compelling argument for more directed research. Especially when the counterarguments are so specious.

Targeted research projects *can* lead to abuses, as the Special Virus Cancer Program dramatically demonstrated. But when managed carefully and directed by the top scientists in the field, as was done by the National Foundation, they can accelerate the search for answers. Notably, Norton Zinder, the researcher who chaired the committee that shellacked the Special Virus Cancer Program, later headed a committee that organized the highly targeted Human Genome Project—which also met significant resistance. "The program was attacked by the entire community," Zinder told me. "It was going to destroy science."

Investigator-initiated, basic research *does* excel at producing knowledge and creative new strategies for uncovering nature's mysteries. But pitting basic against applied research is a dated construct: basic research is no better than other forms of funding at advancing a field, and it often contributes little to the discovery of new drugs and vaccines. As Maurice Hilleman told me in 1993, "I'd be the last person to say that all of this [basic research] isn't needed, but the AIDS problem is so devastating that what you want is answers now and understanding later."

Finally, the NIH's new and improved Office of AIDS Research *did*, eventually, better organize the field. But the OAR did not stage a concerted, all-out attack on the biggest obstacles facing AIDS vaccine researchers. And the office's overhaul, by creating the impression that the organizational problems had been solved, led to the early demise of bolder ideas, such as the Accelerated AIDS Research Initiative, that likely would have done more good than harm.

CHAPTER 10

The Dairymaids of AIDS

When the 38-year-old Glen Redman joined the Los Angeles Men's Study on August 14, 1984, he believed, with great certainty, that he already had been infected with human immunodeficiency virus, which scientists had conclusively proven three months earlier was the cause of AIDS. As the blond and compact Redman explained that day to an interviewer at the Gay and Lesbian Community Services Center in Hollywood, California, he had had sex with between 2,000 and 3,000 men. He hoped the L.A. Men's Study would give him access both to frontline AIDS researchers and the leading experimental treatments, improving his odds in the battle he surely would face against HIV.

Launched earlier in 1984 by researchers at the University of California at Los Angeles, the L.A. Men's Study aimed to track and analyze the spread of AIDS through gay and bisexual men in that city. Redman, a clerk-typist at the Los Angeles Department of Water and Power who had risen through the ranks to become an administrator, would become one of the study's 1,637 volunteers. Every six months he would donate blood samples, which the researchers would scrutinize for HIV, immunologic damage, and any clues they might find to help slow the course of disease and its spread. To assure confidentiality, the study gave Redman a number, 41001.

In August of 1984, no test yet had come to market that could detect an HIV infection, so the researchers could not tell patient 41001 or any of the other healthy men in the study whether they carried the virus. But based on the sexual history Redman reported, they concluded patient 41001 was at "high risk" for developing AIDS. They noted in his records that patient 41001 frequently visited sex clubs, where Redman said he "swallowed semen like a fish." They also recorded that for five years, two to four times each month, patient 41001 had enjoyed with a longtime lover—a man named

Jeff—what AIDS researchers antiseptically refer to as "unprotected, receptive anal intercourse."

Patient 41001's medical records did not tell the half of it. Redman was a randy 23 year old during the signal event that birthed the gay liberation movement: the 1969 raid by New York City police of a gay bar, the Stonewall Inn, that provoked an infamous riot. At the time of Stonewall, Redman often took advantage of bargain night at the Corral Club in Studio City, a bathhouse where the fair-haired, smooth-skinned young man would rendezvous with a swarthy, hairy man. "We got off on each other, do our number, and everybody would stand around and watch and appreciate," recalled Redman years later. "And then, you know, we'd split up and go do our business and hook up again later on. It was a whole different time."

The clubs became racier in the seventies. Once a week or so, Redman would visit Basic Plumbing on L.A.'s Fairfax Avenue, a neighborhood heavily populated by Orthodox Jews. Legend had it that Basic Plumbing required men seeking entrance to first lift their shirts to prove they had trim physiques. This was never an issue with Redman, who typically would visit one of the club's two dozen closet-sized rooms and take advantage of the holes in the walls to give or get an anonymous blow job. He suffered many attacks of diarrhea in those years, requiring him to take a medication so vile that he thought it would kill him. At first, he believed that swallowing massive quantities of ejaculate caused the diarrhea, but later he concluded that he had consumed fecal bacteria from going down on men who had just had anal sex and only did a cursory job of cleaning themselves. Redman stopped visiting Basic Plumbing.

Manspace, another favorite haunt back then, occupied the second floor of a dilapidated row of shops in Hollywood. Manspace specialized in what Redman called assplay. "It was infamous as a mecca for fist-fuckers and fuckees, regular buttfuck enthusiasts, rimmers, and scat devotees," remembered Redman. Nicknamed "Slip 'n' Slide" because of all the Crisco that had melted into the floorboards, Manspace to Redman was full of character—"sleazy, exotic, and fun," he said. Redman stopped visiting the club because it burned down. "Of course, the fire was more spectacular than it might otherwise have been, mostly because the Crisco in the floors and walls reportedly fueled the flames for hours," said Redman.

In May of 1984, scientists published the first compelling evidence that they had found the cause of AIDS, the virus now known as HIV. By the time Redman joined the L.A. Men's Study that summer, AIDS was hitting closer and closer to home, and the popularity of the bathhouses and sex clubs—which doctors long had argued were spreading the disease—was fading. Redman had broken up with Jeff, who in turn had moved back to his hometown in the Midwest. But the two men stayed in close contact. In the fall of 1984, Jeff complained to Redman of flu symptoms that would not go away.

In March 1985, when a blood test first came on the market that allowed the L.A. Men's Study to check participants for HIV antibodies, Redman opted not to learn his status because he "just assumed" that he too had the virus. And by then, AIDS clearly was killing Jeff.

Redman was openly homosexual and his elderly parents knew Jeff. Redman frequently relayed news to them about Jeff's health, chronicling in many letters how HIV was breaking down his former partner's body and mind. Jeff had "a regular (and bothersome) afternoon fever" one day and then all-night bouts of diarrhea two weeks later. He lost 10 pounds in two days and ended up in the hospital. He had "something on his arm removed for biopsy." Redman worried that Jeff concentrated too much on "worst-possible-scenario instead of spending his psychic energy on positive, recuperative thought," adding, "I don't think it's healthy to contemplate and prepare for one's own demise until it's at hand."

As Jeff's health declined, so did Redman's resolve not to contemplate his own future. "I continue to have a difficult time adjusting to his illness and probably should join a support group," Redman wrote his parents on August 27, 1985, adding a story about a "jerk" who came to his office that day. The jerk, visiting a training coordinator in the cubicle next to Redman's, asked, Do you know what gay means? "The answer was 'Got AIDS Yet!' Nice, huh?"

On Saturday night, February 8, 1986, Jeff died from the complications of AIDS. Although AIDS had by then killed nearly 9,000 people in the United States alone, the "gay plague" still carried great stigma with it, and some of Jeff's closest friends and family had abandoned him. Redman flew the next week to Jeff's Midwest home, where he attended a memorial service.

Not long after Redman returned to L.A., his new lover, Max, became ill. Max, also a participant in the L.A. Men's Study, asked for his HIV status. He was positive. Max suggested that Redman check his blood test results. Redman's blood harbored no sign of HIV. But he dismissed the results. "I just couldn't believe that I wasn't infected," he said.

• • •

IN DECEMBER 1997, Glen Redman was 52, gym solid, tanned, and but for a sniffle on the unusually crisp and clear L.A. day, the picture of health. Sitting in the antique-decorated living room of his plaster-walled Spanish home in the old Los Angeles neighborhood of Eagle Rock, he removed his shirt to show off the many tattoos that adorned his torso. One, on his right arm, resembled a tombstone. It simply said MAX, in lettering that Redman described as a blend of "Old English and East L.A.," framed by the years 1940 and 1991. Max, too, died from AIDS.

Like a veteran who returns from a bloody war theater physically unscathed, Redman had no explanation for his fate, and his blue eyes opened wide when he pondered *why me?*—or, more precisely, *why not me?* "I was absolutely floored when they first told me I was negative," said Redman. "I was unbelieving then, and I still am amazed."

But Redman, after subjecting his body to an intensive analysis by the UCLA researchers running the L.A. Men's Study, had finally accepted that the tests were accurate.

Over the years, the L.A. Men's Study found that HIV behaved differently in different people. The virus usually killed a person in about 12 years, but some people died within a few years of becoming infected, while others lived with HIV for more than a decade with no obvious damage done. Still others, despite having practiced "high-risk" sex, never tested positive. Redman and other EUs—shorthand for "exposed, uninfecteds"—by 1997 had become a hot research topic for the UCLA investigators, who wanted to understand how these men had dodged the bullet.

The 31 EUs being studied by the UCLA group in 1997 belonged to one of a dozen such cohorts around the world that researchers had begun to analyze. EUs by then had been found in children of infected mothers, heterosexual spouses of infected partners, hemophiliacs, injecting drug users, healthcare workers, and prostitutes in Kenya, Gambia, and the Congo.

Two possibilities explained the phenomenon. One was genetics, and, as several labs revealed in 1996, some EUs clearly carried mutant genes that deadbolted the cellular doors HIV must pass through to establish an infection. But genetic invulnerability, so far, had explained only a minority of the protection seen. A more likely explanation for the majority was immunologic. Maybe the route by which the virus first entered their bodies—say, oral sex—did not allow HIV to establish an infection but properly trained their immune systems how to fight it. Maybe their immune systems safely learned this lesson by initially being exposed to "subinfectious" low doses of virus or to defective, noninfectious HIV. Maybe their genes predisposed their immune systems to tilt one way or another. Maybe it was a combination of route, dose, and genetics.

Teasing out which specific immune responses succeed and fail against HIV is a messy business. But if history is any guide, it is well worth the trouble to study in great detail people who defy the odds and resist an infection, for they can reveal the secret to making an effective vaccine.

• • •

EDWARD JENNER was the first researcher to develop a vaccine and prove that it worked.

Jenner found the vaccine with the help of Sarah Nelmes, a British dairymaid who, while pulling the teats of her master's cow one day in May 1796, contracted cowpox. The cowpox virus presumably traveled from the udder of the cow to a cut on Nelmes's hand, the result of an earlier run-in with a thorn. But in 1796, no one even knew what a virus was or, for that matter, that microbes could cause disease.

As the cowpox took its course, Nelmes suffered the typical chills, headaches, and vomiting. But after a few days, a large bluish pustule on her hand and a few smaller ones on her wrist remained the only signs of the ailment.

Cowpox was a common malady in Berkeley, cow country situated 150 miles northwest of London in Gloucestershire County, and the case of Sarah Nelmes was nothing exceptional. But Edward Jenner, a country doctor, lived down the road from Nelmes and took a keen interest in her pustules.

The residents of English farm country had long believed that people who became infected with the relatively harmless cowpox would remain forever free of smallpox, a disease that killed at least 20% of the people whom it infected and often left survivors blind and with facial craters. Jenner wanted to put the farm lore to a test, and the large sore Nelmes had on her hand offered just what he needed.

Jenner did not originate the idea that a person could become resistant to a disease. Thucydides, the fifth-century-B.C. historian from Athens, wrote that Grecians who had survived the plague realized that they never suffered a second attack, and therefore could safely serve as caretakers of those dying from the disease.[1] As early as the sixth-century-A.D., the Chinese attempted to create smallpox resistance by artificially causing a mild form of smallpox itself. In a procedure called "variolation," the Chinese dried and powdered scabs scraped from sufferers of mild cases of smallpox, which they then packed into the noses of healthy people. In India, they variolated by rubbing pus from an infected person onto the skin of a healthy one.

Variolation also was popular in the Ottoman Empire, and the wife of the British ambassador to Constantinople imported the technique to England in the early 1700s. (Edward Jenner himself had been variolated as a schoolboy.) The Reverend Cotton Mather of Boston caught word of the practice from one of his African slaves and compelled a doctor friend in 1721 to use variolation to stop a smallpox outbreak in Boston. The practice became so popular in the New World that General George Washington mandated the variolation of all of his troops. Though the treatment worked some of the time, a few percent of its recipients died of smallpox caused by the variolation. The English and the Americans eventually outlawed the procedure.[2]

The beauty of cowpox, then, was that it potentially offered a safe means of creating the same miraculous results as variolation. In what one contemporary admirer called "the most curious experiment that ever was recorded in the history of physiology," Jenner took viscous "matter" from the large pustule Nelmes had on her hand. On May 14, 1796, Jenner inserted the diseased liquid into two, half-inch incisions made in the arm of a healthy eight-year-old boy, James Phipps. July 1, Jenner inoculated Phipps with matter taken from a ripe smallpox pustule to challenge the protection theoretically offered by the cowpox. No disease followed. Several months later, Jenner again challenged the boy with a smallpox inoculation. No disease. Cowpox, then known medically as variolae vaccinae, apparently had protected Phipps from smallpox.

Within a few years, Jenner had performed the experiment on more than 6,000 people and had repeatedly inoculated "poor Phipps," as he called the boy, with smallpox. The smallpox inoculations, reported Jenner, had no effect on any of these patients.

Though Jenner did not know why or how cowpox prevented smallpox, he approached his work scientifically, even going so far as to praise his legions of doubting colleagues. "The scepticism that appeared, even among the most enlightened of medical men when my sentiments on the important subject of the cow-pox were first promulgated, was highly laudable," wrote Jenner in the 1800 landmark paper, *A Continuation of Facts and Observations Relative to the Variolae Vaccinae, or Cow-Pox*:

> To have admitted the truth of a doctrine, at once so novel and so unlike any thing that ever had appeared in the annals of medicine, without the

> test of the most rigid scrutiny, would have bordered upon temerity; but now, when that scrutiny has taken place, not only among ourselves, but in the first professional circles in Europe, and when it has been uniformly found in such abundant instances of the genuine cow-pox in the way that has been described, is never afterwards at any period of its existence assailable by the smallpox, may I not with perfect confidence congratulate my country and society at large on their beholding, of extirpating from the earth a disease which is every hour devouring its victims; a disease that has ever been considered as the severest scourge of the human race![3]

It was heady and premature for Jenner to forecast that his technique would wipe smallpox from the earth. Yet time vindicated his proclamation. By 1977, smallpox had been eradicated from the globe. In 2000, the only viral samples known to exist—save the ones in the hands of bioterrorists—were in laboratory freezers at the Centers for Disease Control and Prevention in Atlanta, Georgia, and at the Center of Virology and Biotechnology in Novosibirsk, Russia.[4]

• • •

FROM THE START OF THE AIDS VACCINE SEARCH, the hunt for people who naturally develop immunity against a disease should have been high on the list of priorities. "Natural protection," after all, potentially can point the way to a vaccine, just as dairymaids had led Edward Jenner to discover an effective way to prevent smallpox. And it is not as though no one had raised the idea early on: David Nalin, the experienced vaccine maker at Merck, emphatically had suggested this at a well-attended 1987 AIDS vaccine meeting held by the Institute of Medicine (see Chapter 4). Yet no scientific journal carried a report about naturally protected people until 1992.[5] The field had such little interest in the topic, in fact, that by the time the *Journal of Infectious Diseases* published this paper,[6] coauthor Janis Giorgi, an immunologist at UCLA who had recruited five exposed but uninfected men to the study, no longer was investigating EUs. "It was so difficult to get that paper accepted," Giorgi complained to me at the time. "I haven't had the opportunity to continue. I can't survive. I don't have direct funding."

Ed Salinas, one of the five men Giorgi recruited to her study, had a story every bit as remarkable as Glen Redman's. A 41-year-old gay man who lived in Los Angeles, Salinas said he once tallied the number of his sexual partners, and it came to more than 1,000 men, most of whom he had met at bathhouses. "The sex becomes so addictive," said Salinas, who asked that I not use his real last name. "It starts out one night, then you're to two nights and then three and before you know it, the whole week's gone and you're there every night. And the part about it is you have sex with one partner and you're still not satisfied, then you'd go on." Yet Salinas, who had even had a longtime lover die from AIDS, tested negative for HIV.

Giorgi's paper said blood taken from Salinas and four other exposed, uninfected gay men showed that they had developed immune responses against HIV.

Discovering immune responses to HIV in EUs certainly required state-of-the-art, sophisticated immunology. It also was possible that, rather than immunologic factors protecting EUs, their cells, for genetic reasons, might have been less vulnerable to infection by the virus. Neither of those facts, though, satisfactorily accounted for the degree of skepticism and disinterest in the pioneering work done by Giorgi and her main collaborators, Gene Shearer and Mario Clerici of the National Cancer Institute. These were, without question, sophisticated immunologists who had developed state-of-the-art techniques. And no genetic explanation for protection had yet been discovered. What, then, best explains the field's disinterest? Scientific culture. From the time of HIV's discovery, virologists dominated AIDS research, and study after study examined how HIV caused disease—drowning out the immunologists who studied how the body protected itself from the virus. "We've discussed the cause of AIDS," Jonas Salk, who was enthralled by the EU data, told me in June 1992. "But we haven't discussed the cause of immunity."

Frustration with the limits of the prevailing culture, however, had led to a sea change—as the publication of Giorgi's paper indicated. Soon, several labs around the world began sifting through the blood of EUs in search of explanations for their apparent protection. Salk joined forces with Shearer and Clerici, popularizing the importance of the data. Even Giorgi, in time, won major funding from the National Institutes of Health to conduct the detailed analyses of Glen Redman and 30 other EUs.[7]

AIDS researchers in 1992 ventured into other uncharted immunologic terrain that held the promise of providing clues to vaccine developers. One of the most intriguing of these *Fantastic Voyage*–like journeys through the immune system investigated "long-term nonprogressors"—HIV-infected people who defied the odds and remained healthy more than a decade after becoming infected.

A common theme ran through all of the studies of EUs and long-term nonprogressors. At that point, the leading AIDS vaccines under development—the recombinant gp120 preparations made by Genentech and Chiron—aimed to trigger the immune system's production of antibodies against HIV, largely based on the rationale that antibodies explained the success of most, if not all, human vaccines on the market.[8] With stunning unanimity, these new investigations found that the often-neglected cell-mediated arm of the immune system appeared, by itself, able to protect people from infection by HIV and slow its relentless destruction of CD4 cells. "The whole AIDS thing was a virologist's game," said the NCI's Shearer, a lanky Kentucky native who spoke with a slight drawl. "Everybody went after antibodies because the assays were easy and available." Cell-mediated immunity, in contrast, took much more work to evaluate, and, said Shearer, "there was just a reluctance to look at it."

As Shearer, Salk, Daniel Zagury, and other immunologically oriented researchers had argued for years, cell-mediated immunity theoretically could mount the most effective defense against HIV. While antibodies primarily prevent HIV from infecting cells, the so-called killer cells—formally known as cytotoxic T lymphocytes, or CTLs—of the cell-mediated immune army can actually identify and obliterate cells that the virus has managed to infect. Now human data backed the theory. The findings also led researchers to probe the cell-mediated immune system's elaborate network of chemical messengers that control everything from the magnitude of the CTL response to the ability of an HIV-infected cell to produce new viruses.

The new insights energized the field. They offered exciting, fresh ways to think about HIV immunity. They challenged the dogma about natural protection and suggested that infection itself need not be a death sentence. They led to collaborations between researchers pursuing far-flung interests. They attracted first-rate immunologists to the field.

Best of all, novel hypotheses soon would emerge, and researchers could test them, build on leads, discover the dead ends, and, finally, decide whether the ideas held merit. But no such drive took place to address the questions raised by Ed Salinas, Glen Redman, and other people whose immune systems effectively handled HIV. Instead, disorganization reigned, with no one ever conducting definitive experiments that would transform the most critical questions into answers. And the once exhilarating hypotheses were relegated to the scientific retirement home, where, except for occasional visits from a few devoted researchers, they were all but forgotten.

• • •

THE CONTROLS OF AN EXPERIMENT often can teach an observant researcher more than the subject of the study. Gene Shearer first learned this lesson after flunking out of graduate school at the University of Tennessee and landing a job as a technician at the nearby Oak Ridge National Laboratory. The can-do Kentucky farm boy, who eventually earned his Ph.D. after seven years of night school and Saturday classes, would apply it many years later when he first entered the world of AIDS vaccine research.

By 1989, Shearer, then 52, had established himself as a specialist in cell-mediated immunity and had his own lab at the National Cancer Institute. Shearer had a Batman-like quality, a regular Joe in blue jeans who loved muscle cars and mountain bikes and would never be mistaken for a scientist—until the talk turned to immunology and he would articulate his novel theories, one of which almost won him a Nobel Prize.[9] He also had found a Robin to his Batman, Mario "Mago" Clerici, a young M.D. from Italy whose father was a prominent immunologist and an old friend of Shearer.

Shearer and Clerici became involved with AIDS vaccine research by helping researchers at the University of Maryland analyze data from their trial of the MicroGeneSys gp160 vaccine in uninfected people. The NCI scientists had the job of measuring cell-mediated immunity against HIV in blood samples taken from the vaccinated participants. Rather than hunting through these blood samples for killer cells that could recognize HIV—a cumbersome and relatively insensitive process—Shearer and Clerici looked upstream for the immune signals that ordered the killing. So instead of counting troops dispatched to a given location, Shearer and Clerici used test-tube experiments to monitor the immune system's generals that gave the orders. The generals in this case were none other than the T "helper" lymphocytes—the white blood cells studded with CD4 receptors that HIV steadily decimates.

When confronted with HIV, CD4s turn on cell-mediated immunity in two ways. First, they copy themselves, a process known as "proliferation." Second, they bark out commands by secreting chemical messengers, called cytokines.

After taking blood from people who received the vaccine and isolating their white blood cells, the researchers tickled the cells with pieces of HIV and then measured whether they proliferated. Shearer and Clerici also developed a more sensitive test that assessed whether HIV-tickled CD4s produced interleukin-2, a cytokine that specifically mobilizes killer cells.[10] As the researchers had hoped, the vaccinated people's CD4s tested positive on both the proliferation and the IL-2 assays.[11] This did not prove that the vaccine worked, but it documented that it did trigger a cell-mediated immune response.

More provocative still was the reaction of one man in the control group who had *not* received the vaccine.

When this 40-year-old gay man joined the study, he said that although he had been promiscuous in the past, he had been in a monogamous relationship with an uninfected man for the preceding eight months. He said he did not shoot up drugs, had not had a blood transfusion, and had not had any sexually transmitted diseases. In other words, he had not—recently at least—placed himself at high risk of contracting an HIV infection. Blood tests confirmed that he had no antibodies or cell-mediated immunity to HIV, nor did he have any evidence of the virus itself when the researchers searched for it with most sensitive test available, the polymerase chain reaction (PCR) assay.

Five months into the study, Shearer and Clerici surprisingly found that when they took a blood sample from the man and again goosed his cells with the HIV pieces, the cells secreted IL-2. The man still had no HIV antibodies, and the PCR assay could not detect any virus in his blood either. At 13 months, the same thing happened. Shearer worried that the IL-2 assay had a glitch: people in the control group, by definition, should test negative on it. Three months later, the results held steady. "I said, 'Maybe this is real,' " recalled Shearer.

At the man's next visit three months later, he had HIV antibodies, and a PCR test found evidence of the virus.[12] Maybe he had been exposed to HIV repeatedly during the study, but for some reason—cellular immunity?—he had resisted infection. "That really caught my attention," said Shearer. "Geez, you can have cellular immunity without antibodies. I got really excited and said, 'How many people like that are really out there?' "

• • •

DURING THE SAME PERIOD that Clerici and Shearer mused over the strange control in their experiment, UCLA epidemiologist Roger Detels asked Janis Giorgi and other colleagues to help him figure out why some gay men who practiced "very high risk" sex had no antibodies against HIV. Detels keyed into these cases because he led the Los Angeles Men's Study.

The researchers first isolated HIV, using a traditional assay, in 27 very high risk men who had no HIV antibodies. They then subjected a few stored samples to the more sensitive PCR test, which revealed that HIV could lurk undetected by the immune system for nearly three years. The implications of the paper, published in the June 1, 1989, issue of the *New England Journal of Medicine*, both baffled and frightened AIDS researchers.

Blood banks used an antibody test to screen for HIV-contaminated donations. Although the test did not catch every batch of tainted blood—it typically takes from a few weeks to a few months after being infected to develop the antibodies—this relatively cheap screen nearly eliminated the risk of HIV infection from a transfusion. But if HIV could "silently infect" people for a few years, as seemed to have happened with these men, then the entire blood supply faced a significant risk. Silently infected people also could unwittingly infect sexual partners. "The results are very shocking," David Imagawa, the UCLA researcher who isolated the virus from the men's blood, told the *Los Angeles Times*.

Despite the study having passed the peer-review process of a topnotch journal, it took a pasting from many AIDS researchers. In its page 1 story, the *L.A. Times* even made the unusual decision to include an anonymous quote from a "leading AIDS researcher" who questioned the study's accuracy.[13] Many suspected that the lab had suffered a contamination, which commonly happens with PCR. "I don't think there was anybody who believed that the study was true," said Giorgi. "Now Gene Shearer was different than that."

Early one morning at an immunology meeting in New Orleans, Giorgi bumped into Shearer and Clerici at the back of the room. Shearer wanted to know if, rather than being silently infected, some of the men in what became known as the Imagawa study had developed natural immunity to HIV. The three researchers quickly began talking about the work of Australian immunologist Chris Parish. "Up until then, nobody else who I'd talked to had ever heard of Chris Parish before," Giorgi said.

In 1971, the year before Giorgi began her graduate studies in immunology at the University of New Mexico School of Medicine, Parish published a dazzling study about the immune system. Parish showed in rats that antibodies and cell-mediated immunity had a seesaw relationship: if an animal made a lot of antibodies, it dampened the cell-mediated response, and vice versa. Parish additionally demonstrated that injecting a rat with a low dose of a foreign substance led to cell-mediated immunity, while a high dose triggered an antibody response.[14] Giorgi's doctoral thesis built on this observation, exploring how mice dispatched their antibody and cell-mediated forces.

Giorgi and Shearer independently had concluded that Parish's old work might help explain the phenomenon both of them had observed. What if a low dose of HIV immunized people by triggering a state of cell-mediated immunity that protected them? And what if stimulating antibody production actually worked against that mechanism of protection? "It's so basic to what I did as a graduate student," said Giorgi. "It just seemed like it would be the explanation for the exposed uninfected." They decided to collaborate, with Giorgi providing blood from high-risk men in the L.A. Men's Study, while Shearer and Clerici contributed their IL-2 assay. "We were both just delighted with the idea that we could do this," said Giorgi.

When Giorgi returned home, she ran an ad in the L.A. Men's Study newsletter to recruit HIV-uninfected men who said they had had unprotected, receptive anal intercourse twice or more in the preceding nine months with an HIV-infected partner. "It was right after Christmas, so there were plenty of people who had had recent high-risk exposure," said Giorgi.

Ed Salinas was of the men who responded to the ad.

Salinas had joined the L.A. Men's Study in 1984 for a simple reason: fear. "All of my friends, everybody was terrified that they were going to catch this and just die," said Salinas, a stocky, short man who acquired a passion for bathhouses while traveling in Europe as a bank auditor. "I felt the same because I had been pretty wild with the sex all over the place." After watching a close friend with AIDS return to his mother's home to die, Salinas decided that would never be his fate. As soon as he learned he was HIV-positive, he imagined, he would "just go on and end it right away." Salinas, who kept his homosexuality hidden, said he suspected that his family would have accepted that he was gay, "but it was like you'd have to be a child back at home, and I just didn't want to do that." But to his continuing surprise, Salinas kept testing negative for HIV, and he never had to make this decision.

Ed said sex for him was "mainly blow jobs"—raising the intriguing possibility that he first was exposed to a subinfectious, low dose of virus—and that he shied away from anally receptive intercourse because "if you do that, you're really, really gay." He acknowledged how odd this distinction sounded for a man with his past, but explained that he had a difficult time accepting his homosexuality. Still, he would "just go with it" to please some partners, and that is what led him to sign up for Giorgi's study.

Giorgi first presented data about the five men recruited for the study at NIAID's annual AIDS vaccine meeting in 1990. All had tested positive on the IL-2 assay, meaning their CD4s had recognized HIV and appropriately launched a cell-mediated immune attack. A group of men in the L.A. Men's Study who said they had not recently had high-risk sex served as controls, and, as expected, HIV did not lead their CD4s to secrete IL-2.

The preliminary data intrigued some of the few hundred scientists at the meeting, but it left open many questions. Two months into the study, one of the five men developed HIV antibodies. Maybe he had been silently infected, like the men in the Imagawa study. Although the other four men remained antibody free, maybe they would soon prove to be infected too. And Giorgi had yet to screen their blood for HIV with the PCR assay, the most direct and sensitive way to determine whether they harbored the virus.

Giorgi, Shearer, and Clerici kept at it, hoping to describe their work at the international AIDS conference in Florence, Italy, where more than 10,000 researchers would congregate for one week in June 1991. For AIDS researchers, winning a slot to give a talk at the then annual international conference indicated a stamp of approval that made it easier to find funding, attract collaborators, and publish the work. In all, the conference organizers received condensed versions, or abstracts, of 4,716 scientific papers and axed 1,299 of these. The organizers awarded a lucky 468 scientists "oral presentations," where they could narrate slide shows in halls lined with giant TV screens that projected their data. Another 2,949 abstracts earned an honorable mention to become "poster" presentations, which allowed the researchers to tack their work onto the cloth-covered corkboards that commonly divide office workers from each other. The study of the exposed, uninfected men—which now included PCR data showing that they indeed had no detectable HIV—warranted only a poster.[15]

The conference organizers did, however, choose Clerici to speak about the odd control they had found in the vaccine study, the man who tested positive on their IL-2 assay for more than a year before he showed any evidence of being infected. At the end of Clerici's talk, as is tradition, scientists in the audience lined up behind microphones set up around the room to ask questions. "You mean to tell me that we came all the way over here from the United States to hear this ton of garbage?" asked one.[16]

A few months later, the *NEJM* published a letter from Imagawa and Detels that offered more details about their earlier, controversial finding of 27 "silently infected" men. The researchers said they did not find evidence of a contamination, but they also had revised their hypothesis. Further testing, they wrote in their October 24, 1991, letter, revealed that only one of the 27 men had evidence of an HIV infection.[17] Perhaps, they suggested, the men had transiently been infected with HIV, and the original analysis had caught these men while they were in the process of aborting the virus. "I'm very suspicious that some people are at least resistant to HIV, but I'm very hesitant to say they're immune," Detels told me. "My feeling is it's real. What it represents is the question."

To Detels, who said he personally had taken much flak "since the day we published this damn thing," the data represented a promising means to slow the spread of HIV. Monkey and chimp studies unequivocally had shown that infection depended greatly on how much virus the animal received, the strain of the virus, and whether it entered via a vein, the anus, or the vagina. If these men truly could resist HIV, maybe they could handle the virus in most, but not all, situations, reducing their susceptibility. Would that a vaccine could do as much. But rather than seeing the Imagawa study as potentially leading the way to uncovering the biological factors that allowed these men to resist the virus, the field by and large thought it, too, best belonged in the circular file.

• • •

JONAS SALK hurried over to Peter Bretscher after hearing him speak about immune regulation at a symposium held March 27, 1992, at the La Jolla, California, institute that bore the famous polio vaccine developer's name.[18] Bretscher, a researcher at Canada's University of Saskatchewan and onetime collaborator with Chris Parish, had worked at the Salk Institute for Biological Studies two decades earlier, and the two men knew each other. Salk, adopting a seer's voice that he much enjoyed, told Bretscher that "some of the things said here could influence things in the future."

Bretscher's work in mice with a parasite that causes serious skin lesions particularly thrilled Salk. Bretscher showed how injections of low doses of the parasite "locked in" a cell-mediated immune response such that the mice subsequently resisted infection when challenged with high doses of the same parasite.[19] "I'd felt for years that people in AIDS [research] were not looking at cell-mediated immunity half enough," said Bretscher.

Salk then connected the dots to the work Giorgi had done with Clerici and Shearer. "When I discovered the Shearer/Bretscher stuff, I said, this is it," Salk told me during an excited phone call in June 1992, the same month the paper appeared about Ed Salinas and the four other EU men.

The addition of a high-profile and controversial scientist like Salk to the small cadre of researchers studying EUs brought immediate attention to this struggling subfield. Heightening the buzz, the EU data fit beautifully with new insights by Bretscher and other hard-core immunologists—a group who largely had avoided AIDS research—about the specific mechanisms that the immune system used to mount an antibody versus a cell-mediated response.

According to data from both mice and humans, two different types of T helper, or Th, cells regulate these immune responses. Th1 cells secrete IL-2 and other cytokines that turn on the cell-mediated arm of the immune system. Th2 cells, in contrast, pump out cytokines, such as IL-4 and IL-10, that gear up production of antibodies. And these two families of cytokines are "cross-regulatory," such that high levels of a Th1 cytokine can shut down the ability of Th2 cells to secrete their cytokines and vice versa.

At the annual international AIDS conference held in Amsterdam during July 1992, the then obscure Th1/Th2 theory and its relationship to EUs might have received scant attention—but not with Jonas Salk promoting it and 1,000 journalists covering the meeting.[20] While a researcher like Janis Giorgi would struggle to win a speaking slot at an international gathering, Salk's fame often afforded him special access, as occurred in Amsterdam, where the organizers added him to a "late-breaker" session to discuss data that Shearer would cover in an oral presentation later and that other researchers would detail in a poster.

A talk by Salk, however, had little in common with standard scientific presentations. This often led to fierce complaints, such as the famous gripe by a polio colleague in 1954 that Salk should "show us, not tell us" (see Chapter 2). Salk, though, saw himself as a breed apart. "I prefer to know what nature thinks, not what my colleagues think," he told me a few weeks before this talk. "I'm interested in patterns in nature. That's why I refer to myself as an artist of science."

Salk began his Amsterdam talk—momentously entitled "Reconceptualization of Requirements for Induction of HIV Immunity by Vaccination"—by explaining why he decided to abandon the whole, killed approach that he had earlier championed as the most promising strategy for a preventive vaccine. He described a new monkey challenge experiment, detailed in a conference poster, that he and his collaborators in France had conducted to test two different formulations of whole, killed SIV vaccines.[21] One of the vaccines stimulated mediocre antibody responses in the monkeys, but no detectable cell-mediated immunity. Monkeys injected with the other vaccine, however, produced both high levels of antibodies that could neutralize SIV and respectable cell-mediated immune responses. The researchers then challenged both groups of monkeys with SIV grown in monkey blood cells, and all of the animals became infected. This result led Salk to the conclusion that more was not necessarily more: Maybe, as he had observed in his own monkey studies decades before, antibody production dampened cell-mediated immunity.[22] In short order, he weaved in Shearer and Clerici's EU work, Bretscher's studies in mice, and Th1/Th2 theory.

At Shearer's presentation, he offered hard new data about EUs. Shearer and Clerici—who received direct funding from the NCI and hence, unlike Giorgi, did not have to win grants—had moved well beyond their first examination of five gay men, to study CD4 cells from many different groups of uninfected people who likely had been exposed to HIV. The researchers found that their HIV pieces triggered positive IL-2 responses in 20 of 32 gay men, 10 of 22 injecting drug users, 11 of 21 babies born to infected mothers, and 8 of 9 health-care workers who had jabbed themselves with HIV-contaminated needles. Positive responses for these uninfected people, then, ranged from 45% to 89%. As a control, they analyzed blood from 155 presumed unexposed people, and only 10, or 6.5%, reacted.

Data unveiled at the conference from other labs added to the momentum for the link between Th1/Th2 theory and EUs. Frank Plummer of the University of Manitoba and coworkers since 1985 had been following a group of prostitutes in Nairobi, Kenya. As they reported in a poster at the meeting, 29 of the women they were studying had remained HIV-antibody negative for more than two years, despite having similar risk factors as the 1,200 other women in the study: an average of five sexual partners a day, about 25% of whom were HIV infected. Although the investigators had only used PCR to hunt for HIV in blood samples from 13 of the women, all of these had tested negative.

Th1/Th2 theory intrigued many researchers, but the field as a whole still had deep reservations. The theory not only challenged AIDS vaccine researchers to think about the problem differently; it also suggested that they might have chosen a dangerous path. This point came up at a press conference at the Amsterdam meeting—a less genteel forum for questions than the scientific sessions—during which Martin Delaney of Project Inform confronted Salk with the garden-party skunk. "Would you discourage government groups from their plans for large-scale testing of preventive vaccines that are based on an antibody reaction?" asked Delaney.

If Th1/Th2 theory held true for HIV, then vaccines that primarily aimed to stimulate antibodies carried serious, unappreciated risks: they might actually thwart effective immune responses and increase the likelihood that someone would become infected. Salk played coy. "I'm not inclined to discourage anything, but rather to encourage some things more than others," he said. "In this instance, since this remains to be demonstrated, I think we have to learn by doing. The [monkey] experiment that we did that pushed things to the limit and said that's not the way to go is what caused me to stop, look, listen. I'm hoping the same thing will be true for all of the other work that other people are doing, that they will come to their own conclusion."

Adding to the misgivings that some researchers had about the theory, Shearer, during his Amsterdam talk, extended it to explain why people who become infected by HIV progress to disease. In this scenario, infected people contained the virus as long as they maintained a Th1 state, which meant secreting an abundance of IL-2 and other Th1 cytokines. But Shearer contended that a person who had "shifted" into a Th2 state would progress to disease. This hypothesis intriguingly suggested that analyzing people's cytokine profiles and classifying them as Th1 or Th2 could better predict their vulnerability to AIDS-related diseases than could their CD4 counts; in other words, an HIV-infected person who had a relatively low CD4 count but was in a Th1 state potentially was better off than a person in a Th2 state who had a higher CD4 level.

Other studies bolstered the idea that cell-mediated immunity could control HIV in people who did become infected. In particular, Jay Levy from the University of California at San Francisco had evidence that an elusive chemical secreted by CD8 cells—cellular siblings of CD4s—could suppress the ability of HIV to copy itself. In Amsterdam, he said that his studies of HIV-infected people who remained unharmed by the virus for a decade or more—the long-term nonprogressors—produced this "CD8 antiviral factor," or CAF, at a higher level than HIV-infected people who did suffer immunologic damage and disease.[23]

The next March, Shearer and Clerici published two papers backing the theory that the immune systems of HIV-infected people, over time, shifted from a Th1 to a Th2 state. Specifically, they analyzed 45 HIV-infected, but healthy, people over 15 months and found that production of IL-2 decreased, while IL-4, a Th2 cytokine, increased. Buttressing the thesis that a Th1 tilt protected HIV-infected people, 92% of the people who had strong IL-2 responses did not develop AIDS over a three-year period. In comparison, nearly half the people with weak IL-2 responses developed AIDS. "Whatever the mechanism, our test is predicative of who would develop AIDS," said Shearer.[24]

Critics of this thesis came rushing out, their guns loaded for bear. "This Th1/Th2 stuff, it's not totally wrong, but the package being pushed forward could be a real problem," Duke University's Dani Bolognesi warned me early on. "I've talked with the experts. The suggestion is, it's wildly misinterpreted. Antibodies versus cell-mediated immunity doesn't divide that way."

One of the experts was Sergio Romagnani, an immunologist at the University of Florence who first demonstrated in 1991 that humans, like mice, had discrete populations of Th1 and Th2 cells. Romagnani assailed Shearer and Clerici's theory about a shift and progression to AIDS. "This Th1/Th2 story has been considered by a lot of people in too much of a simple way," Romagnani told me. "This is a very confused situation, and we need to do other experiments."

In contrast to the NCI researchers, when Romagnani analyzed 50 HIV-infected people whose symptoms ranged from mild to severe, he did not find increasing levels of IL-4 or other Th2 cytokines. An analysis of Th1 and Th2 cells from infected people also did not support the idea that infected people shifted to a Th2 state as disease progressed. Romagnani insisted, too, that the Th1 and Th2 divisions are not absolute: Th1 cells, for example, do trigger production of antibodies, just not as effectively as Th2 cells do.

Similar data—and criticisms—came from another formidable force, NIAID director Anthony Fauci. "I think Shearer and Clerici have made an important contribution in focusing on profiles of cytokines they've seen, but they got a step ahead of their data," Fauci told me. "I'm not saying these people are not seeing these different cytokine profiles. It just may not have a damn thing to do with Th1/Th2." Rather, suggested Fauci, the supposed shift likely just reflected what happens when an infected person's CD4 cells decline.

Shearer, true to his unflappable demeanor, sloughed off the criticisms.[25] "The question becomes: How much right are we and how much wrong?" he said. "We all have to wait and see if it's been oversold."

When it came to Th1/Th2 theory and EUs, Romagnani had no comment. Fauci's only objection centered on the point raised by Martin Delaney at the Amsterdam press conference. "I don't know how Clerici and Shearer got caught up in antibodies 'bad' and cell-mediated immunity 'good.' " said Fauci. "That's absolutely nonsense. They're mutually compatible. It isn't that one is good and one is bad."

Data steadily mounted from AIDS researchers around the world that backed the link between Th1/Th2 theory and protection from HIV. But Salk's Amsterdam expectation that AIDS vaccine researchers would find it compelling and shift strategies was quaint. While the debunkers of Th1/Th2 theory concentrated on how it applied to infected people—not EUs—that detail escaped most researchers. Leading scientists, after all, had attacked the theory, and in the world of science, that meant the topic now had a taint. And not only did the message of Th1/Th2 theory cause discomfort in many researchers, so, too, did its main messenger, Jonas Salk.

• • •

HAD JONAS SALK'S LIFE UNFOLDED ACCORDING to his own script, he would have coached the AIDS vaccine field across the finish line, healing the world one more time, reinforcing his ideas about focusing on the Big Picture, and simultaneously deflating the Pooh-Bahs of science who had excluded him for most of his career from what he saw as their "cabal." Salk well recognized that many of his colleagues had little patience for his ideas and prophecies, altogether ignoring what he believed were the hottest leads. "It's as if I continue to be invisible," he told me shortly before the Amsterdam gathering. "I'm just misread. A number of times I've gotten up and spoken about cell-mediated immunity at these meetings and there's been dead silence." Still, he kept hoping beyond hope that his detractors, when reviewing the data from no less of an authority than nature itself, would at long last abandon their biases, stop their professional jockeying, and join him in solving the problem.

It was a romantic vision for a most romantic of men, a Don Quixote armored in a V-neck sweater and an ascot, a syringe rather than a lance tucked under his arm. "I want to communicate for those who are interested in reflecting rather than just collecting data," he explained at an AIDS vaccine meeting in 1993, offering one of his many sing-songy aphorisms. "I'm not trying to insist. I'm trying to suggest a way of thinking. I put this all in a testable hypothesis. Nature will have the last word. Nature has already done the experiments."[26] He thrilled to the attacks by the likes of Fauci and Romagnani, whom he saw as hopeless reductionists. "Will you break open my brain and show me which cell has memory for what?" he parried. "It's part of a system. And what we're seeing is a systemic effect."

In his own eyes, Salk was an accident of evolution, a mutant who, he liked to joke, had a strange birth defect: the ability to see the world differently. "I see things in an inverted way," he confided to me, noting that Leonardo da Vinci wrote from right to left. "I see things upside down or backwards." And because of his special vision, said Salk, he often saw things before others did. "I've suffered from this all my life," he said. Salk time and again would recount the complaint of a polio colleague,[27] who said, "You came to the right conclusions based on inadequate data."

Of course, if Salk had special access to a crystal ball—if he had the wisdom, as he liked to say, "to make retroactive judgments prospectively"—why, then, had he so ardently championed the whole, killed HIV approach? I asked him this shortly after his Th1/Th2 conversion, noting that had the field listened to him and moved quickly into human trials with a whole, killed preparation, the vaccine likely would have triggered production of antibodies that—if the new theory panned out—might have left people more vulnerable to HIV infection. The vaccine probably would have "disadvantaged" a person's cell-mediated immune response, Salk allowed. "We would have fallen short of what we would have liked to have seen," he said. But he doubted that his colleagues would have ostracized him for passionately pushing the whole, killed vaccine. "They'll forgive you for being wrong," Salk said, quoting his deceased friend and Manhattan Project physicist, Leo Szilard. "They'll never forgive you for being right." Anyway, he said, he realized the limitations of the whole, killed HIV vaccine strategy before a human trial took place. "It didn't happen because I covered my bases," he maintained. I later pushed Salk on this point again. "Look, you can't take it away from Columbus that he discovered America," he said to me testily. "And he wasn't looking for it."

The difficulty that Salk had acknowledging his limitations was his blessing and his curse. The very force that drove Salk to move faster than his more cautious colleagues alienated him from them—and not simply, as he liked to believe, because he had proved them wrong time and again. "Although he does not profess infallibility, he submits plausible, praiseworthy reasons for everything he has done," wrote Richard Carter in *Breakthrough*, his 1965 biography of Salk. "Some who know him, including some who love him, think that some of his inability to condone his frailties may have generated a protective inability to recognize them." This "gift of retroactive infallibility," as one person described it to Carter, also chipped away at Salk's credibility and, consequently, his ability to play head coach for a scientific field.[28]

For a few years, exciting data quelled the wariness many researchers felt toward Jonas Salk and the concerns they had about the criticisms of Fauci and Romagnani. Shearer, Clerici, and Plummer expanded their EU work, finding more and more evidence that the phenomenon was real and linked to cell-mediated immunity. And a torrent of similar reports came from labs around the world. In New York City, researchers studied "discordant" heterosexual couples, where one is infected and the other is not; in many of the 80 uninfected partners—who, on average, said they had had unprotected sex 500 times—their CD4 cells began dividing when confronted with HIV.[29] Israeli investigators described how CD8 cells from high-risk gay men secreted a substance similar to the CAF that Jay Levy had described in long-term nonprogressors.[30] At Tulane University in New Orleans, scientists analyzed blood from a small group of uninfected hemophiliacs who had received HIV-tainted clotting factor before tests existed to screen for the virus, and discovered that their CD8 cells also seemed trained to prevent an infection.[31] Immunologists in Oxford, England, went a step further, detecting in an uninfected newborn of HIV-infected parents a more specific form of cell-mediated immunity: killer cells that had been primed to recognize and eliminate HIV-infected cells.[32] The Oxford scientists later found HIV-specific killer cells in five uninfected Gambian prostitutes.[33] A French group reported that discordant, heterosexual couples had evidence of HIV-specific killer cells, too.[34]

Studies of exposed, uninfected people suddenly were hot. NIH held a small, three-day conference on the topic.[35] Leading biomedical journals devoted prominent attention to Th1/Th2 theory too. *Science* ran a bold perspective coauthored by Salk, Bretscher, Shearer, and Clerici that cautioned colleagues about antibody-based approaches, warning that "an attempt to have the best of both worlds may well be at the expense of the more important one."[36] The *Journal of the American Medical Association* published a paper by Shearer and Clerici that described cell-mediated immune responses to HIV in uninfected health-care workers who had been exposed to virus.[37] Articles discussing this once ignored topic began to appear in the *New York Times*, the *Washington Post*, the *Los Angeles Times*, and other major daily newspapers.[38]

With a groundswell of enthusiasm building, the next logical step would have been to test the theory in animals. Salk, with the help of Bretscher and the Immune Response Corporation, appeared poised to do just that.

Salk first wanted to find a vaccine that, given the appropriate low dose, could lock mice into a Th1 state. (Although mice cannot be infected with HIV, they do develop immune responses to the virus.) At Bretscher's suggestion, Salk thought they could best achieve this by stitching HIV genes into BCG, the tuberculosis vaccine, which contains live mycobacteria that would copy the HIV proteins. Their clever, if fanciful, idea was that an injection of BCG would automatically determine the proper low dose in each person: as soon as cell-mediated immunity kicked in, it would kill the mycobacteria, stopping the production of HIV proteins. Once they had established that this worked in mice, the researchers hoped to give monkeys an SIV version of the vaccine and then challenge them.

But by 1994, Salk's plan began to falter as his relationship with IRC deteriorated. From Salk's perspective, the company had trudged forward with the therapeutic vaccine trials, shamelessly shilling their results to investors, and fighting with its corporate partner, Rhône-Poulenc Rorer; Salk also felt that the company had made a particular fool of him the preceding summer at the international AIDS meeting in Berlin, where IRC held a press conference for journalists and awkwardly patched in investor analysts via satellite phone.[39] "I've gone along with them, and they've messed up everything, including me," he groused when we spoke that February. "The public I represent is depending on me and expecting that I'll behave in a responsible fashion. . . . I have things to rethink. I'm not going to allow myself to be connected to that kind of behavior." Two months later, Salk's frustration boiled over. "I'm looking at starting a new company," he told me. "This has just gone on too long."

Other researchers in 1993 and 1994 did conduct monkey experiments to test the thesis. Clerici, working with primate researchers at the NCI and the University of Washington, studied two monkeys that researchers had injected with SIV as part of an experiment to determine the dose necessary to cause infection. Both of these animals received extremely low doses of the virus, and neither became infected. Sixteen months later, researchers reused these animals in another such "dose-ranging" study, but this time, they gave each animal relatively high doses of SIV via the rectum. Again, neither monkey became persistently infected. Clerici ran the proliferation assay on blood from these animals and found that both reacted to SIV, meaning that they had cell-mediated immunity against the virus. In contrast, the researchers gave two other monkeys that had never been exposed to SIV the same intrarectal doses, and both readily became infected and developed disease.[40]

In an independent study, David Pauza and colleagues at the University of Wisconsin, Madison, had similar results, with three of four exposed monkeys resisting an intrarectal challenge. But when the researchers rechallenged the protected monkeys with an intravenous dose of SIV, all became infected. This outcome raised the intriguing possibility that a low dose immunization might work against sexual transmission but not dirty needles.[41]

The data from these monkey experiments, provocative as they were, stirred little interest. One problem was that these experiments had not so much tested vaccines as a potential correlate of immunity. A true vaccine would contain a form of the virus that could, without causing disease, teach an immune system how to fight the virus. The researchers here used the fully lethal virus itself at a low dose, a strategy that no one ever contemplated for humans.

The makeshift designs of these studies, which salvaged animals from other experiments, introduced another snag. When testing a new vaccine, researchers traditionally give it to an animal three or four times, spaced over a few months, before a challenge. This schedule boosts the immune system as much as possible, optimizing the system such that the vaccine has the best chance of working. If it does work, even to some degree, then scientists can build on the success. In these experiments, the scientists either did not boost the animals or they reexposed them to the virus without analyzing the booster's impact on the immune system.

Pauza, for one, noted that he would have liked to have explored the question of low-dose immunization more carefully. "We struggled very hard to even get the initial studies published," he told me in an e-mail. "I did not believe that in 1993–4, we could survive as a lab by pursuing the low-dose immunization approach."

Salk became increasingly disenchanted with the entire AIDS vaccine enterprise. "There are none so blind as those who *will* to not see," he would complain repeatedly. "They're groping," he said to me in July 1994. "They're feeling their way through it rather than thinking their way through it." He had decided not to attend the international AIDS conference to be held in Yokohama, Japan, the next month, the first one he had missed since Paris in 1986, the year he became involved with the disease. "I'd only poison the place," he said. "Whatever I bring is bad." Industry's only concern was how much product it could sell to how many people, he said. "They've come up with the value of a human," he said incredulously. He no longer had any faith whatsoever in the scientists or the institutions leading the search. "Those who are presumably the leaders do not really understand the nature of the problem," he said. "In lieu of any one person you need a minion. We're dealing with institutionology, and the institutions that presently exist are just inappropriate. Just adding money reinforces." By October, he told me he no longer even wanted to pursue his mouse and monkey experiments. "I'm not going to do animal studies," he proclaimed. "What did I do with flu and polio? Go into people. We've spent enough time on this."

In March 1995, Salk phoned me to vent about the slow pace of progress. "It's come to the point where next week I'm supposed to give a talk," he said.

"What the hell for?" He had particular disdain for the drive, launched by Bernie Fields's *Nature* editorial the year before, to encourage more basic research. "I'd like to see anybody dissect a squirrel to find out why it climbs a tree," he said. "Back to basics is a scam. It's an excuse for not finding out whether something works. Damn it all, I'm going to make it happen."

On the morning of June 23, 1995, Salk woke up feeling short of breath. Earlier in the week, he had undergone heart surgery to open a clogged artery, so he returned to the Green Hospital of Scripps Clinic, a few miles away from his La Jolla home at 8:30 that morning. Four hours later, his heart failed, and the 80-year-old Jonas Salk died.

• • •

WITH THE PASSING OF JONAS SALK, AIDS vaccine researchers lost what little interest they still had in Th1/Th2 theory, and the question of whether a low dose of an AIDS vaccine could stimulate cell-mediated immunity and prevent infection remained open.

A few primate researchers did continue to investigate the possibilities, but their experiments served to confuse more than to clarify. In 1995, Gerhard Hunsmann, a primate researcher in Gottingen, Germany, recycled five monkeys from an experiment that failed to infect them with a low-dose, intravenous injection of SIV. The animals neither had killer cells that recognized SIV nor did they produce Th1-like cytokines, but they did show evidence of cell-mediated immunity in the less convincing proliferation assay. Hunsmann and his coworkers then boosted this so-called proliferative response by injecting the animals with another low dose of SIV. Upon challenge, however, all of the animals became infected.[42]

The next year, a group in Sweden headed by veteran AIDS vaccine tester Gunnel Biberfeld recycled animals that scientists earlier had exposed to subinfectious intrarectal doses in other experiments. Biberfeld and colleagues challenged four uninfected monkeys that 18 to 40 months earlier had been intrarectally exposed to HIV-2, the less common form of the human AIDS virus and a much closer relative of SIV. Two of the animals completely resisted infection, which the researchers attributed to their strong killer cell responses to SIV, and a third became infected but powerfully suppressed the virus, suggesting that its immune system probably could stave off disease longer than an "unvaccinated" animal. The investigators had an odd result, though, when they repeated the experiment with uninfected monkeys that had been exposed to SIV rather than HIV-2: all three animals had cell-mediated immunity to SIV—two even had CTLs—but the challenge, done at 10 months, infected all of the animals.[43]

This jumble of primate data yielded one clear message: Monkeys exposed to a low dose of the AIDS virus, like people, can develop cell-mediated immunity, and this immunity can, in the absence of antibodies, protect the animals from SIV put into their rectums. After that, nothing is clear. No one knows why a low-dose rectal exposure would protect while an intravenous one would not. No one knows why HIV-2 primed the animals differently than SIV. No one knows whether repeated boosting with a low dose of a vaccine could drive up levels of cell-mediated immunity even higher. No one knows whether turning up antibody production turns down cell-mediated immunity in uninfected monkeys. And, most disconcerting of all, no one in 2000 was even investigating these and a long list of other unknowns first raised by the Th1/Th2 theory eight years before.

The most obvious unasked question was whether a vaccine given at a low dose could protect as powerfully as some of these low-dose exposures had. Beyond that fundamental query were others: How does low-dose vaccination compare with other strategies? Does it make sense to combine a low-dose vaccination with another vaccine approach? How about combining a low-dose vaccine with specific Th1 cytokines? Can a low-dose AIDS vaccine "lock in" an animal to a cell-mediated state?

Studies continue with exposed, uninfected humans. But, as Salk lamented, scientists mainly seek answers to the most basic questions—dissecting the squirrel to see why it climbs the tree—and there is no detectable push to translate those data into experimental vaccines that researchers can test in animals. The field also has made no effort to address, as Shearer put it, how many people like that are out there. This information could prove critical in evaluating the AIDS vaccines that have been developed. Consider: To determine whether an AIDS vaccine works, a trial must explicitly recruit people who are at high risk and HIV-negative—the very population where researchers find EUs. But, to date, no trial in a high-risk population has, at its start, explicitly checked the cell-mediated immune status of the participants. If, as Shearer and Clerici's studies suggest, EUs are relatively common in high-risk populations, this could hopelessly muddle the data from a trial, making it impossible to distinguish between protection conferred by a vaccine and by natural immunity.[44]

Shearer and Clerici long ago concluded that their colleagues are like a herd of buffalo on the prairie running in the wrong direction. "It's difficult to stop them and make them turn," said Clerici. "Maybe after almost 20 years of failing to develop a vaccine, it would be time to stop and search for other ways to fight the disease."

Clerici was not saying, with the certainty of a Jonas Salk, "I am right." He was saying that, because the field had not mustered its forces to address the issue, he did not know whether he was wrong. The AIDS vaccine field simply had too short an attention span and was too easy to distract, so it did not follow through on much of anything, just as you might expect of a teenage enterprise that had no strong guardians leading it forward.

CHAPTER 11

Perpetual Uncertainty

On October 31, 1993, Halloween, several laboratories spooked AIDS vaccine researchers with troubling new data that immediately put on hold the plans to stage large-scale efficacy trials of the leading preparations, setting in motion a series of events that dramatically reshaped the field. Enmity between the reductionists and the empiricists would grow to the point where they divided into two warring camps. Industry's wary view of its partnership with the National Institutes of Health would change to one of outright distrust. And rich and poor countries, which until then were working side by side, would face the discomfiting reality that they used different criteria to assess a vaccine's potential.

The disturbing Halloween news surfaced in Alexandria, Virginia, at the annual AIDS vaccine conference sponsored by the National Institute of Allergy and Infectious Diseases.[1] More than a year before this meeting, NIAID had announced that it planned to launch large-scale efficacy trials in the United States by the end of 1993 with the most promising AIDS vaccines.[2] NIAID's two favorites for this trial were the recombinant gp120 vaccines made by Genentech and, separately, by Biocine—the joint venture between the biotech firm Chiron and the pharmaceutical giant Ciba-Geigy. In small-scaled human trials, both of these vaccines had proven safe and able to teach the immune system how to make antibodies that could neutralize HIV. Yet at the Virginia conference that Halloween, three researchers gave back-to-back presentations that questioned whether the supposed neutralizing antibodies found in vaccinated people could truly disable the virus.

The neutralizing antibody dilemma that roiled the Virginia meeting came down to the credibility of an assay used by every laboratory. While many antibodies can bind to HIV, only neutralizing antibodies can prevent the virus from infecting cells. To assess whether a vaccine has trained the immune system to produce neutralizing antibodies against HIV, researchers run a laboratory test that mixes blood from a vaccinated person with the virus. If neutralizing antibodies are in the blood, HIV will be locked out of healthy cells.

Back then, everyone who ran the neutralizing assay grew the HIV they used for the test in immortalized cell lines, assuming that if antibodies could defeat this lab-adapted virus, they would also work against a real-world strain. But studies by David Ho first published in 1990 suggested that real-world strains of HIV, so-called primary isolates, behaved much differently from their lab-adapted cousins.[3] To leave no stone unturned, laboratories at Duke University, the Walter Reed Army Institute of Research, and Chiron independently ran the neutralizing antibody test with HIV that researchers had freshly harvested from patients. None of the labs—which analyzed blood from people who had received the vaccines (also called "immunogens") made by Genentech, Biocine, and four other companies—found antibodies that could neutralize these primary isolates of the virus.[4] "The obvious implication is that the immunogens used to date don't induce the breadth of response needed to neutralize primary isolates," Duke University's Tom Matthews said at the end of his presentation, adding with dark humor: "We hope we don't know what this means."

In workshops and hallway talk, researchers tried to digest the bad news. They debated the vagaries of the neutralizing antibody assays, and noted that the differences between the lab-adapted and primary isolates might have more to do with laboratory artifacts or different conditions used in the different tests. They also noted that one study presented at the meeting contained a smidgen of success: United Biomedical Inc., a Long Island, New York, company that had just started small trials of its AIDS vaccine in the United States and China, showed that blood from one vaccinated person neutralized a primary isolate. Still, few researchers could dismiss the negative data. "If it turns out that the data are real, any of the products that are presently in clinical trials are not going to be effective," predicted UBI's Wayne Koff, the former head of NIAID's AIDS vaccine branch.

The data sent NIAID reeling. "It certainly does make me anxious about going forward with large-scale efficacy trials," NIAID Director Anthony Fauci told me a few days after the meeting. Fauci and his staff decided to reevaluate the question in the spring, by which time they hoped to have a clearer sense of the importance of the neutralizing assay data. If it turned out that, as some suspected, the assay used to measure neutralizing antibodies of primary isolates suffered from technical problems, said Fauci, "You just wipe your brow and say, 'phew, thank God.'"

Following an article I wrote for *Science*, "Jitters Jeopardize AIDS Vaccine Trials," newspapers from coast to coast weighed in with overinflated, doom-and-gloom headlines. "AIDS Vaccines Seen Ineffective Outside the Lab," claimed the *New York Times*. "Anti-AIDS vaccines a bust," barked the blunter *New York Daily News*. "AIDS Vaccines May Be Too Risky for Human Tests," the *Arizona Republic* inaccurately warned.[5]

NIAID's reaction incensed representatives from Genentech and Chiron. Jack Obijeski, who ran the AIDS vaccine project at Genentech, complained that his company had manufactured 200,000 doses of HIV vaccine because it thought NIAID had made a commitment to moving forward. "To leave that vaccine on the shelf, something that might help someone—we think that's ridiculous," said Obijeski, who did not believe the primary isolate question should override other positive animal and human data from experiments with Genentech's vaccine. "If that's the case, this is a monumental disincentive for Genentech." And at Genentech, he cautioned, many other projects competed with the AIDS vaccine work for company resources. "What needs to be forthcoming is for the NIH not to dwaddle about, one step forward, one step back. That's what makes CEOs nervous."

Chiron's Dino Dina faced a similar crux with the executives at Biocine. But unlike Genentech, Biocine would have to build a plant before it could even produce the quantity of vaccine needed to stage an efficacy trial. "To move in that direction without some assurance that the vaccine will be tested is unthinkable," said Dina.

Many AIDS vaccine researchers worried, too, that the debate about the neutralization assay would derail the push to stage AIDS vaccine trials in poorer countries. By 1993, save for Daniel Zagury's ill-fated experiments in Zaire, no developing country that had been hard hit by AIDS had taken part in a vaccine test. The World Health Organization had, however, agreed to help Uganda, Rwanda, Brazil, and Thailand lay the foundation necessary to stage everything from small phase I safety studies to full-bore, phase III efficacy tests. Working in concert with the NIH and the U.S. military, the WHO had established collaborations among scientists in developed and developing countries. The same troika also had begun to characterize populations that were at high risk for infection and likely to participate in an efficacy trial when the time came. And WHO representatives had briefed nervous government officials, providing assurance that any trials that did take place would not exploit their people and would adhere to the most rigorous ethical and scientific standards. Yet one important player still was missing: industry.

Every AIDS vaccine then being tested was made from the strain of HIV that was most common in the United States and Europe. Theoretically, antibodies against this strain, known as subtype B, would have little if any impact on the half dozen other subtypes of the virus circulating around the world. So to poorer countries, the debate about the neutralization assay had an absurd quality: they simply wanted a vaccine to trigger antibodies that matched the strain of HIV in their country. Anything else was gravy. "If I had a vaccine made from a strain that was Ugandan, and knowing the state of the epidemic in my country, I'd do everything I could to tell my authorities, 'Look here, with so many dying and getting infected, we should be trying to get into efficacy trials,' " Edward Mbidde, director of the Uganda Cancer Institute, told me at the Virginia conference.

Chiron's Dina made it plain at the meeting that his company would not invest in making a vaccine based on a Ugandan strain if the NIH did not support efficacy trials of the gp120 vaccine designed for the United States and Europe. "If we want to move from type B viruses to other areas, we must work together," said Dina. Donald Burke, head of the U.S. military's AIDS vaccine program, underscored to the conference attendees that when it came to tailor-making an AIDS vaccine for a poor country, Chiron's reluctance hardly was an anomaly. For more than a year, Burke said, he had been trying to find a company willing to make a vaccine for the unique strain of virus circulating in Northern Thailand, where he was helping to lay the groundwork with Thai officials for trials. "I don't want to close the door, but none of the manufacturers have made a commitment up to this point," said Burke.

NIAID's backpedaling had another, less obvious divisive effect. Ideally, rich and poor countries would conduct, in parallel, tests of vaccines deemed by the scientific community to have the most promise, with the WHO and the NIH working in lockstep. Now it suddenly became apparent that different countries had different standards, as did the WHO and the NIH. "NIH can make a decision about the United States, but they cannot make a decision about the rest of the world," José Esparza, the WHO's chief of AIDS vaccine development, told me at the meeting. And the impassioned Esparza, a native of Venezuela who sported a broom mustache and a perpetually rumpled suit, stressed that vaccine development was an empirical business, rendering the neutralization of primary isolates little more than an academic dispute. "For every point you try to prove there's a counterpoint," said Esparza. "I think a trial will give you more information than 1,000 lab experiments."

So the commotion about the neutralization of primary isolates extended far beyond determining whether the Genentech and Biocine vaccines merited large-scale efficacy trials. As would become clear during the coming months, the issue ultimately centered on NIAID—which ran the best-organized, best-funded AIDS vaccine effort in the world—and whether it merited its role as leader of a field that, to more and more observers, appeared to be falling apart at the seams.

• • •

BEING A GOVERNMENT ORGANIZATION, NIAID could not make big moves without consulting outside advisers, who sat on various committees. These committees fit together like the wooden Russian doll that has inside it a smaller replica of itself, which has inside of it a smaller replica of itself, and so on. When it came to staging efficacy trials of AIDS vaccines, NIAID consulted with no fewer than three committees of advisers. NIAID established one of these committees, the HIV Vaccine Working Group, in 1992 largely to help the institute move AIDS vaccines into efficacy trials, and the 31 members all knew the issues intimately. The congressionally mandated AIDS Research Advisory Committee, or ARAC, had 15 members on a subcommittee that explicitly specialized in therapeutics, and 10 others who focused on vaccines; of these 10, five had little, if any, expert knowledge about AIDS vaccine research. The third group, the six-member AIDS subcommittee of NIAID's advisory council, had only two scientists involved with AIDS vaccine R&D.

Three times a year, NIAID held a joint meeting between the ARAC and the AIDS subcommittee of its advisory council. On March 2, 1994, the joint meeting spent much of its time exploring the AIDS vaccine efficacy trial conundrum. Ashley Haase, a retrovirologist at the University of Minnesota whose Buddy Holly glasses gave him a faintly hip air, chaired the meeting. Haase opened the discussion by announcing that, at the next joint meeting in June, these advisers would be asked to make a decision, which he described as "fateful" and "awesome," about whether NIAID should stage the risky and expensive real-world test of the gp120 vaccines. "The state is really one of confusion," said Haase.

As the advisers would hear during the discussion, a complex mix of scientific, financial, and ethical factors had pushed the gp120 vaccines made by Genentech and Biocine to the front of the field. None of these factors, when isolated, made a satisfying argument that these two vaccines held substantially more promise than the other ones being developed. Still, together they had managed to nudge the vaccines all the way to the final stage of testing, phase III, the efficacy trial, an experiment that would cost tens of millions of dollars and would require up to 10,000 volunteers who were at high risk of becoming infected.

Duke's Dani Bolognesi recounted that the main scientific argument in favor of the gp120 vaccines had, from the start, been about antibodies. It was a most mechanistic, reductionist perspective. The surface of HIV is studded with gp120, a molecule that can dock onto CD4 receptors, the first step in the infection process. If injecting gp120 into a person led to antibodies that could gum up that binding process, infection, in theory, would not occur. Genentech also had shown in the chimpanzee trial published by the journal *Nature* in 1990 that gp120 could protect these animals from an HIV challenge, while a vaccine made from gp160—the complete envelope protein, which MicroGeneSys, Immuno, Oncogen, and Daniel Zagury used—could not.[6] Genentech and Biocine had another putative scientific advantage: They made their vaccines from supposedly common strains of HIV—MN and SF2—while their competitors all used the rare strain known as LAI/3B. The antibodies stimulated by HIV_{MN} and HIV_{SF2}, the thinking went, logically would work best against similar strains, so the more common they were, the better the vaccines derived from them would be at protecting a population.

But powerful counterarguments beset these vaccines. For starters, scientists now had evidence that the antibodies spurred by the vaccines—which were made from lab-adapted strains of HIV—had no effect on primary isolates of the virus. "The bottom dropped out of the bucket," Bolognesi said of the finding. Maybe gp120 antibodies, as the reductionists maintained, could foil HIV, but relying on these particular gp120 antibodies to do the job appeared as foolhardy as fighting a shark with bare fists.

The success with chimpanzees withered upon close examination, too—and highlighted, once again, how disorganization impeded the search for an AIDS vaccine. Genentech made much ado about its chimp challenge in part because it had no other animal data to promote: monkey researchers who genetically engineered the SIV analog of the gp120 vaccine routinely failed to protect challenged animals.[7] And the chimp challenge itself had important limitations. Not only did Genentech use so few chimps that the results lacked statistical significance, the researchers, as was customary, gave the vaccine its best chance of working by waiting for the antibody levels in the animals to peak before challenging them. The route of the challenge, an intravenous injection, also might not reflect what happens during sexual exposure. A real-world vaccine additionally would need to work against a variety of strains of HIV, and this experiment had only tested gp120 made from HIV_{3B}—the strain Genentech used in its first generation preparation—against HIV_{3B}. "The call for more animal studies is I think a conclusion that I would make here, too," said Bolognesi. "The difficulty has been that we simply have not had the challenge stocks for the chimpanzees to do these studies."

Challenge stock is a batch of virus that scientists have characterized by determining the precise dose needed to infect an animal. Although the NIH had made a challenge stock of HIV_{3B}, Genentech did not want to test this against its MN-based vaccine because the result might be misleading: Even if the vaccine failed to protect against the distantly related 3B, the preparation still might have an impact in parts of the world where MN—and other MN-like strains—predominated. So, ideally, the company first wanted to test its vaccine against an MN challenge stock. But no one had characterized HIV_{MN} in chimps.

The AIDS vaccine enterprise failed at this task for a variety of reasons. One was expense. To make challenge stock, scientists must test various doses of the virus in several chimps to determine the amount needed to cause an infection, an experiment in the early 1990s that could cost as much as $500,000. Coupled with the daunting price tag, companies that needed challenge stock had to find someone with chimps who would run the experiment. In 1991, Jorg Eichberg, a veterinarian at the Southwest Foundation for Biomedical Research in San Antonio, Texas, who managed one of the largest private chimp colonies in the United States, told me that he had turned down several requests to make challenge stock as it was "scientifically unexciting work and you're really burning up animals." Eichberg, who then had performed more HIV chimp challenges than any researcher in the world, agreed with many of his colleagues that the federal government should do this type of work.[8] The NIH, however, was notoriously slow on the draw. MN had been identified in 1989 as a prevalent isolate of HIV, but the NIH did not try to make a chimp challenge stock from it until 1991. The ill-fated project crashed and burned in 1992 when Bolognesi and his coworkers, who were contracted to help with the project, realized that the MN they used in the experiment had been contaminated with 3B.[9]

Like Genentech, Biocine wanted to test its SF2-based vaccine against SF2 challenge stock before subjecting it to a challenge with a more divergent strain like 3B. The NIH agreed to make this challenge stock, but it did not become available until June 1993, and, as was first reported at this meeting,

Biocine had not challenged the animals until November 1993 and still, in March 1994, had no data.

To make matters worse, scientists had designated MN and SF2 as common strains based on a misleading classification scheme that likely had little meaning when it came to vaccines. An influential analysis conducted by a powerhouse team—which included Bolognesi, Scott Putney from Repligen, and Merck's Emilio Emini—classified the various strains of HIV based on the differences between the strings of amino acids that formed the small region of gp120 known as the V3 loop. By examining the 40 amino acids that make up the tip of the infamous loop taken from 245 different isolates, the researchers arrived at a "consensus sequence" made up of the most common amino acid in each of the 40 slots. They then ranked the prevalence of the 245 isolates by comparing each one's sequence with the consensus sequence. This exercise showed that MN was more prevalent than SF2, but that both were more prevalent than 3B.[10]

There was a method to this madness. Most AIDS researchers at the time believed that antibodies directed against the V3 loop could powerfully neutralize HIV. From a reductionist vantage, it followed, then, that if the HIV gp120 in a vaccine had a common V3 loop, the antibodies directed against it would work against the more common strains of HIV circulating in that population.

But this analysis had a major flaw: the scientists had based the comparison of V3 loops on laboratory-grown viruses. So even if the lab-adapted MN and SF2 had common V3 loops, the comparison only described how they stacked up against other lab-adapted isolates. Common *primary* isolates, the ones that caused infections, might have entirely unrelated V3 loops. It was as though researchers had determined the most common plants in a region by visiting nurseries there rather than taking samples from the wild.

On top of all these negatives, the Genentech and Biocine gp120 vaccines hardly stimulated cell-mediated immunity at all. Indeed, their emphasis on antibodies created an immune state that was precisely opposite to the one seen in exposed, uninfected people. While most AIDS researchers did not ascribe to the Th1/Th2 theory that such an antibody-only state would cripple the cell-mediated response, the utter lack of cell-mediated immunity in vaccinated people gave them great pause.

If scientific issues alone determined the fate of the Biocine and Genentech gp120 vaccines, they would have stood little chance of moving into efficacy trials that had the financial support and the imprimatur of the NIH. Mike

Saag, a University of Alabama clinician on the ARAC, minced no words. Saag said he was reminded of the David Crosby lyric "Beneath the surface of the mud / there is more mud." "There is a lot of data that goes in different directions simultaneously," said Saag. "Before we jump into complexity to create more mud, we need to have some momentum or some vector, some direction of data that is consistent, that is telling us that, empirically even, this is the way to go."

But science was just one part of the equation. Fauci and others at the NIH well realized that industry largely had avoided the AIDS vaccine field, leaving it wide open to biotech companies that had little experience, shallow pockets, and a shaky commitment to scientific experiments that aggressively evaluated their potential products. Chiron and Genentech, however, were different from the likes of Repligen, MicroGeneSys, and Viral Technologies. These two San Francisco Bay–area companies both had solid scientific reputations and had earnestly pursued the problem. Scientists from both companies also had integrated themselves into the AIDS vaccine community, collaborating with leaders in the field, regularly publishing papers in the best journals, and openly discussing their work at scientific meetings, including negative data like the primary isolate study done by Chiron. So NIAID officials and staffers did not want to alienate the most attractive industrial friends they had.

ARAC member Daniel Kisner starkly described the dilemma. Kisner, a former National Cancer Institute higher-up, knew from where he spoke. Kisner left the NCI to become an executive at a major pharmaceutical firm, Abbott Laboratories, and had since moved on to help run Isis, a San Diego biotech company. Kisner cautioned that the biotechnology investment community would watch NIAID's every move in this arena. "If they get the sense that this is not an area of fertile investment, I think you are going to see a lot of little companies disappear, and a source of potential future vaccines disappear," said Kisner. "That's not a threat, I think it's just a fact."

Kisner further emphasized that biotech companies did not have the resilience of pharmaceutical houses. "The difference between the large companies and small companies is very simple: most of the small companies have nothing to sell but good news, and bad news doesn't sell stock, and bad news causes damage to those companies," said Kisner. "So you are going to be faced with a biotechnology community that will react I think with some fear and panic with regards to decisions made here, should they be sort of consistently negative."

The relentless spread of HIV further increased the sense within the trenches at NIAID that they had to take their best shot today—even if it was in the dark. The institute's own calculations, as Division of AIDS head Dan Hoth had articulated at the Amsterdam conference in 1992, showed that a mediocre vaccine today could do more for a population than a great vaccine that comes along five years later. For some, this translated into an ethical argument. They had potential vaccines in hand that had proved safe and able to stimulate the immune system. How could they *not* test them in efficacy trials? So what that researchers had yet to identify the correlates of immunity for HIV. So what that envelope vaccines did not work in the monkey model. Scientists had produced evidence, skimpy as it was, that a gp120 vaccine had protected chimps. And vaccines for other diseases in which there was no "good" animal model and no known correlates of immunity had moved into efficacy trials: rotavirus, typhoid fever, malaria, pertussis. When efficacy trials failed, other vaccines moved forward. What was the big deal? Why was AIDS being held to a different standard? And if these vaccines did not enter efficacy trials, it would take a few years at a minimum before other vaccines made it through phase I and phase II testing.

Compelling counterarguments challenged these financial and ethical rationales. If Biocine and Genentech believed so strongly in their vaccines, they could fund their own efficacy trials. And when it came to ethics, staging efficacy trials did carry unique risks with a disease that was behaviorally, rather than casually, transmitted. Counselors certainly would advise people who joined vaccine trials not to assume that the injections they received offered them any protection, but what if the participants ignored this message, developed a false sense of security, and decided that they no longer had to use condoms or clean needles?

The arguments and counterarguments went round and round and round. But come the spring of 1994, the wheel would at long last stop spinning.

• • •

ON APRIL 21 AND 22, 1994, NIAID summoned the members of its third band of advisers, the HIV Vaccine Working Group, to a hotel near Virginia's Dulles Airport to help the institute decide whether to move the gp120 vaccines into efficacy trials. In addition to NIH staffers, the 31 members of the working group included vaccine makers, vaccine testers, virologists, immunologists, behavioralists, and AIDS activists. The diverse group had a much better handle on AIDS vaccine issues than the ARAC or the NIAID advisory council's subcommittee, which between them had only a handful of scientists who worked in the field. Yet the Vaccine Working Group's opinion carried little official weight, and whatever it recommended would simply be forwarded to the body that would offer Fauci a final verdict: the joint meeting of the other committees in June.

All of this advice, however, was just that: Fauci, in the end, could take it or leave it.

NIAID closed the meeting to the public. But I obtained official "confidential" minutes from the meeting and spoke with several of the participants, who, in addition to the Vaccine Working Group, included representatives from the Walter Reed Army Institute of Research, the Food and Drug Administration, the White House's AIDS Policy office, the WHO, and three different universities.

The meeting began with everyone stating their conflicts of interest, a result of Genentech's complaint that some members of the Vaccine Working Group had close ties to their competitors. Indeed, the chair of the group, Dani Bolognesi, had such a pronounced conflict of interest that he did not attend the meeting at all, which Ashley Haase agreed to chair.

Bolognesi's lab at Duke had joined forces with Gallo's lab at the NCI and Virogenetics, a New York biotechnology company that was 85% owned by the pharmaceutical giant Pasteur Mérieux Connaught.[11] The three groups were developing a vaccine by inserting HIV genes into a weakened form of canarypox, a cousin of smallpox that is harmless in humans; because this vector system produced HIV proteins in the human body, it theoretically would lead to a robust killer cell response. Human trials had by then begun in both the United States and France, showing that up to 40% of vaccinated people did produce HIV CTLs. NIAID believed that priming with this vaccine, combined with booster shots of an engineered envelope protein, had a better chance of defeating HIV than the envelope protein alone. This prime-boost strategy was nowhere near ready for an efficacy trial—NIAID had just started the phase I human trials—but in the development pipeline, it was widely seen as the next best alternative to gp120 alone.

Mary Lou Clements, who ran the AIDS vaccine trials at Johns Hopkins University and previously had worked on the WHO's smallpox eradication effort, gave the first presentation at the meeting. Clements recounted the limits of animal models, listing the many vaccines that had moved into efficacy trials without those data. She also highlighted the catch-22 of the demand that researchers identify correlates of immunity before launching efficacy trials. "Valid correlates of protective immunity against human disease are most often determined in a retrospective analysis of data from a clinical trial in which the vaccine conferred partial protection against naturally acquired disease," said Clements.[12] She ended by urging her colleagues to see that even if a vaccine proves worthless, an efficacy trial has great value as it can help researchers design new and improved preparations.

Next came the Biocine and Genentech scientists, who powwowed ahead of time and coordinated their talks. "We didn't want to get to the meeting and start peeing on each other's leg," remembers Jack Obijeski, who then headed Genentech's vaccine program. Although scientists from the two companies had cordial relations, their unusual united front for this meeting signified how deeply troubled they were by the NIH's 11th-hour misgivings about staging efficacy trials. "NIH was flabbergasted that we did a joint presentation," recalled Phillip Berman, the lead Genentech scientist. "I sensed incredulity."

Both companies conceded that antibodies from people given their vaccines could not neutralize primary isolates. But Berman questioned the validity of the assay. Fights about assays are legion in biomedical science and often become vitriolic. This one was true to form. Berman contended that the assay likely could not predict whether the Genentech or Chiron vaccines would protect humans. He gave a detailed description of the primary neutralization assay, first popularized by Vaccine Working Group member David Ho, saying it required scientists to idealize conditions for viral growth—conditions that did not exist in the human body. "It would be impossible to give the virus more of an advantage," said Berman, years later. "David Ho had a kind of hurt look on his face that I was challenging his assay." Told of Berman's memory, Ho said it was not his assay, and, anyway, HIV grows in "exactly" the same conditions in the body. "It is sad that he is still in denial," said Ho.

The Vaccine Working Group next heard an update from a phase II, placebo-controlled trial of both vaccines that involved 296 people, 60% of whom were deemed at high risk of becoming infected. The trial so far had found no serious adverse affects. Two of the people in this trial had become infected, but that held little meaning: The trial was still "blinded" such that no one yet knew whether these people had received placebo or a vaccine; if it turned out that they had been vaccinated, then the vaccines were not 100% effective, a level of efficacy no one expected anyway.

Finally, and most importantly, the companies had new data from chimp studies that used the recently made SF2 challenge stock. Biocine challenged two chimps that earlier had received its SF2-based, gp120 vaccine. One animal completely resisted the infection, while the other became infected briefly, but then appeared to clear the virus. Two control chimps readily became infected. Genentech challenged three chimps given its MN vaccine with the SF2 virus—which had 17% different amino acids in its gp120—to see whether the immunity against one strain could protect against another. None of the animals became infected while the one control did. This was the first time any vaccine had protected chimps from a strain of the virus that differed from the one in the vaccine. Tellingly, antibodies taken from the vaccinated chimps could not neutralize what the researchers said was a primary isolate of SF2.[13] So the inability of vaccine-stimulated antibodies to neutralize primary isolates in test-tube studies did not accurately predict what would happen in the challenge study. If the same held true in humans, then the jitters that afflicted AIDS vaccine researchers in the fall might have been for naught.

NIAID's Division of AIDS staffers came next. They had a strong self-interest in pushing these two vaccines up the development ladder. They had helped move the vaccines this far, establishing the AIDS Vaccine Evaluation Group, which funded centralized labs, a statistics operation, and a network of academic testing sites, known as AIDS Vaccine Evaluation Units, that would conduct the human trials. They also had built the HIV Vaccine Efficacy Trials Network, HIVNET, to prepare sites for efficacy trials by identifying the potential cohorts—groups of high-risk, uninfected volunteers—and establishing the baseline rate at which HIV was spreading. They had strong bonds with the scientists at Genentech and Chiron. So they, rightly, saw the decision about whether the gp120 vaccines deserved a phase III trial as a decision about whether their leadership had been sound.

After staffers overviewed the AIDS vaccine data from animal and human studies, epidemiologist Sten Vermund, chief of the clinical trials branch of NIAID's Division of AIDS, made a pragmatic suggestion. Since so much uncertainty surrounded the gp120 vaccines, Vermund raised the possibility of a midsized efficacy trial. "The meeting sentiment was that the [gp120] products did not look like 'grand slam home runs,' but that there was a need to study them further to see if they might be a 'single,' " remembered Vermund.

The great benefit of a "lite" efficacy trial was cost. Estimates suggested that trials, regardless of size, cost between $1,000 and $2,000 per participant per year. The full-scale trial, which Vermund designed to assess whether either of the two vaccines could protect 60% to 80% of vaccinated people, would probably take three years and 8,000 to 10,000 people. The cost, then, would total at least $24 million and perhaps as much as $60 million. The scaled-down trial called for enrolling about 4,500 people for three years.

A shortcoming of the smaller trial was that it would not have the statistical power to prove whether the preparations worked unless the vaccines had an efficacy of 80% to 90%. This presented a paradox: logically, the less faith scientists had in a vaccine, the larger a trial should be in order to detect even small differences between the control and vaccinated groups. Vermund, though, argued that the midsized trial could reveal hints of efficacy, if they existed, and then researchers would have a strong rationale for proceeding with full-scale phase III tests.

At the end of the presentation, the Vaccine Working Group hashed out the pros and cons, adding novel arguments to the debate. On the pro side, they speculated that even if the gp120 antibodies had a relatively small impact on HIV, they might give the immune system enough of a head start to mount an effective cell-mediated response and clear the virus. Failing that, maybe the improved immune response in vaccinated people would allow them to better handle an infection, slowing the course of disease.

The Vaccine Working Group members agreed that running more laboratory or animal challenge studies designed to predict efficacy was "futile without more clinical data," because regardless of what they found, scientists would still debate the meaning of the results. The group also noted that even if the vaccines failed to protect many people, much could be learned about correlates of immunity by comparing those who became infected with those who did not. They acknowledged, too, that launching efficacy trials with these vaccines did not preclude staging new efficacy trials as other preparations moved through the development pipeline; researchers ran nine efficacy trials involving hundreds of thousands of children while testing different *Haemophilus influenzae* type b vaccines.[14] Finally, they recognized that standing still might not only lead Biocine and Genentech to jump ship, but "could have a chilling effect on continued vaccine research and development by the pharmaceutical and biotechnology industries."

When it came to the cons, the group warned about the provocative possibility that proving the worth of mediocre vaccines carried great risks. What if a trial showed that gp120 vaccines had, say, 50% efficacy? If those vaccines won approval, the next-generation preparation that came along would have to be compared with them. Those future trials, then, would have to recruit "inordinately large cohorts to detect incremental increases in efficacy," the group noted. Future efficacy trials might also suffer if the vaccine failed completely, as researchers might face an ever tougher time recruiting volunteers.

At the meeting's end, Fauci, who attended, asked the Vaccine Working Group members whether NIAID should organize this midsized efficacy trial of the gp120 vaccines. "People were really sweating," said Pat Fast, chief of clinical R&D at the Division of AIDS. "They were really thinking. They really squirmed. They had to come up with a sensible decision, and it's not easy." Merck's Maurice Hilleman captured the sentiment of the vast majority at the meeting who, despite having a long list of reservations, wanted to move the gp120 concept forward. "It's not a bad enough vaccine not to consider it," said Hilleman. "I can't see any damn good reason not to do the trial." Only Harvard primate researcher Norman Letvin emphatically objected to the trial, because the vaccine did not stimulate killer cells. Australian immunologist Gordon Ada, who sat at the other end of the table, then changed his vote because, he said, "I feel badly for Norm." So it was nearly unanimous: hold your nose and go.[15]

The overwhelming support for staging a midsized efficacy trial left Fast and other NIAID staffers stunned—and elated. Their excitement, though, soon would change to exasperation.

• • •

ASIDE FROM THE TWO AIDS ACTIVISTS who were part of NIAID's Vaccine Working Group, Derek Hodel and Robert Vázquez, gay men and other communities that had been hard hit by HIV largely had avoided this issue. AIDS activists, powerful as they were, had concentrated on speeding the search for treatments, attacking discrimination, and protecting the privacy of infected people. But on May 9 and 10, 1994, Hodel, the treatment-issues director of a lobbying group called the AIDS Action Council, organized a meeting at the swank Mayflower Hotel in Washington, D.C., that, for the first time, created a public dialog about preventive AIDS vaccines among community advocates—the politically correct name for activists—as well as scientists, ethicists, and vaccine makers.[16]

One of the first speakers at the gathering was Vázquez, a gay Puerto Rican with AIDS who ran the Minority AIDS Coalition in Philadelphia. After making a crack about the U.S. government's 1984 AIDS vaccine predictions—"We won't be seeing Margaret Heckler on the Psychic Friends Network very soon"—Vázquez lamented his own community's indifference toward the research. "I don't know whether I should start laughing or screaming," Vázquez told the 300 attendees. "For the last few years, I've been talking to community people about preventive vaccines. . . . It's too complicated and they're too frightened to deal with it. Vaccines occupy the same wistful space as the dream of a cure. The vaccine is just part of AIDS mythology."

David Gold of Gay Men's Health Crisis later gave a talk that similarly emphasized the lack of community interest in the issue and tried to make sense of it. "If you look at all the groups involved in the AIDS advocacy, I don't know of one full-time position that is devoted solely to monitoring and advocating for effective HIV vaccine research," said Gold. "When you think of what the stakes are on a worldwide basis, that's a very telling figure."

Gold, a lapsed real estate lawyer, contended that HIV vaccine advocacy and treatment advocacy could work against each other. "We need to discuss openly the potential conflicts that we may personally feel," said Gold. "What happens in the unlikely possibility that there is a safe and effective vaccine? What does that do to treatment research? Dr. [Joep] Lange, of the World Health Organization, said, and I quote, last month, 'A successful AIDS vaccine would inevitably delay research for an outright cure.' If you look at polio research, once a vaccine was discovered, research into a cure effectively dried up."

Now, as the question of efficacy trials had moved to the fore, AIDS advocates recognized that they had to weigh in about HIV vaccine research—and their opinions would add another level of complexity to the go/no-go debate. A central unknown was whether researchers could recruit thousands of at-risk people in the United States to join an efficacy trial. In one study presented at the meeting, the Centers for Disease Control and Prevention surveyed 2,043 gay and bisexual men in San Francisco, Denver, and Chicago who reported either having had a sexually transmitted disease or anal sex without a condom. Only 6.6% said they were not willing to participate in an efficacy trial, with 37% reporting that they were "definitely willing," 36.5% "probably," and 20% "might be." A separate study done at Johns Hopkins University of injecting drug users found that 84.3% of the 375 people were "likely" to enroll in an efficacy trial, nearly half of whom said they were "very likely."[17]

On the surface, it appeared that recruiting qualified volunteers would not present a major stumbling block. But a closer analysis raised many quandaries. People given the gp120 vaccines could test positive on an HIV blood test, which looked for antibodies, not the virus itself. Most vaccinated people would have such a weak and transient antibody response that, realistically, this would not pose a major problem, but, still, it could lead to cases of discrimination—and researchers would have to warn people participating in a trial of this potential outcome. In the Hopkins study, 57% of the people surveyed said they would be "much less willing to participate" if they subsequently would test positive on the HIV blood test.

A behavioral researcher at Columbia University, Robert Fullilove, added the disquieting truth that scientists conducting the efficacy trials would have to do "a tremendous amount of trust building" to attract another pool of potential participants in the United States: poor minorities, living in the inner cities, who had high rates of HIV infection. "These trials are being considered at a time when there are tremendously high levels of distrust about what the government's intentions may be in affected communities where we might consider conducting these kinds of trials," said Fullilove, who also worked at Harlem Hospital. "What is it going to take in all of this to convince people that are more insecure that we're really interested in a protection experiment, not in attempting to do a redo of Tuskegee?"[18]

NIAID director Fauci said the lack of faith that scientists had in the gp120 vaccines further complicated the recruitment of volunteers. "Should you tell people that you don't think there's a great chance that this is going to be an effective vaccine, or should you tell them you think it's going to work?" wondered Fauci. "Because people will make the assumption that if the NIH and the groups that have come together have felt that we should go ahead, [then] they must really believe that it's going to work." Fauci said the only solution to this ethical dilemma was to make the trial not simply a vaccine trial, but one that attempted to prevent infection both by vaccination and by intensively counseling the participants about condoms, clean needles, and other ways to dodge HIV. "It's become apparent to me that if we're going to move ahead with this, we have to link it."

The coupling of counseling and vaccination, however, created its own quagmire. As Fauci pointed out, if counseling had a significant impact on reducing infection in both the placebo and the vaccinated groups, researchers might find it impossible to determine whether the vaccines had any worth. "If we are successful, the [prevention] trial will have been successful and the vaccine trial will have failed, something that is very difficult for people to comprehend," said Fauci.

Don Francis, who finished his 20-year tour of duty with the CDC in 1992 and the next year joined Genentech's AIDS vaccine program, had little patience for what he saw as a lot of hand-wringing by people who had unwarranted fears about moving forward. Declarations about gp120's worth were not only premature, complained Francis in his talk ("HIV Vaccine Efficacy Trials: Is it Now or Never?"), but foolhardy. "The reason for these empirical efficacy studies is that we really do not understand how vaccines work," said Francis. "Often we have some laboratory test that we think correlates with the protection, and we target our vaccine to maximize that testable response. But in reality, the immune response in humans is far more complex than can be determined by some simple laboratory test. Indeed, to say that we understand protective immunity I think is scientific arrogance."

Francis assailed Fauci's idea that the only ethical way to justify efficacy trials with what appeared to be second-rate vaccines was simultaneously to evaluate a risk-reduction program. "We are testing the efficacy of a vaccine," declared Francis. "We are not testing the efficacy of behavioral interventions. The counseling provided in an efficacy trial should be that which is standard in the community in which the trial is being conducted, no more, no less. If the standards of the clinics in the area are to dispense free condoms and needles, the vaccine clinic should do it. If not, it should not."

Never one to shy from striking moral high tones, Francis crescendoed with an FDR-ish plea. "We should neither fear failure nor fear success, and we certainly must not just sit around and fear everything," thundered Francis. "Momentum, courage, and action—that's what we need."

Chiron's Anne-Marie Duliège, chair of the session at which Francis spoke, had a smirk on her face. "Thank you for this, uh, speech," said Duliège. The room burst into laughter.

The introduction of Francis to the corporate side of the efficacy trial debate was no laughing matter though. Francis—main protagonist in *And the Band Played On*, a principal investigator in the efficacy trials of the hepatitis B vaccine, nemesis of Robert Gallo, former head of the CDC's AIDS lab, and a veteran of the smallpox eradication program—relished a good fight and was a media darling. Though some derided him as St. Francis, he'd flip a hank of his stringy, dirty blond hair out of his face and smile a toothy, defiant grin at his detractors. Yes, he had the religion: He believed his mission was to protect the health of the world. If outsiders carped that his corporate ties to Genentech muddied his motivations, well, to hell with them.

As he had hoped, Francis had riled the room. NIAID's Vermund came to the microphone and suggested that providing "prevention intervention" that matched the community standard was "almost Tuskegee-like," as many places did nothing. Francis volleyed that his proposal had nothing to do with Tuskegee, as he wasn't suggesting that researchers intentionally put trial participants at risk. "We've got to be very careful not to destroy efficacy trials with Cadillac, impractical behavior-change programs," said Francis. Vermund countered that the behavioral interventions also should not be a Yugo, a reference to the notorious Yugoslovian junker car. An AIDS activist from San Francisco next lit into Francis. "You suggested it's moral cowardice to not move forward, and that's outrageous," he said. "To not move forward in this context is of concern for the harm this vaccine might do to populations. And that's where folks are coming from."[19] A lawyer from the American Association of Physicians for Human Rights then advised Francis that his position might bite him back. "It's not very smart business for a company that has a profit incentive to create the impression that it is giving short shrift to the best prevention efforts," he said. "That is going to keep people away from participating in those trials."

At the end of the meeting, I spoke at length with Fauci.

All of the brambly issues that surrounded recruiting people into efficacy trials added to the great uncertainty that Fauci personally felt about his institution putting its reputation and resources behind two vaccines that he believed "quite frankly really don't indicate they're going to have a high probability of success."

On the one hand, the working group's deliberations in April had much impressed him. "It was extraordinary how unanimous the feeling was that, in some manner or form, we have to do something," Fauci told me. He also allowed that the new chimp challenge data offered "more fortification to move ahead."

Still, the proposal to stage midsized efficacy trials made Fauci particularly uneasy. An inherent illogic existed: upon learning bad news about the neutralization of primary isolates, his staff proposed staging trials that could detect only high levels of success. "That's why I'm having considerable trouble with that concept," Fauci said. He also had to confront the growing sentiment, crystallized by Bernard Fields's manifesto published in that week's *Nature*, that AIDS funders needed to boost basic research. "The focus on drugs and vaccines made sense a decade ago, but it is time to acknowledge that our best hunches have not paid off and are not likely to do so," wrote Harvard's Fields, a widely respected voice on weighty scientific matters.[20]

I asked Fauci whether he was leaning toward conducting efficacy trials with the gp120 vaccines or whether he would rather wait until something more promising came along. "To be honest with you," he said, "I don't know." One of his frustrated staffers described him as a tank with one tread.

• • •

ACROSS THE COUNTRY IN SAN FRANCISCO, a man in a phase II trial of the Biocine gp120 vaccine received the requisite four doses and then became infected. That did not surprise NIAID staffers: no vaccine protects every person who receives it, and a handful of other "breakthrough" infections had occurred in AIDS vaccine trials. But this man, it appeared, rapidly was losing CD4 cells. And that detail tripped alarm bells within NIAID because one of the other breakthrough cases, a woman whose steady partner was HIV-infected, had in a short 15 months seen her CD4s plummet to below 200; a healthy person has between 600 to 1,200 CD4s, which after infection by HIV depletes by about 60 per year.

A central ethical tenet of any clinical trial holds that researchers must fully inform all volunteers about the potential risks. NIAID helped each of the six academic centers that it designated as AIDS Vaccine Evaluation Units to write their informed consent forms. These highlighted that the vaccines did not contain infectious HIV and therefore could not cause an infection, that they could cause temporary side effects like pain from the injection, and that they were experimental and offered no direct benefit to a volunteer. But Jack Killen, director of the Division of AIDS, became concerned because the informed consent did not explicitly state that people who became infected by HIV after receiving the vaccine might, for some unknown reason, develop disease more quickly than unvaccinated people. Killen decided to halt enrollment in every trial. "Jack got very, very anxious about the breakthrough business," remembered Fast. Although she believed "there was legitimate reason to be concerned," she thought stopping enrollment was overkill. "I think it was done for show," she said. "It created all of this excitement and noise."

Late on the Friday afternoon of May 20, Fast faxed a notice to the AIDS Vaccine Evaluation Units that said all trials should stop enrolling volunteers until they had changed their consent forms. NIAID offered a model consent form, which noted that, as of May 1994, its sites had enrolled more than 1,400 uninfected people in 17 trials of 12 vaccines. Seven adults in these trials—some of whom had received the vaccine, others the placebo—had become infected through sex, the form explained. One vaccinated person, it noted, had had a steep decline in CD4s. (The San Francisco case proved to be a false alarm.) "If you do get infected, we do not know what effect the vaccine will have on the disease, if any," the form concluded. "The disease course might be the same or better, but you might become sick with AIDS even more quickly than expected. THEREFORE, PLEASE PROTECT YOURSELF AGAINST ANY EXPOSURE TO HIV THROUGH SEX OR INJECTION."

The next Tuesday, NIAID held a conference call with the AIDS Vaccine Data and Safety Monitoring Board, an independent group set up to oversee the trials and protect participants. In all, NIAID had now identified nine people in various trials that had received AIDS vaccines and later become infected. The board resolved that the informed consent forms should be changed, but to lift the hold on enrolling people in trials.

What would normally have been a minor incident exploded into a crisis the next Sunday, May 29, when the *Chicago Tribune* ran a front-page story by investigative reporter John Crewdson about the breakthrough cases: "AIDS Vaccine Study in Peril." The story, which mentioned only five of the cases, quoted Killen as saying they contributed to doubts "that this is going to be a really good vaccine."

During the next two days, Crewdson's story spread widely, and the facts became badly exaggerated. "5 Get Vaccine, Then AIDS Virus," stated the *New York Daily News* headline, which wrongly implied that participants had been infected by the vaccines. The usually august *New York Times* made a similar mistake, heading the wire story it ran with "5 in Vaccine Trials Develop AIDS Virus." *USA Today* had the biggest offender: "Volunteers for Vaccines Get HIV."

The media coverage made many AIDS vaccine researchers seethe. Barney Graham, who headed the AIDS Vaccine Evaluation Unit at Vanderbilt University, complained to me that the "twisted" stories had done harm to his program.[21] "It creates the general impression in the community that something is wrong with these vaccines," complained Graham, who even had to reassure physicians and staff at his hospital. Mary Lou Clements, director of the same testing program at Johns Hopkins, similarly was livid. "I'm so appalled by this feeding frenzy with half-baked data," said Clements.

The press, complained Clements, had not explained that many of the vaccinated people had yet to receive their full series of shots. In all, six of the 10 breakthrough cases were of people who became infected before receiving the minimum three injections typically needed for a vaccine to prime the immune system. "If they don't get all the shots, you can't expect a vaccine to work," said Clements. The press also failed to note that the breakthroughs involved five different vaccines.[22] And after checking around with epidemiologists, NIAID's Fast learned that rapid declines in CD4 cells happen in 3% to 10% of infected people, calming fears about the one such case they had observed.

Fauci acknowledged that, scientifically, the data from the breakthrough cases had little meaning. "It tells you for sure you aren't looking at a 100% effective vaccine," he said. "It doesn't tell you anything that has statistical significance or is even approaching statistical significance."

But Clements, Graham, and others feared that the press about the breakthrough cases would encourage NIAID's advisers to vote against the efficacy trials at their June meeting. "Perception is part of the decision tree, and the way this has been distorted puts a very negative perception on this," said Clements. Graham was certain it had influenced the political side of the equation. "Political decisions are mixed up with emotions," said Graham. "But emotions don't necessarily line up with the data."

Fauci conceded that the breakthrough cases were not "irrelevant" and said the coverage "absolutely tilts the political framework" for deciding whether the NIH should launch the efficacy trials.

ARAC member Nancy Haigwood, who left the AIDS vaccine program at Chiron to join the star-crossed effort at Bristol-Myers Squibb, hinted that these fears were well grounded. "Many people on the committee have been hoping to find positive data," said Haigwood. "This isn't positive."

• • •

Before and after. The epidemic would cleave lives in two, the way a great war or depression presents a commonly understood point of reference around which an entire society defines itself.

—RANDY SHILTS, *And the Band Played On*

FOR AIDS VACCINE RESEARCHERS, the June 17, 1994, meeting that once again brought together the vaccine subcommittee of the NIAID advisory council and the AIDS Research Advisory Committee became a defining moment, a point of reference, the before and after event.

The ARAC meeting, as it became called, attracted more public attention than any AIDS vaccine issue ever had. Held in the aptly named Crystal Ballroom of the Bethesda Hyatt located down the block from the NIH campus, the daylong gathering attracted what the *New York Times Magazine* rightly noted was "an extraordinary 55 reporters and 15 camera crews, hoping for scandal."[23] Representatives from Genentech and Biocine had come out in force, as had NIAID staffers, academic researchers, and AIDS activists. At the front of the room, seated at tables that had been arranged to form a giant rectangle, were 21 members of the ARAC and another seven from NIAID's advisory council.

The ARAC meeting had parallels with the surreal events that unfolded on the Los Angeles freeways that very day, where O. J. Simpson crouched in the backseat of his Bronco, a gun to his head, while TV cameras recorded the crowds of onlookers who stood on the overpasses, observing the low-speed chase. The ARAC panelists had a gun to their head: they had to make a momentous decision by the end of that day. TV crews, not one of which had ever attended an ARAC meeting, now had come out in force, while the sidelines were lined with a more discreet collection of onlookers observing a different type of low-speed event. And no one knew where the proceedings would end.

A few moments after Fauci began his introductory remarks, he had a problem with his microphone, which a TV cameraman walked over and fixed. "You guys make vaccines?" Fauci cracked. Waves of laughter rolled through the room.

Fauci's eloquent opening talk, however, attempted mightily to conceal that he had but the slightest of hope for the gp120 vaccines. "People of good faith and integrity can and have articulated cogently on either side of this argument," he said. After ticking off the pros and cons, Fauci said he would gladly answer clarifying questions during the day, but he had come to observe, not to participate. "Good luck to all of us," he concluded.

Ashley Haase, once again the meeting chair, ran through the deliberations of the April meeting of the HIV Vaccine Working Group. Haase explained that the Vaccine Working Group evaluated whether the basic science called for advancing the gp120 vaccines into efficacy trials. Other concerns, Haase said, were "really somewhat beyond the scope" of that group. "What is different this morning is that it is ARAC that has the responsibility of taking all of these issues into account—the science, the social, the ethical and the other dimensions." This was a curious distinction—the working group meeting had included intense philosophical discussions about social, ethical, and other dimensions. None of the five scientists on the panel who had been at the Vaccine Working Group meeting challenged him on this point, however. This created the false notion that the ARAC members would more thoroughly discuss the issue, which, supposedly, they could better evaluate in all of its complexity.

The presentations that followed Haase's opening talk closely mirrored the ones made to the working group. NIAID staffers once again overviewed the data from animal and human studies, different potential designs for efficacy trials, and the feasibility of recruiting people. Derek Hodel, as he had at the working group, discussed the concerns of potential volunteers. Genentech and Chiron scientists recounted their rationales for proceeding.

A few important new details did surface. In particular, NIAID's statisticians had worked out the specifics of staging a large efficacy trial versus a midsized one.

To design an efficacy trial, researchers juggle four variables: the number of people participating, the trial's "endpoint," the length of the trial, and the rate of new infections in that population. Changing any one of these impacts the others.

In this case, the trial designers chose to assess the endpoint of sterilizing immunity—complete protection from infection—a standard, interestingly, that most primate researchers by then had decided was pie-in-the-sky. Yet if the efficacy trial designers had set a more relaxed endpoint and assessed, say, whether the vaccine could prevent or delay disease, they would either have had to follow the trial participants for many more years, enroll many more people, or test the vaccine in a population that had a terrifically high new infection rate.

Because this theoretical trial would take place in the United States, the designers assumed a new infection rate of 2% per year—roughly what researchers observed in gay men living in U.S. cities. The large efficacy trial thus would need to enroll 9,000 people, with equal numbers receiving the Biocine vaccine, the Genentech product, and a placebo. This trial would take 3.5 years, and have the statistical power to detect with great certainty a vaccine that "worked"—that is, created sterilizing immunity—in 40% or more of the people who received it. Using cost estimates of $1,000 to $2,000 per participant per year, the total price tag for this trial ran from a low of $31.5 million to a high of $63 million.

By comparison, the smaller trial would involve only 4,500 people and take two years, costing between $9 million and $18 million—less, incidentally, than the amount the Congress appropriated for an efficacy trial of the therapeutic MicroGeneSys vaccine. But the smaller trial could detect efficacy, with certainty, only if the vaccine worked 60% of the time or more. The trial could give glimmers of efficacy in the 30% to 60% range, but, to prove this convincingly—and to get licensure from the FDA—researchers still would have to follow up with a larger trial.

In his talk, Derek Hodel insightfully explored the risks faced by people who might join a vaccine trial. "We sometimes reach for pristine data before subjecting volunteers to vaccine trials, as though there exists no risk for those not yet infected," Hodel said. "For those of us who are HIV-negative, it is important to remember that the risk for HIV infection, although a statistical one, is hardly neutral."

Genentech's Don Francis came next. Prophet, scold, public health crusader, voice of industry, ex-government insider, battle-scarred vaccine tester—he stacked all of these hats on his head. No more could be learned in the laboratory, he reiterated, and it was high time to take to the field. "You are talking about 750,000 dead Americans from the HIV epidemic, if it stopped today," implored Francis. "Adding up all the war mortalities of the Civil War, World War I, World War II, the Korean War, and Vietnam War, they come up to only a portion of the deaths that already we can expect from HIV infection, unless there is some dramatic intervention that we can apply." He assailed the nail-biters who agonized over whether enough people would volunteer for the trial. "To say that these can't be done, I think, is naïve," chided Francis. "We should expect chaos, and they can be done." Keep the momentum going, he pleaded. Genentech had 15,000 vials of gp120 vaccine sitting on the shelf and had engineered enough of the protein to make 300,000 more doses. He ended his oratory with a 1968 quote from no less than a former NIH director, Donald Frederickson:

> Field trials are indispensable. They will continue to be an ordeal. They lack glamor, they strain our resources and our patience, and they protract the moment of truth to excruciating limits. Still, they are among the most challenging tests of our skills. I have no doubt that when the problem is well chosen, the study is appropriately designed, and that when all the populations concerned are made aware of the route and the goal, the reward can be commensurate with the effort. If, in major medical dilemmas, the alternative is to pay the cost of perpetual uncertainty, have we really any choice?[24]

Chiron scientists followed with, as usual, a much cooler presentation, and, in a startling last-minute move, they even attempted to separate themselves from Genentech. "We recognize that fundamentally, probably, we only need to test one gp120 vaccine and not necessarily a series of them, especially if they are perceived as being indistinguishable or very similar," said Anne-Marie Duliège.

Potentially, Genentech and Biocine vaccines had an important difference: their gp120s were mixed with different adjuvants, the immune-stimulating chemicals used to boost the wallop of the viral protein. But the message came through clearly that the days of one-for-all and all-for-one were over. A free-for-all had begun, with every party protecting its own interests.

The next portion of the meeting, which opened the microphones for public comment, revealed even more plainly how divisive this issue had become. Don Burke, head of the U.S. military's AIDS research program, spoke first, urging that the panel state that its decision applied only to the United States. In particular, U.S. Army researchers had worked with scientists in Thailand for 35 years to combat diseases, and they had begun gearing up to test vaccines against HIV, which was ravaging that country. Because of the severity of the epidemic in Thailand, said Burke, his colleagues there used a different calculus to weigh the science and ethics of testing the gp120 vaccines, and he read from a letter written by them to Fauci that made this point.[25]

Next came a surprising barrage of negativity from a group that had, until then, rarely said anything about AIDS vaccine R&D: AIDS activists. "You shouldn't take lightly the fact that there is growing opposition from communities both in the United States and outside of the United States to the trials as they are planned now," warned a representative from ACT UP's New York branch. Indeed, ACT UP New York distributed a flyer at the meeting that threatened "a massive boycott" of these "extremely unethical and dangerous" tests. Robert Vázquez, the Vaccine Working Group member, went to the mike to warn about the "very large fear" in the HIV-infected community, which he was part of, that a vaccine would quash treatment research. "It would be terrible to see vaccine trial sites, you know, with [people with AIDS] demonstrating in front of them," said Vázquez. "That is a real possibility." More cautionary words came from a member of the influential Treatment Action Group, an ACT UP Golden Gate representative, and a gay, former Boston City Council member.[26]

For the next several hours, the panelists wrestled with the slippery beast that had been brought before them. Many questioned the Vaccine Working Group's conclusion that the only way to learn more about these vaccines was to stage efficacy trials. "I think it is ridiculous to say we can't learn more about these things by studying them further without going into an efficacy trial," said Bolognesi, who, although absent from the Vaccine Working Group meeting because of a conflict of interest, participated in this meeting as an ARAC member.[27] "I think that is absolute nonsense."

Several panel members suggested that they might be persuaded to give a green light if more chimpanzee experiments had been done to test the durability and breadth of the immunity induced by the vaccines. Alan Schultz, head of NIAID's AIDS vaccine branch at the Division of AIDS, reminded the gathering why the working group had determined that nothing more could be learned from animal studies: Researchers would argue endlessly whether the model, be it a chimp or a monkey, truly reflects what happens in humans. And the ever practical Schultz, who viewed the AIDS vaccine world with a wry detachment, turned the table on the panelists. "If in fact this group or some decision-making group would say, 'If you did this experiment and if you got that result, we will do the trial,' then, in fact, I think maybe those experiments would be done," said Schultz. As diplomatically as he could, Schultz explained that the real message coming from this panel was, "Well, go ahead and spend a half million on a couple of chimpanzees, show us the results, and we'll think about it."

Other panel members asserted that the ongoing phase II trials in humans would likely inform the efficacy trial decision. Studies of the breakthrough infections, for example, promised to uncover clues about the worth of the vaccines. If the vaccinated person had high levels of anti-gp120 antibodies at the time of the infection, this would add to the sense that they had meager protective power. And a comparison of the viral strain that infected a person and the strain used in the vaccine would offer insights, too: if they were similar, then this would be additional evidence that the vaccine had little worth.

This line of reasoning made sense, but again, it went against the logic that ruled at the Vaccine Working Group. A phase II trial, no matter how it is sliced and diced, cannot definitively answer the fundamental question, How well do the gp120 vaccines work? Ed Tramont, a member of the NIAID Advisory Council who had attended the Vaccine Working Group meeting as an observer, became particularly agitated on this point when one member of the panel went so far as to state that she wanted to know "what kind of [efficacy rates] you are going to see" in the phase II trial before proceeding to phase III. "What more do you want out of your phase II trial?" snapped Tramont, the University of Maryland researcher who formerly headed the U.S. Army's AIDS research program. "You have shown safety and you have shown immunogenicity as defined by neutralizing antibody to laboratory strains. Now, you want more than that?"

Another NIAID advisory council member, Larry Corey, cut to the quick. "From a statistical point of view, it is a $10 million or $12 million investment that really doesn't tell us what the scientific questions that are to be articulated, and I find the concept of that to be a very difficult one, and one that I am really having a large degree of problem with," said Corey, who had been testing AIDS vaccines at the University of Washington. Corey, then, like many others on the panel, did not believe that the midsized trial could even do what its proponents said it could. Fauci, in his only comment during the entire seven-hour proceeding, noted that NIAID did not even have $10 million budgeted for efficacy trials. "If we do that and it stays within NIAID, which it likely will, we would have to divert money away from other endeavors."

Repeated assurances that other products in the pipeline soon would be ready for efficacy trials bolstered the naysayers. "[T]he very realistic likelihood is [that] within two to three years we would be ready to go with a different concept," said Killen. Corey begged to differ, predicting that an efficacy trial could begin with the canarypox-boost idea in "12 to 14 months from today."

Further off, but even more exciting, was a recently discovered, revolutionary vaccine strategy that potentially offered a simple, safe, efficient, and cheap way to stimulate production of killer cells without using a vector or an attenuated AIDS virus. Called "naked DNA," this vaccine contained nothing more than HIV genes stitched into a circular piece of bacterial DNA called a plasmid.[28] When injected into muscle cells, the naked DNA produced the viral proteins. Naked DNA, also known as genetic immunization or gene vaccines, bowled over AIDS vaccine researchers. "This is one of the most exciting things in modern vaccinology," Hilleman had told me earlier. "It's a new world out there." Duke's Bolognesi agreed, praising naked DNA vaccines as "a tremendous breakthrough" and "on everyone's top 10 list as a possibility."

Yet a half dozen other panel members countered that the midsized study, which one member joked was a phase 2.76 trial, had much to offer. Maybe the trial, empirical as it was, would not end the perpetual uncertainty—but it might. Staging the trial also would bring the whole debate back to earth, demonstrating to the public, as one panelist put it, that they did not have "any grandiose expectations" about these vaccines, and that this was one of what likely would be many such trials. "Doing nothing at this point, I think, would be a mistake," declared Nancy Haigwood, a basic researcher on the

ARAC who had once helped Chiron developed its gp120 vaccine. "I have this sense that somehow, in the midst of looking for the perfect product, we have forgotten how far we have come. . . . I think we may see some surprises. Let's do the experiment."

As the meeting neared its scheduled 5:30 P.M. ending time, the panelists began scrambling for ways to reach consensus. Fauci wanted a vote, to recommend, yes or no, whether NIAID should move these vaccines into larger trials. One panelist who shared Haigwood's belief that a midsized trial might hold surprises and accelerate the field saw the proposal as something of a hedge bet: he imagined that, if the smaller test left questions open, in two years these vaccines and others that had advanced through the pipeline could all be compared in a full-fledged efficacy trial. Another ARAC member asserted that the sticking point was the word "efficacy," and he proposed they avoid that language and do the midsized trial. Yet another member, frustrated by all the incertitude in the room, suggested they postpone the decision and leave it to a new committee that included members from these panels and the Vaccine Working Group.

Mary Lou Clements paced the sidelines, livid. "This," said Clements, "is a disaster."

• • •

IN THE RUSH TO END THE ARAC MEETING, the panel cobbled together a proposed recommendation and voted. It read, in full:

> Continue current gp120 program and development of other candidates currently under study. Proceed with expanded clinical trials and evaluation when another concept and/or compelling data from current studies are available.

Although the language sounded positive, it was anything but. Translated into plain English, it said, The Vaccine Working Group was wrong: do not stage the midsized efficacy trial.

The panel, remarkably, unanimously supported the proposal. Some of the supporters of the midsized trial, like Haigwood, had abstained from voting because they had conflicts of interest. Other supporters, like Susan Zolla-Pazner—an immunologist at New York University who voted for the trial as a Vaccine Working Group adviser and against it as an ARAC—member-later said they did not understand the full implication of the vote. "Being given a day, or even a few hours, of unstructured discussion and quiet reflection might well have changed many votes," Zolla-Pazner told me years later. "It would have changed mine. But none of us on the committee had that luxury—or the presence of mind to ask for it—before casting our votes."

Immediately after the vote, Fauci and the panel chair, Ashley Haase, held a press conference. I asked Haase why they had not voted on the essential question before them. The group, he said, basically could not do it. "That's why the discussion went on and on," Haase said. "We couldn't get any closure on the idea." Fauci added that the point was moot, because the recommendation they did make nixed the midsized trial. "You vote for one, you rule out the other," said Fauci. He then announced that he planned to accept the group's recommendation: no go.

Fauci tried mightily to put a positive spin on the story, insisting that it was neither a "setback" nor a "disappointment." One journalist asked why the public should conclude that this group was right and the Vaccine Working Group was wrong. "I don't think this should be [described as] one group said one thing and another group said another thing," said Fauci. "Each of the groups were asked different questions and they merged into each other." And the Vaccine Working Group, he maintained, had reached a "conclusion" not a "recommendation."

Don Francis said he was "absolutely shocked" by the decision. "It was so wild going from one extreme to the other," said Francis, who did not conceal his disgust. "I don't see how this decision can hold. It's so illogical, and it was taken casually."

The decision did hold, and, over the years, it would be scrutinized incessantly. Some, like the epidemiologist who first proposed the midsized trial, Sten Vermund of the University of Alabama at Birmingham, would accuse Fauci, who did not like the working group's recommendation, of having "orchestrated" the vote at the ARAC meeting through friends on the panel. If you look at the "EXACT WORDING on the ARAC resolution, you'll see what I mean by NIAID's manipulation on that vote," Vermund wrote me in a 1998 e-mail.

Others would point to subsequent events that proved how shortsighted and misinformed the ARAC panel had been. New vaccine candidates would not prove ready for efficacy trials within 3 years, and the canarypox-boost approach, the one NIAID had banked on as the next best hope, lost much of its allure as other ideas moved forward. As many had warned, the no-go decision crippled the AIDS vaccine projects at both Genentech and Chiron;

Genentech, in fact, would bail out of the effort altogether, leading Francis to build an offshoot company, VaxGen, that existed only to raise money to stage the efficacy trials. In 1998, four years after this meeting, VaxGen actually would launch a 5,000-person efficacy trial in the United States of a slightly improved version of the Genentech gp120 vaccine.[29]

A more dispassionate view of the vote argues that confusion, rather than manipulation, carried the day.

In the end, the ARAC decision saved the NIH money that many researchers believed it could have afforded to gamble. And the decision came at a cost that few opponents rarely acknowledged: It further divided an already fractious field. NIAID, as a result, lost what little power it had to steer the agenda. And the decision revealed for all the world to see that the people in charge felt more comfortable with consensus and the perpetual uncertainty it fostered than with taking a bold, controversial position that, scientifically dubious as it was, could answer, with enough finality to satisfy both sides in the debate, whether an approach they once fervently believed in should be scuttled or celebrated.

CHAPTER 12

New World Order

In October 1994, four months after the ARAC decision, the World Health Organization called a meeting in Geneva to evaluate whether efficacy trials with the gp120 vaccines made sense in countries like Thailand that had more severe epidemics than the one in the United States. John Moore, the biochemist from the Aaron Diamond AIDS Research Center who specialized in HIV antibodies, emerged as one of the strongest critics of the vaccines, and on the second morning of the WHO meeting, Don Francis cornered him. Francis had a copy of a note that Moore had written to a few basic researchers at Genentech, offering his condolences about the ARAC decision. How, Francis wanted to know, did Moore reconcile the kind words in his note with his behavior at this meeting?

"Get real," Moore remembered telling Francis. It was true: Moore *had* the previous spring appeared to support the expanded trials of the gp120 vaccines. First, in May, following the HIV Vaccine Working Group decision, he faxed a congratulatory note to Genentech's Phil Berman, a sometime collaborator. "I'm very pleased, and I'm sure you must be relieved that the stress of the past few months has been worthwhile," wrote Moore. "Well done!" Then, as Francis reminded Moore, the ARAC decision led him to send the note offering his condolences. But during that summer, Moore had been invited to attend the WHO meeting and, as a result, studied the issue more closely. "You formulate your views when you have to," Moore later explained to me. Moore concluded that the gp120 vaccines had all the promise of dishwater. He told Francis as much and walked off, incensed by what he saw as Francis's cheek.

By then, October 14, 1994, Moore had a reputation as the bad boy of AIDS vaccine research, a Brit with a wicked sense of humor who described himself as a "fat bastard" and often would give a scientific presentation wearing a sweatshirt that broadcast a smartass message. One of these sweatshirts said, UPON THE ADVICE OF MY ATTORNEY, I HAVE NO COMMENT AT THIS TIME. The joke was that Moore, who grew up in the gritty town of Liverpool before leaving for Cambridge University, rarely held his tongue, especially when he thought scientists had played fast and loose with data, which for years led him to zing MicroGeneSys for hyping the wonders of its therapeutic AIDS vaccine. But infamous as Moore was to AIDS researchers, he had not been anointed into the inner circle that steered policy at the National Institutes of Health or anywhere else, and many colleagues viewed him as a loudmouth, the sidekick of his more famous boss, David Ho. Still, Moore had published extensively on gp120 antibodies, and his influence, as his invitation to the Geneva meeting attested, rapidly was growing—as was his disdain for the gp120 vaccines.

Moore's reservations about the gp120 vaccines had intensified, he said, in part because during that summer, many colleagues he respected voiced their own pessimism, as negative data about the approach kept amassing. The way Moore saw it, the Division of AIDS at the National Institute of Allergy and Infectious Diseases was lousy with mid-level officials who badly wanted to stage the gp120 efficacy trials. Because these officials could influence whether an academic received funding from NIAID—which had the biggest AIDS research war chest in the world—Moore contended that they created "a climate of fear." Basically, he said, no one wanted to risk crossing them on such a critical issue. "After ARAC, a lot of people realized that they could afford to express their opinions without being punished," Moore asserted.

Moore's reservations about Francis had increased too. Following the ARAC decision, Francis told me that Genentech planned to take its case to Congress and the executive branch, which I reported in *Science*.[1] To Moore, this foreshadowed a MicroGeneSys-like scandal. And while Genentech did have lobbyists in D.C., Francis, unlike MicroGeneSys's Frank Volvovitz, knew how to draw attention to the gp120 case without their help. A July 19, 1994, op-ed piece in the *Washington Post* that Francis cowrote proved this point. Written with Donald Kennedy, the former FDA commissioner and Stanford University president, the op-ed was headlined "A Private Sector AIDS Vaccine? Don't Hold Your Breath."[2] The carefully wordsmithed piece diplomatically implied, repeatedly, that the NIH had dropped the ball and suggested that "the government"—which was code for the Centers for Disease Control and Prevention and the U.S. Army—had to do more.

Later that day at the Geneva meeting, Moore publicly struck back at

Francis. Genentech, Francis reminded the group, had manufactured thousands of doses of vaccine. "Are you really proposing that we should keep the vaccine in the freezer for the next three years?" Francis asked.

"No, Don, I'm not suggesting you should leave it in the freezer," replied Moore. "You should turn the damn thing off."

The feud between empiricists like Francis and reductionists like Moore had been simmering for years. But until this pithy dig, public exchanges always had the civilized, genteel tones that marked scientific debates: maybe start with a compliment, acknowledge the other person's point, then beg to differ. Now war had been declared.

What had changed? The ARAC decision shifted the tectonic plates beneath the AIDS vaccine world, fundamentally altering the landscape and the prominence of its inhabitants. Here was the WHO, until then a side player that specialized in establishing ethical guidelines and collaborations between rich- and poor-country scientists, calling a meeting that, in effect, would decide whether the gp120 vaccines should move forward. Here was Thailand replacing the United States as ground zero for AIDS vaccine efficacy trials. Here were the CDC and, separately, the Walter Reed Army Institute of Research, both of which had established programs in Thailand, moving ahead of the NIH as the U.S. government's main advocates of efficacy trials.

And that was just for starters. At the NIH itself, the newly rehabbed Office of AIDS Research soon would eclipse the influence of NIAID and Tony Fauci. A private group sponsored by the Rockefeller Foundation, the International AIDS Vaccine Initiative, or IAVI, would come to life, prodding both industry and governments to do more. A new company, VaxGen, would emerge from the ashes of Genentech's AIDS vaccine program and make as its sole goal the funding of gp120 efficacy trials. A company, Acrogen, would surface that sang the virtues of the whole, killed HIV approach. Activists would launch the AIDS Vaccine Advocacy Coalition, their first organization dedicated to the endeavor. And the White House would form a Presidential Advisory Council on HIV/AIDS, which would aggressively assert its views about how best to proceed, eventually influencing Clinton to take action himself.

Much progress occurred because of this new world order. The gp120 vaccines did advance. Creative plans surfaced to improve the marketplace and spur more companies to devote more attention to the field, including the tailoring of vaccines for developing countries. NIAID seriously searched its soul about the organization of its AIDS vaccine program, which ended up winning much better funding and building stronger partnerships with industry. A drive by NIH's OAR, IAVI, and others explicitly sought out innovative vaccine ideas. The notion of testing risky vaccines in populations that had explosive new infection rates gained currency. Something of a sense of urgency began to build too.

But as earnestly as all these new agenda setters attempted to fill the vacuum of leadership left in the aftermath of the ARAC decision, none fully succeeded. Everything still took too long. Coordination remained the exception rather than the rule. And many of the best ideas to emerge from these yeasty days became mere words on paper, left to yellow on personal library shelves that were beginning to buckle from all the similar trailblazing ideas that had preceded them.

• • •

JOHN MOORE WAS not the only one to run down the gp120 vaccines at the WHO's October 13–14, 1994, meeting, "Scientific and Public Health Rationale for HIV Vaccine Efficacy Trials." Jean-Paul Lévy, director of France's version of the NIH—an important backer of Pasteur Mérieux Connaught's AIDS vaccine program—strongly objected, maintaining that the vaccines not only appeared worthless but could harm people.[3] Lévy worried that the antibodies against gp120 might *enhance* the ability of HIV to establish an infection. Although such enhancing antibodies, which had been made famous by dengue hemorrhagic fever, had led to much hand-wringing by AIDS researchers in the late 1980s, the concern eventually faded because little data supported the possibility.[4] Yet Lévy's had such strong convictions that he made a dramatic exit from the meeting. "How much did Genentech pay you?" Lévy shouted at the WHO's José Esparza, the meeting's organizer.

In the end though, Lévy was the only one of the 26 participants who objected to the group's conclusion that it made sense for poor countries with high rates of HIV infection to test the gp120 vaccines if they so chose.

The decision, if not the meeting itself, required something of a decoder ring to understand. The cash-strapped WHO had no money to fund phase III efficacy trials. Esparza also went to great lengths to frame the meeting around general concepts rather than proposals to test specific vaccines in specific countries. The WHO indeed issued a press release at the meeting's end that made no mention of Genentech, Chiron/Biocine, gp120, or Thailand. But the participants well understood that the WHO had put those variables on the table. And they knew how others would interpret their declaration. "It's a decision to be a little more empirical," plainly explained the U.S. Army's Donald Burke.

A month after the meeting, *Nature* published an editorial by John Moore and another participant, epidemiologist Roy Anderson of the University of Oxford, that explicitly discussed Thailand and the gp120 vaccines and clarified the vagaries of their pronouncement. Their decision was not, they assured, "the result of pressure by biotechnology and pharmaceutical companies who wish to experiment with products deemed unsatisfactory for use in Western countries."[5] Rather, a presentation made at the meeting by Anderson powerfully influenced the deliberations. Anderson developed a model of a vaccine that worked in only 30% of the people. If widely used, Anderson asserted, this vaccine still could offer great relief in communities that had high new infection rates.[6] They further noted that a recent malaria vaccine trial had demonstrated an efficacy of 31% and was deemed "a valuable addition to existing control measures."[7]

Ultimately, the WHO meeting gave skittish officials in Thailand and other hard-hit countries a stamp of approval should they decide to stage scientifically and ethically sound efficacy trials of the gp120 vaccines. "These governments are quite capable of deciding in their own best interests," concluded Moore and Anderson, "so perhaps it is time for us all to leave them to do so."

A detente, then, had been brokered between the empiricists and the reductionists. It would not hold for long.

• • •

A MOST SURPRISING PLAYER soon emerged on the AIDS vaccine scene, questioning whether the NIH's approach to the problem had too narrow of an agenda: President Bill Clinton.

On June 1, 1995, William Fletcher, a fellow Rhodes Scholar during Clinton's Oxford University days, wrote his old friend about a would-be AIDS vaccine maker, Acrogen. Fletcher, who taught law at the University of California at Berkeley and a few months earlier had won Clinton's nomination to become a federal judge, explained in his letter that Acrogen wanted to make a whole, killed HIV vaccine but could not find either public or private funding, as the old-fashioned approach lacked sex appeal and raised substantial liability issues.

Fletcher, who codirected the northern California campaign office for Clinton during his first presidential race, learned about Acrogen's plight from another of his friends, Burton Dorman, the company's president. A Berkeley-trained biophysical chemist who did a post doctoral fellowship in human genetics at Yale, Dorman in the 1980s built and sold a company that made veterinary vaccines using whole, killed techniques. Many of Dorman's employees stayed with him in his new venture, Acrogen, a privately held company based in Oakland, California, that hoped to get off the ground by making diagnostic tests for infectious diseases and, in time, treatments and vaccines.

At the urging of Don Francis, who briefly served as a consultant to Acrogen, Dorman began exploring the possibility of making a whole, killed HIV vaccine. The NIH repeatedly turned down his grant proposals to make a vaccine that would meet the standards for human tests, as did the Walter Reed Army Institute of Research. Dorman similarly had high hopes when he turned to private investors. A group of "well-heeled financial people" took a keen interest in the project, Dorman said.[8] "The responses were wonderful, sort of, 'What took you so long to ask?' " But the group ran for the hills after speaking with an attorney who warned them about the liability risks that Acrogen would face if it marketed a whole, killed HIV vaccine.

Fletcher's letter to Clinton led White House staffers to question the NIH about whether it supported AIDS researchers interested in exploring this age-old vaccine approach. The NIH, in turn, assured the White House that it had this base covered.

I spoke with Dorman a few months later. "As far as I can discern," he told me, "I've not been able to accomplish anything."

• • •

AS PRESIDENT BILL CLINTON took a seat next to Bruce Weniger, the CDC epidemiologist thought to himself that, for once, it paid off to have a last name that began with a letter near the end of the alphabet. It was July 28, 1995, and Weniger had been invited to the White House's Roosevelt Room along with a half dozen other members of the newly formed Presidential Advisory Council on HIV/AIDS, or PACHA. They sat in alphabetical order, which placed Weniger at one end, adjacent to the empty chair that Clinton chose.

PACHA had just held its first meeting, and these representatives, in a highly choreographed presentation, passed on their initial recommendations, which included calling on the President to hold a national summit on AIDS at which he would detail his vision for ending the epidemic. Weniger had not been designated to speak. But Weniger, who did pioneering work in Thailand that detailed the emergence of HIV there, had an urgent message he wanted Clinton to hear: the President could accelerate the search for an AIDS vaccine.

The day before, in his formal self-introduction to PACHA, Weniger made a bold proposal. "A number of respected scientists and public health experts have recently suggested new targeted, goal-oriented, empirical, trial-and-error approaches, similar to what was used to develop the polio vaccine, but which, astonishingly, to this day have not yet been tried for AIDS," said Weniger. The idea, he said, was for the federal government to provide seed money for a targeted effort that would be run by "a consortium of the applied research community and the private sector." The government could help by overseeing the program and dealing with legislative and regulatory problems, but other than that, he argued, it should get "out of the way." The President, though, could play a special role, said Weniger. "I can think of no greater tribute to the memory of Jonas Salk, who spent the final years of his life struggling to develop an AIDS vaccine using the empirical approach that has been suggested, than for our President to declare a national goal for the development of these technologies by the 50th anniversary of the success of Dr. Salk's polio vaccine, which would be the year 2005."

Weniger had the eager-beaver drive of a newcomer to a religion, looked like a mid-forties version of Ernie from *My Three Sons*, and even struck some *supporters* as overbearing. Now, with Clinton sitting next to him, Weniger saw an opportunity that he could not pass up. When the topic turned to visionary leadership, Weniger saw his chance, and interjected that Roosevelt created the National Foundation for Infantile Paralysis and JFK had proclaimed the goal of putting a man on the moon at the end of a decade. Why not set a goal of developing an AIDS vaccine? suggested Weniger, briefly outlining his idea of tying it to the anniversary of Salk's polio vaccine.

The appointment of Weniger to PACHA had raised the eyebrows of many White House insiders, who saw this little-known AIDS researcher as trading on his father's ties to Clinton. During the late 1970s and 1980s, Sidney Weniger, a real estate developer, attracted much attention in Little Rock, Arkansas, where he purchased several high-profile properties, became an influential supporter of Clinton, and, before declaring bankruptcy in 1987, held a famous Christmas bash that the newspaper *Arkansas Business* once noted "grew bigger and better with each passing year."[9] But as little regard as some White House staffers had for the younger Weniger, he had caught the attention of the President, who dutifully took notes on unlined 4-by-5 index cards.

When Weniger finished his pitch, he looked over and read the top card, on which Clinton had written "Salk 2005."

"He had got the message," said Weniger.

Yes, he had. At least that part of it.

• • •

WITH THE gp120 AIDS VACCINE PROGRAM at Genentech barely showing a pulse, Don Francis on August 16, 1995, flew to Seattle to meet with a possible savior, Robert Nowinski. Nine months earlier, Genentech management, disheartened by the ARAC decision, had broken up its AIDS vaccine team, reassigning Phil Berman and the other scientists to new programs. A consultant's study suggested that Genentech either find a corporate partner to develop the vaccine or spin it off to a new outfit. With no partners in sight, Francis hoped that he could convince Nowinski, who had started three biotech companies and had earned the nickname No-lose-ski, to help them launch a spinoff company dedicated to nothing more than staging efficacy trials of the gp120 vaccine.

Francis had never met Nowinksi, so, before making the trip, he did a Nexis/Lexis search to learn what he could about this potential white knight. Nowinski, one of the first wave of scientists to make the leap from academia to biotech, cofounded Genetic Systems in 1980, which itself would spawn Oncogen, the onetime AIDS vaccine maker. A Brooklyn native, Nowinski had "a quick wit, boundless confidence, and a real talent for articulating the romance of science," wrote one business reporter.[10] He also had a knack for making money: In 1986, these two companies sold to Bristol-Myers for nearly $300 million. Nowinski, featured in a 1987 *Fortune* magazine article headlined "Striking It Rich in Biotech," went on to cofound another biotech firm, Icos, which soon attracted Microsoft's Bill Gates as a major investor.[11] But Nowinski by 1995 had left biotech altogether and become an art dealer.

When Francis pulled into Nowinksi's driveway, his eyes went wide. The house, a copper-roofed, Tudor-styled hunting lodge originally owned by the Post cereal family, sat 200 feet above Puget Sound. Water flowed over a stainless-steel staircase sculpture on the lawn. Inside he passed more sculptures and a living room filled with Picasso paper paintings. It was a precious, sunny Seattle day, and Nowinski led Francis outside to a brick path on the edge of a cliff, where four Doric columns framed four steel patio chairs. They sat down for a business talk.

"Don, why are you doing this?" asked Nowinski.

They spoke for four hours, never discussing business.

A few weeks later, Nowinski agreed to chair the board of the company that became known as VaxGen, which aimed to raise $30 million from private investors to stage efficacy trials of the gp120 vaccine in the United States and Thailand.

• • •

EARLY IN THE EVENING OF SEPTEMBER 19, 1995, Bruce Weniger and three dozen of the top brass in the campaign to develop an AIDS vaccine gathered in a meeting room of the Pang Suan Kaew Hotel in Chiang Mai, the largest city in northern Thailand. As a storm brewed outside, they began to lay the groundwork for staging an efficacy trial in Thailand of the Biocine/Chiron gp120 vaccine that had been rejected by the ARAC for trials in the United States.

The participants, most of whom had come to Chiang Mai to attend a major Asian AIDS conference then taking place in the hotel, could have debated the primary neutralization data. They could have sifted through the breakthrough cases. They could have lamented the lack of cell-mediated immunity in vaccinated people. They could have debated the meaning of a chimp challenge experiment that had no statistical significance. They could have noted the growing sentiment that SF2, the viral strain used in the challenge, grew so poorly in chimp blood cells that it might be too wimpy to reflect a real-world infection. They could have made great hay about a recent demonstration that a hybrid virus engineered to combine proteins from HIV's surface and SIV's core could cause disease in monkeys; this so-called SHIV would allow researchers for the first time to test an HIV gp120 vaccine in this much cheaper animal model.[12] But no one in this room saw fit to raise any of these points. They all understood that Thailand urgently needed to try *something*—even something that had only an outside chance of working. HIV already had infected as many as 1 million of Thailand's 60 million residents, mostly through heterosexual sex, and as anyone who traveled through the country could easily see, the disease was destroying families and entire communities.

And so, as the rain fell and the wind rose in the dark outside the Chiang Mai hotel, Chiron's Anne-Marie Duliège faced no hostile questions. Rather, the gathered officials from the Thai government, the WHO, and the U.S. military earnestly discussed the possibility of Chiron manufacturing a gp120 from the subtype E virus circulating in northern Thailand. They talked about the demographics of potential volunteers who were at high risk of infection. They even proposed time lines for a full-scale efficacy trial. Then, suddenly, there was a clap of thunder and the room went black. "Just keep talking," someone said in the darkness.

And they did.

No decision was made at the meeting that night to stage the efficacy trial. But everyone at the table made it clear that they backed the idea of moving forward, as quickly as possible, with large-scale human tests of this vaccine—even if they were in the dark.

For the Thais, deciding to test vaccines that the United States had rejected caused great anxiety. "They like consensus—they don't like to go against the current," said the WHO's José Esparza, a participant at the Chiang Mai gathering. "Taking a different position from the United States is not easy for a small country like Thailand. They admire U.S. science and can't just ignore it."

Which explains why the Thais ultimately felt deeply stung by the ARAC vote and Fauci's decision. "You have the enemy around here and you give me the pistol," explained Prasert Thongcharoen, head of an AIDS vaccine group of researchers called THAIVEG that wrote Fauci on the eve of the decision. "One day you tell me, don't fire that pistol! It will burst and will harm you. . . . It's unfortunate for us. We prepared everything." Prayura Kunasol, who at the time headed Thailand's Department of Communicable Disease Control, was equally beside himself about the decision. "I was not happy with that," said Prayura. "It demonstrated selfishness."

Yet, in the end, the NIH decision fortified the Thais' resolve to go it their own. "We Thai people have to do for our Thai people," said Prayura. "We have to be self-reliant."

The short history of the Thai AIDS epidemic, which Weniger and his Thai coworkers had exhaustively documented, accounted for the heightened sense of urgency felt by Prayura, Prasert, and other Thai officials.[13] In 1985, when the AIDS blood test became available, Thailand began wide-scale screening for HIV. By 1987, fewer than 100 HIV-infected people had surfaced out of 200,000 tests. But in 1988, the numbers began to change. In one methadone clinic, for example, the HIV rate jumped from 1% at the beginning of the year to more than 32% by September. Come 1990, HIV had infected 44% of the injection drug users.

A second wave of infection began in 1989, with rates among female prostitutes jumping from 1% to more than 40% in some places. And in the far larger population of heterosexual men who visited clinics for sexually transmitted diseases, a third wave struck: the rate went from 0.2% in 1988 to 5% in 1991—a prevalence 50 times higher than that among heterosexual men in the United States. Surveys of Thai military recruits in the north explained this explosion. A whopping 96.5% of the HIV-infected recruits reported having had sex with a "commercial sex worker," as did 79% of the uninfected men.[14]

But simple numbers alone only hint at the devastation. Pratoom Thajorn, a nurse and social worker, knew it intimately. Pratoom worked in San Patong, a district 20 kilometers outside of Chiang Mai that some AIDS researchers believed in 1995 had higher rates of HIV infection than anywhere else in the country. I joined her on another rainy September day as she traveled from village to village, where HIV had assaulted Thai families from every angle, killing so many people that, legend had it, a local crematorium broke.

Pratoom first visited the home of a 68-year-old man who had died two days before from AIDS. A widower who tended water buffalo, the man was infected by a blood transfusion, Pratoom said. "He didn't have any money for prostitutes or for any risky behavior," she explained. Pratoom sat on a woven mat in his wood shack on stilts, a bare lightbulb hanging from the ceiling and idyllic posters of mansions in bucolic settings tacked to the walls. As his disease progressed, she said, he had to sell his animals to pay for medical care. He died broke, and his family, with less than $7 to their name, could not afford to have him embalmed, let alone to put on a proper funeral with a traditional meal for the mourners. But as his body sat packed on ice to keep it from rotting in the sweltering heat, the villagers chipped in nearly $800, and yesterday his family fêted 300 people who came to his funeral. Now it was Pratoom's turn to help: she had brought along another social worker who specifically cares for elderly people with AIDS, and he handed the family an envelope stuffed with *baht*.

Pratoom then drove over rutted muddy roads that cut through rice fields to reach another clutch of shacks on stilts. Again, the client was a 68-year-old man, but this one was not HIV-infected. Rather, he spent his days caring for his 30-year-old daughter and his 25-year-old son, both of whom were infected and had the thin limbs and sunken faces of people whose diseases have progressed to full-blown AIDS. The father, who had pustules on his limbs from his own battle with disseminated tuberculosis, explained that both of his children moved in with him during the past six months because their spouses had died from AIDS. The son, who had a five-year-old boy himself, appeared to have dementia and wandered aimlessly about the little village.

The daughter said that though her 10-year-old son is uninfected, AIDS was taking a heavy toll on him. "He doesn't want to get married because he's seen both his uncle and father get sick from AIDS," the woman said.

Pratoom ended her home visits with a stop at the more upscale shack of a petite 63-year-old woman. The woman, who said she weighed 28 kilos (61.6 pounds), sat in the shade found underneath her stilted house, a cozy shelter from the heat and downpours. Nine neighboring women sat on a bench or on the ground around her, three of whom said they had relatives with AIDS, and watched. The woman, who had slight cataracts and often looked in between people when she spoke, told about the slow death of her husband of 23 years. "He was sick with a lot of things," she said. Two years after his death she too, became ill. When the doctor told her she had AIDS, she was dumbfounded—but then her son revealed that a doctor earlier had confided that his father had died from the disease.

Little wonder, then, that Thailand wanted to test vaccines deemed worthless in the United States: AIDS here had only the faintest resemblance to AIDS in the Western world. This point was dramatically underscored that evening, when an American epidemiologist who lived in Chiang Mai and was helping Thai researchers ready the country for vaccine trials took me on a tour of brothels. The most startling "commercial sex worker" I spoke with that night was Anek—a 22-year-old man. Anek plied his trade from a gay bar in Chiang Mai. "I don't like men," Anek, who said he was a college student, told me. Do you like women? I asked. "*Pu chai*," he said, which means "I am a man." Anek, it turned out, hoped to marry and have kids, but, for the time being, was shying away from women. "I worry about HIV with women," said Anek, "but not so much with men."[15]

William Heyward, a CDC epidemiologist detailed to the WHO, came to Thailand specifically to help the country lay the groundwork for AIDS vaccine efficacy trials, and he stressed that Thai scientists and officials well understood the scientific arguments for and against the gp120 vaccines. But Heyward, who once headed the largest AIDS research project in Africa, Projet SIDA in Zaire, said outsiders often did not appreciate that Thailand had a long history of aggressively staging human tests of vaccines. "Hepatitis A, hepatitis B, malaria, dengue, Japanese B encephalitis—it just goes on and on and on," said Heyward. "They know that vaccine development is a long and arduous process. They're committed for the long haul."

Peter Piot, a Belgian epidemiologist who cofounded Projet SIDA, spoke about the Thais' resolve at the opening of the Chiang Mai AIDS conference.

"The Thais are not looking to others to test these vaccines for them," said Piot, who headed the newly formed UNAIDS, a program that combined the WHO's Global Programme on AIDS with other UN AIDS projects.[16] "They are doing it for themselves—and, I might add, for the sake of the whole world."

Thailand soon decided to conduct small-scale trials of the gp120 vaccines to assess their safety and ability to trigger immune responses in Thais. Biocine thought that, if all went well, it could start efficacy trials there by the end of 1997. Scientists from Genentech, who had yet to start VaxGen, were more optimistic, predicting that they would have the required preliminary data by December 1996. As it happened, an efficacy trial would not begin in Thailand until 1999.

• • •

ON THE FINAL MORNING OF THE five-day Chiang Mai AIDS conference, Seth Berkley from the Rockefeller Foundation announced the birth of the first nonprofit, nongovernmental organization dedicated to speeding the search for an AIDS vaccine.[17] Berkley said that the International AIDS Vaccine Initiative, which had taken two years to organize, would work "full stop" to produce an AIDS vaccine. "The AIDS vaccine effort is foundering," warned Berkley. "We can't let the effort die."

A lanky, curly-haired physician, pilot, and long-distance runner, Berkley had turned in his white coat to work in public health, and he approached it with an impatience that distinguished many successful field epidemiologists. Berkley grew up a lower-middle-class, New York City kid, the son of a Midwest boxing champ turned proofreader and a mother who ran a travel agency for gay men called Fairy Travels. After medical school, Berkley did a stint at the CDC, which included an investigation of the Sudan famine in 1985 that required him to travel by camel behind rebel lines with an AK-47–toting support team. He next hooked up with the Carter Center, President Jimmy Carter's attempt to build a healthier and more peaceful world, which sent him to Uganda, where he quickly began assessing the spread of HIV.

At the Thai meeting, Berkley came out swinging. He called for more directed research. He said IAVI would push for exploring more vaccine strategies, including the killed and attenuated approaches. He lamented the dearth of vaccines made from strains of HIV afflicting developing countries. He noted that the few companies involved in HIV vaccine R&D were abandoning their efforts right and left. "In the current environment, it makes intrinsic sense for industry not to invest in AIDS vaccines," he said, describing the lack of incentives as a "market failure" that IAVI would strive to address.

Others had attempted to organize nonprofits that promised to spur the AIDS vaccine field. Jonas Salk had tried with his ill-fated World FAAIR. In 1993, a former pharmaceutical executive, H. R. Shepherd, founded the more general Albert B. Sabin Vaccine Foundation, which promised to accelerate development of vaccines against AIDS and other diseases. Similarly, Wayne Koff, the former head of AIDS vaccines at NIAID, in 1994 established the short-lived Global Vaccine Development Foundation. But IAVI, unlike the rest of this pack, came into being with a clear sense of purpose that had been hammered out by diverse groups of experts during in-depth meetings devoted to the issue.

The idea for IAVI came from talks that began in 1993 between Don Francis and Bruce Decker, the former head of California's AIDS Task Force. A gay, HIV-infected man, Decker was becoming increasingly ill and wanted to make a difference before he died. Francis suggested that they join forces and tackle the problem of promoting AIDS vaccine development for poor countries.

They sent a memo to 45 people, outlining their ideas for what they then called the World AIDS Vaccine Initiative and inviting them to attend a meeting during the international AIDS conference to be held in Berlin that June. Their immediate goal was to have leading public health and AIDS vaccine experts brief the "largest multi-national funding entities" about the problem. "We will demonstrate to them that, if they provide sufficient funding to stimulate research and administration, an AIDS vaccine can be expeditiously developed for use in the Third World," they wrote. In the long term, they hoped to establish low-cost vaccine factories in poor countries, and create a "purchase fund" that could buy vaccines. "[W]e can no longer sit on the sidelines, helplessly watching our less fortunate neighbors struggle with little hope on the horizon," they wrote. "We believe that we have finally reached the point where we can break through entrenched thinking and have reasonable vision for a better future."

Had Decker and Francis merely announced their grandiose goals, wrapped in familiar public health invocations, their initiative might have gone the way of the many other well-meaning projects that said a lot and did little. But the Berlin get-together led to another in London, which, with Seth Berkley's connections, spawned a novel gathering at the Rockefeller Foundation's villa in Bellagio, Italy.

Reserved for think tanks and monthlong residencies of scholars, the Bellagio center is set on Lake Como, with the Italian Alps as a backdrop, and features buildings that date back to the 17th century graced by 50 acres of gardens. In these lush environs, Don Francis, Seth Berkley, and 22 others spent March 7–11, 1994—five days—detailing the problems with AIDS vaccine R&D. The participants mixed familiar vaccine overseers like the WHO's Esparza, France's Jean-Paul Lévy, the Army's Burke, and NIAID's Peggy Johnston with a former Levi Strauss president, an official of the United Nations Development Programme, a technical adviser to the Wellcome Foundation Ltd. (the U.K. parent to Burroughs Wellcome, the U.S. maker of AZT), a struggling biotech firm's CEO, and a World Bank representative. Other scientists came from Thailand, Uganda, Japan, Sweden, and Amsterdam. (No media were invited.) It was the single most diverse, intense AIDS vaccine think tank ever held.

A report from the Bellagio meeting starkly spelled out the problem. Although over 90% of new HIV infections were taking place in developing countries, the report warned, "current efforts are directed almost exclusively towards vaccine products catering to the needs of the developed world."[18]

All told, according to the report, governments, industry and philanthropic organizations spent $160 million on AIDS vaccine R&D in 1993—less than 10% of the amount spent on general AIDS-related R&D. The report then rightly noted that most of this $160 million came from governments, which primarily funded investigator-initiated research. "Although this is an excellent way to make new scientific breakthroughs, it is not necessarily the most effective way to develop a particular product," the report contended. The limited resources also meant that AIDS vaccine strategies progressed one at a time: try gp160; if that fails, try gp120; if that fails, try a prime boost; etc. "Given the many scientific uncertainties remaining, the development and testing of multiple empirical approaches in a parallel fashion, rather than sequentially, will be a faster route to the development of safe, effective, and inexpensive vaccines appropriate for widespread use."

The meeting participants did not scold industry for feebly responding to this public health emergency. Rather, they articulated the reason why the potential AIDS vaccine market had failed to excite companies and investors. The report smartly divided industry into biotechs and big pharmaceuticals. Biotech firms relied on venture capitalists, who wanted to make money as quickly as possible. By 1994, these investors, many of whom had poured money into the likes of Repligen, had discovered faster ways to make back their investments than betting on AIDS vaccines. Pharmaceutical houses not only recoiled at the long list of scientific unknowns and the staggering liability issues; they also saw a limited market in wealthy countries. Indeed the report estimated that developed countries had fewer than 5 million sexually active homosexuals and injecting drug users, a far cry from the 19 million that a Shearson Lehman AIDS vaccine market analysis once calculated for the same populations in the United States alone (see Chapter 6). And because of these negatives, the companies that did have AIDS vaccine projects stuck with the safest, least speculative approaches.

A new initiative, the Bellagio report concluded, could reduce obstacles and fill gaps. "I had assumed—and I feel like a fool to say this—that the vaccine effort was taken care of and that everything was going great," Berkley told me after the Bellagio meeting. "But as I began to look at it closely, I saw that the vaccine effort was in trouble. And the situation was getting worse, not better, in terms of incentives for industry and the attention paid to the developing world."

After two more meetings over the next year and a half with new, diverse groups of scientific and financial experts, IAVI had a plan for incorporating itself, a scientific agenda, and original ideas about how to stimulate the field.[19] (Don Francis, concerned about a conflict of interest because of his work at Genentech, stopped attending meetings after Bellagio.) To push the field, IAVI promised to create a "highly targeted applied vaccine development effort" that would promote vaccines others had ignored. Over seven years and at an estimated cost of $569 million, this program would explore 10 new vaccine approaches, fund the development of six vaccines made from developing world subtypes of the virus, and test four preparations in efficacy trials.

To "pull" the field, IAVI hoped to establish—and this was a mind-blowing idea—a "credible market" in developing countries. One possible way to do this, IAVI envisioned, was a "guaranteed purchase" plan in which a group of developing countries agreed to buy a certain number of doses of a safe and effective AIDS vaccine at a given price. The World Bank would play a key role in this scheme, offering these countries a line of credit, up front, in the same way banks prequalify potential home buyers.

The world simply was not set up to design and test an AIDS vaccine as quickly as possible, Berkley argued. Imagine, he suggested, if you were to "reaggregate" the entire planet into one community of 5 billion people.

"Alright, we've got this problem, it's spreading, and costing us $5 billion to $9 billion a year," said Berkley. "We know from history that vaccines are the only way to stop it. There's no question we'd get it together."

But the world is not, try as the Seth Berkleys might to change it, one people, one planet. It is a fractious place, populated by groups that have a difficult time seeing beyond their own borders, let alone their continents. And that is a reality that would relentlessly confront IAVI, which long would remain idea rich and cash poor.

• • •

OUTSIDE PARIS sits a storied chateau known as Saint-Cloud. Once home to Louis XIII and later Napoleon himself, the chateau later had an annex built to house the 100 guards who protected Napoleon III. Cent Gardes, as the building became known, eventually became a summer residence and lab for Louis Pasteur, who died there. A century later, Cent Gardes had become synonymous with AIDS vaccines by serving as the annual gathering place for a three-day meeting each October of the world's leading AIDS vaccine researchers.

On October 20, 1995, John Moore revealed to the Cent Gardes crowd a much-anticipated analysis of the cases of breakthrough infections that occurred during the earlier trials of the Genentech and Biocine/Chiron recombinant gp120 vaccines. As part of a star-studded team of scientists from six institutions that analyzed the cases, Moore reported immunologic and virologic data from 12 such breakthroughs.[20] Not only did all but one of the vaccinees have high levels of antibodies prior to becoming infected; the researchers also found that antibody responses induced by the vaccines quickly waned, suggesting that the antibodies were neither effective nor long-lasting. No one had antibodies against primary isolates, and test-tube experiments with lab-grown strains of the virus further showed that the antibodies they did have worked poorly against strains that differed from the ones used to make the vaccines. The vaccines also did not seem to impact how well people who became infected controlled the virus: the amounts of HIV in their bodies roughly matched the viral loads of people who had not received an AIDS vaccine.

Moore, appropriately, cautioned that the data still did not prove the vaccines worthless. "It's not final proof of anything," Moore told me. "It just reinforces what I previously believed. Nobody can say the data were encouraging. It's all smoke and mirrors. There's nothing there."

The AIDS vaccine research community reacted to the data as predictably as any polarized group deals with new information: both sides became more entrenched. Supporters of the gp120 vaccines stressed that these inconclusive data proved that the vaccines deserved an efficacy trial. Detractors saw the thorough study as a knockout punch and contended that this debate should be retired once and for all.

Ronald Desrosiers, the AIDS vaccine developer from the New England Regional Primate Research Center, had a more dispassionate view than most. "There's no way you can look at that data and not be discouraged," said Desrosiers. "But I'm very enthusiastic about seeing these go forward. It could be that the vaccines are 70% effective and that would be consistent with the breakthroughs. You're never going to know. I don't think they're going to work, but I don't see anything bright over the horizon. And miracle of miracles, they might work. Then we'll all be kicking ourselves and saying we should have been starting these trials earlier."

• • •

ON NOVEMBER 6, 1995, John Moore hurled a missive at the research committee of the Presidential Advisory Council on HIV/AIDS that sparked another battle royale in the AIDS vaccine arena. The resultant firefight, which took place via e-mail and fax, offered one of the clearest written delineations of the differences between the empirical and reductionist points of view. Yet it also served to polarize the two camps even further.

The 10-member PACHA committee—only two of whom were researchers—had solicited public views about AIDS research problems that it might encourage the White House to solve. Moore, in a letter addressed to the committee's chair, urged PACHA to work closely with the NIH's Office of AIDS Research.[21] The OAR, with help from the country's leading AIDS academics and activists, was in the midst of a massive review of the NIH's entire AIDS research program. Moore noted that he sat on the OAR committee that was analyzing AIDS vaccine research, and, as such, he took particular exception to some of the comments that Bruce Weniger had made the preceding July in his formal self-introduction to PACHA. "I disagree wholeheartedly with the research philosophies he expresses," wrote Moore. "Indeed, I could scarcely imagine a strategy less likely to succeed in creating an AIDS vaccine than the one Dr. Weniger proposes, for the empirical, trial-and-error, private sector–led approaches to an AIDS vaccine have been a monumental failure to date."

The OAR vaccine committee, wrote Moore, finally would set AIDS vaccine research "onto a rational, coordinated path" that would increase fundamental knowledge about the immune system's relationship to HIV. "To adopt an alternative approach would be a catastrophic blunder, for it would sacrifice long-term success for short-term expediency," wrote Moore.

Moore went far beyond chanting the familiar basic research mantra, however, thrashing Weniger's contention that the government had to do more to involve industry in the vaccine search. "[N]ot only has the involvement of the private sector in AIDS vaccine development not been a success, I believe it has actually hindered progress to an AIDS vaccine, by causing finite Federal research funds to be spent on paths leading nowhere," snapped Moore. "Companies can indeed develop vaccine concepts, but when majority scientific opinion deems these concepts to be failures, the companies cling to their investments and cause considerable turmoil and wastes of energy trying to get Federal decisions rescinded."

Moore then cited the MicroGeneSys fiasco, and warned that Genentech and Chiron/Biocine might use PACHA to do a similar political end run around the scientific community. Indeed, Moore wrote, his own studies of breakthrough cases had made him "more convinced than ever that NIAID made the right decision in postponing efficacy trials of these products."

In closing, Moore allowed that a national program to develop an AIDS vaccine could help, especially if the government contracted with industry to produce vaccines that researchers wanted to test. "Such a system could allow a more objective, rational and rapid evaluation of product performance than has been possible with company-owned products," concluded Moore. "Rejection of a poorly performing concept must become a routine matter, not a political drama that unfolds like a glorified soap-opera."

Two days later, Weniger sent an e-mail to 17 people he had grouped together into his HIV Vaccine R&D Interest Group, asking them to write PACHA about "the state of federal efforts to develop vaccines for HIV/AIDS." As he noted in his e-mail, he planned to fax Moore's submission to everyone in the group he could.[22]

Later that day, Moore got word of Weniger's e-mail and sent him a wicked one-line e-mail. "This e.mail-friendly text of my letter to Dr. Levine should save you some fax paper," wrote Moore, who included an electronic version of the letter.

Weniger e-mailed Moore back, thanking him and noting that he "certainly had started the discussion off with some substantive issues to debate." But Weniger attacked Moore's criticism of industry, writing that most vaccines in use came about because of the private sector. He then tried to make common cause with Moore about the idea of a goal-oriented vaccine program. "A main reason for this is to take the direction of it out of the purview of our competitive, peer-reviewed, researcher-initiated, science funding mechanisms, which are philosophically and administratively not well suited for developing products as rapidly as possible, as NASA did for the Apollo program and as Los Alamos did for the Manhattan Project," Weniger said in closing.

Moore immediately volleyed back. "That the private sector succeeded in the past is no guarantee that it will do so in the future," Moore wrote in his e-mail. "There are many failed vaccines, after all." He went on to declare that industry's contributions to AIDS vaccine research "is good in principle, but bad in practice." The essential problem, he argued, is that the private sector had relied on "massive injections of Federal funds," which gave the government a say in the product's future, "but companies have never accepted this point, at least not with any grace." Then Moore eloquently highlighted how the bottom line in the debate was the bottom line. "In a world of infinite resources, this would not matter, but this is the real world, and there must be a time of choices, however hard."

In a parting shot, Moore acknowledged that what he called the "rational and empiricist schools of thought" both should be included in a structured AIDS vaccine program, but that Weniger only seemed to embrace the empirical one. Moore took particular exception to the interest group listed on the e-mail routing—which Moore wrongly concluded was the PACHA research committee—as it had no NIH scientists, no member of the OAR review's vaccine committee, and few if any basic researchers. "Instead you have filled the committee with corporate scientists, old friends of the corporate scientists, and clinical scientists who have long-lasting relationships with the corporate scientists," wrote Moore. "You will not be having a debate on your committee, you will be mutually applauding one another's views."

Had the Moore-Weniger showdown remained between the two of them, it would have been no more than a spat. But Weniger not only sent out Moore's fax to his interest group, he also e-mailed this entire back-and-forth to them. Some subsequently wrote PACHA impassioned letters, including one that sarcastically said if the council aimed to promote funding for HIV/AIDS research, "then a view such as Dr. Moor's [*sic*] should be greatly valued." But if the goal was to develop an AIDS vaccine, this letter advised, listen to the biotech industry and vaccine manufacturers. "The measurable end-point for vaccine development should be a vaccine product, not the number of publications in scientific journals."[23]

An equal number of scientists who shared views with Moore wrote PACHA too. "I do not believe any of the current vaccines will work," wrote Arthur Ammann, the director of research at the Pediatric AIDS Foundation who noted that he previously had worked for five years at Genentech on the gp120 project. "Much of the current rhetoric represents the last gasps of individuals who have had their vaccine programs abandoned by their own scientific review boards and now are looking for government support. The President should not become involved in this debate."

When John Moore framed the debate as a struggle about how best to spend limited resources, he perfectly captured the essence of what the reductionist camp believed. Bruce Weniger did the same for the empiricists when he asserted that peer-reviewed, investigator-initiated research was ill suited to the goal of developing products as quickly as possible. But Moore and Weniger never squarely addressed each other's points. This fostered the mistaken sense that if one was right, the other was wrong and that the agenda setters faced an either/or choice.

While both camps made convincing arguments, the heat waves emanating from the debate obscured the truth—and exaggerated the distance between the opponents. Once again, basic and applied research had been pitted against each other, and in this paradigm, more generously feeding one took food out of the mouth of the other. In the world of the NIH, as Fauci made clear at the ARAC meeting, there indeed was a limited pie to share. But if the goal was to find a working AIDS vaccine as quickly as possible, that was too provincial a perspective. The real question was this: if money were devoted to a new, targeted research effort for AIDS vaccines—and NIH-sponsored, investigator-initiated research did not take any financial hit whatsoever—would it speed the search? As the OAR's review of the AIDS vaccine effort soon would make clear, the prevailing view, whether researchers tilted toward empiricism or reductionism, was yes. Logically, then, the next step for the AIDS vaccine agenda setters would have been to put aside their differences and, following IAVI's lead, outline what this targeted program would look like, how much it would cost, and who would pay for it. But the acridity between the empiricist and reductionist camps had intensified to the point that logic did not stand a chance.

• • •

FOR SEVERAL YEARS, NIAID held its annual AIDS vaccine meeting, the premier gathering of its kind, in secluded, even romantic, enclaves like beachfront hotels in the Florida Everglades or colonial-style resorts in the Virginia woods. The world's top AIDS vaccine researchers would flock to these resorts and hang out together for a week, gorging on data and debates. You could make a name for yourself at the meeting or shoot your mouth off and damage your reputation. It was a must-attend, from beginning to end. But in February 1996, NIAID held the meeting in an auditorium on the campus of the National Institutes of Health in Bethesda, Maryland.[24] Dani Bolognesi, whose role had changed from the dean of AIDS vaccines to the field's therapist, made only a brief appearance, as did Tony Fauci. John Moore did not even show. The primate researchers who once dominated this meeting gave no talks at all. "The AIDS vaccine search is in disarray," said Wayne Koff, the erstwhile organizer of this meeting who left NIAID to work on AIDS vaccines at United Biomedical Inc. "It's a wheel that's spinning and needs to be stopped."

The real problem with the meeting, though, had nothing to do with where it was held or who spoke or even who attended. The problem was that there was so little new work to talk about. Once, the AIDS vaccine search seemed logical, linear, as though it had urgency and purpose. Several vaccines had entered clinical trials, and several more were in the wings. The monkey model had the power to move researchers' emotions from despair to delight. Passions flared right and left. Wacky ideas seemed plausible and worth exploring because the people backing them, people like Jonas Salk, had such fires in their bellies to solve the problem that only the jaded could ignore them.

The meeting agenda itself provided a powerful symbol of how much had changed. In past years, the gathering always included an entire session devoted to the results of initial human tests of candidate vaccines. Now the organizers deemed that only one early vaccine strategy had produced clinical results sufficiently promising to merit an oral presentation. "There's not a great deal out there, and that's a concern," said Patricia Fast, head of NIAID's AIDS vaccine division. Dani Bolognesi similarly was worried. The pipeline of new AIDS vaccines, he said, "is almost shut down."

Adding to the sense that the vaccine field had lost its way, other areas of

AIDS research recently had announced findings that promised to alter everything from fundamental understandings of how HIV caused disease to treatment strategies. On the basic research front, Robert Gallo's lab stunned the field when it reported that CD8 cells secreted chemokines—a family of immune system chemical messengers that few AIDS researchers had even heard of—that could prevent HIV from infecting cells. This raised new possibilities for treatment and, potentially, a new correlate of protection that vaccine researchers could assess. AIDS researchers testing experimental treatments had more immediately practical news: A much-hoped-for synergy occurred when infected people took a new class of drugs called protease inhibitors along with already marketed compounds like AZT and 3TC (lamivudine) that attacked HIV's reverse transcriptase enzyme. Although these so-called triple combo cocktails had not cured anyone, they blasted HIV so effectively that the most sensitive tests had trouble even finding the virus in the blood cells of treated people, holding out the first real hope that drugs could delay disease and death for years rather than months—and, as would later become evident, complicating the evaluation of vaccines.

At the NIAID AIDS vaccine meeting, the short list of promising candidate vaccines provided a subtext for much of the discussion. NIAID indeed unveiled an ambitious new plan that, in the wake of the ARAC decision, aimed to forge stronger ties between the institution and vaccine developers. "The importance of industrial partners cannot be overemphasized," said Fauci when he described the new "development plans."

To act like a true business partner, NIAID decided to sit down with companies and negotiate specific criteria that a vaccine must meet before it could move from a small human trial to a medium-sized one and then to a large efficacy test. If companies knew what they had to do to win NIAID funding for an efficacy trial, the rationale went, they might have more resolve to remain in the business. "It's essential that we don't have a moving target," said Fauci, who allowed that he took "considerable heat" for not staging the gp120 efficacy trials.

NIAID and the two companies involved in the only human tests whose results were presented at the meeting, Pasteur Mérieux Connaught and Biocine, had decided to draft just such a contract. NIAID had just begun phase I trials of Pasteur Mérieux's vaccine, which contained one or more HIV genes stitched into a harmless canarypox virus, followed by a boost with Biocine's gp120 vaccine. Results presented at the meeting showed that the vaccines easily stimulated production of neutralizing antibodies—although these antibodies still could not neutralize primary isolates—and, more importantly, triggered killer cells in anywhere from 12% to 44% of the people vaccinated. Before these vaccines could move on to the next phase, the companies and NIAID agreed that at least 90% of the people who received both vaccines must produce neutralizing antibodies and at least 30% must produce CTLs. If the vaccines met those milestones and worked in either monkey or chimp challenge experiments, they could move into efficacy trials by the middle of 1998.

Immediately, leading researchers began questioning the setting of specific criteria. "It concerns a number of people that we don't know what's going on," said the University of Minnesota's Ashley Haase, who presided over the fateful April 1994 HIV Vaccine Working Group meeting and the ARAC decision. "They're essentially setting up numbers arbitrarily."

Fauci conceded that they had set these criteria in a "semi-arbitrary" manner and had based them on the results already seen. But they need some criteria, Fauci maintained. "I made the leadership decision we want to deal in good faith," said Fauci. "Either we're going to have a vaccine program or not."

NIAID badly wanted to put the field back on track and recapture the momentum it once had. But hammering out business plans, well meaning as they were, based on "semi-arbitrary" criteria, would prove unable to either keep companies in the game or reliably predict when efficacy trials would begin.

• • •

AT 9 P.M. ON THE EVENING OF FEBRUARY 15, 1996, less than five hours after the close of NIAID's meeting, the "working group" that ran the OAR review of the NIH's $1.4 billion AIDS research portfolio met down the block at Bethesda's Hyatt Regency Hotel to discuss the recommendations that they would soon issue. David Ho brought up NIAID's new proclamation of criteria for efficacy trials. "I want to know the structure and process behind the current decision," said Ho, one of 16 members on the working group.

This working group, known as the "Levine committee" (it was cochaired by Princeton University molecular biologist Arnold Levine), relied heavily on input from teams of academic scientists organized into six "area review panels" that each analyzed specific research efforts. So David Ho's question fell to Dani Bolognesi, who, as chair of the vaccine panel, had been invited to this meeting. "I don't know what the process was," said Bolognesi.

Mark Harrington, one of two AIDS activists on the working group, was aghast. "If Dani doesn't know the process, the process stinks," said Harrington.

This exchange evinced how crippled NIAID's AIDS vaccine leadership had become: no sooner had Fauci announced what he viewed as a dramatic plan to restart the stalled AIDS vaccine R&D engine, than another, more powerful agenda-setting body had begun to criticize it. And when the members of this working group released their report a few weeks later, it became clear that they viewed improving the interactions between NIH and industry as a minor part of a real solution, anyway. "The entire AIDS vaccine research effort of the NIH should be restructured," their report declared.

The Levine committee report, which ran more than 400 pages, articulated the thinking of 100 scientists and activists who sat on the area review panels and the working group during the yearlong investigation. Their unsparing, frank, and remarkably thorough analysis covered vaccines, drug discovery, epidemiology, behavioral science, pathogenesis, clinical trials, alternative medicine, primate centers, and repositories for reagents. The lists of participants read like a Who's Who in academic AIDS research coupled with such leading scientific voices as Nobel laureates David Baltimore and Phillip Sharp, former Merck CEO P. Roy Vagelos, and *Science* editor Floyd Bloom.[25]

The working group's broad overview of the field called for doubling the amount spent on investigator-initiated research, noting that "there is no better way to enhance the diversity and productivity of research approaches." But the report otherwise had few predictable rallying cries, especially when it came to vaccines. The working group had several bold vaccine-specific proposals of its own. Rather than each institute running an independent AIDS vaccine R&D program, a new committee made up of "distinguished, non-Government scientists" should set up and oversee a new "trans-NIH" effort. In addition to this so-called AIDS Vaccine Research Committee, the group called for a new, White House–level National AIDS Vaccine Task Force that would mesh the workings of "all U.S. Government agencies and coordinate them with those of pharmaceutical and biotechnology organizations, private agencies, other nations, and international organizations."

The section of the Levine committee report written by the vaccine area review panel similarly avoided platitudes and bloodless prose. "The major theme of this report is that HIV vaccine research and development is in crisis," the panel reported. Creative new vaccine concepts had trouble winning funding from the peer-review groups, called study sections, that evaluated grant proposals sent to the NIH. Primate researchers had not done head-to-head comparisons of different vaccine strategies and their work had only weak links to human studies. Capable, creative immunologists had avoided the AIDS vaccine field. Industry's interest in AIDS vaccine R&D was "waning" and might "vanish altogether." Of the $112.9 million that the NIH supposedly spent on HIV vaccine research in fiscal year 1994, $18 million supported work that the panel deemed only "tangentially related" to vaccines and another $6 million funded studies that had nothing to do with AIDS. The National Cancer Institute additionally spent $16 million a year on various in-house AIDS vaccine researchers, although the panel found that the NCI made "no attempt to coordinate them towards a common goal." And the panel assailed HIVNET, a $16 million network of researchers set up mainly to lay the groundwork for HIV vaccine efficacy trials.

In what amounted to a manifesto, the panel members offered a raft of imaginative, smart recommendations. Create a "culture" that would attract top-notch immunologists and support creative vaccine ideas, they urged. To do this, they advised that rather than sending grant proposals to peer-review groups that specialized in such general subjects as virology or immunology, the NIH should establish a study section dedicated to vaccine research. "At present, no single study section has the essential combination or depth of talents in microbiology, host defense mechanisms, immunology, and chemistry that is pertinent to vaccine biology, especially when considering an agent as complex as HIV-1," they noted. Cell-mediated immunity especially needed attention, noting that their analysis of grants revealed "an overemphasis" on antibodies. Other ideas to improve the culture included making blood samples from human vaccine experiments more widely available, and allowing academics who did not work at or with primate centers to have more access to monkeys.

A second excellent recommendation of the OAR's Levine committee was that the NIH do more targeted AIDS vaccine research. According to the panel, "the NIH must be prepared to go beyond its traditional role, for the discovery and development of a vaccine demands more than just the acquisition of fundamental knowledge; it requires that the information be applied and resultant vaccine strategies appropriately evaluated." If the NIH had "a strong, targeted vaccine initiative," the panel predicted, it could provide the "currently missing bridge between the NIH-sponsored basic research and commercial product development." As part of this initiative, the panel recommended that the NIH see to it that promising vaccine strategies being ignored by industry move forward. Specifically, the NIH or DoD could offer their facilities to make small lots of vaccines that were of high enough quality to test in phase I human studies. Or the NIH could contract with the private sector to make vaccines. More radical still, the panel advised the NIH "to consider cooperative or public ownership arrangements for AIDS vaccine products, especially for riskier approaches, since private industry has largely been reluctant to enter this area at levels required if there is to be an active program."

This beefed-up targeted program should not cannibalize funds for basic research, the panel insisted. "Because of the importance of protecting the Nation and potentially other areas of the world against the expansion of the AIDS epidemic, the budget for these activities needs to be provided from new funds," the panel stated. In all, these experts suggested that $25 million to $50 million each year should go to this targeted effort, which a new task force, made up of NIH and outside scientists, would oversee. This recommendation resembled the targeted plan laid out by IAVI, as did the suggestion that the OAR should make it "a high priority" to analyze the economic disincentives facing industry and to offer solutions.

When it came to primate studies, the panel "felt that a better return could have been received from the enormous investment that has been made," noting that the British group led by Jim Stott had "yielded a high return from a relatively small investment of targeted funds." Unlike the British, primate researchers in the United States had been conducting challenge experiments with a wide variety of conditions that did not allow one lab to compare its results to those in another, a situation that could be remedied by "a transition from independent investigator-initiated trials to a venue of centralized coordination, evaluation, and problem solving." Researchers also had to use more animals in their studies to reach definitive conclusions. "Too often, nondefinitive end points are reached with small animal studies that are ultimately a wasted effort or that turn out to mislead the field, causing even greater losses." As for chimpanzees, the panel said NIH-funded chimp research should focus "almost exclusively" on developing a challenge strain of HIV that would replicate vigorously in these animals and cause disease.

Finally, the panel urged the NIH to overhaul the way it classified research as vaccine-related, gut the NCI's internal AIDS vaccine program, and carefully scrutinize HIVNET.

Although some panel members told me they remained far from convinced that the NIH had the backbone to follow through on their recommendations, the widely respected OAR director, immunologist William Paul, assured me they were wrong. Said Paul: "There's no good having a report if we don't do anything about it."

• • •

BRUCE WENIGER STRONGLY DOUBTED that the OAR reforms would happen, as became clear in the "draft recommendations" for President Clinton that he asked his fellow members of PACHA's research committee to review. The OAR reforms, Weniger asserted in this April 8th draft—which only reflected his personal views—"will likely come too slowly, be incomplete, and will not address the inherent weaknesses of the agency in fostering the targeted, applied AIDS vaccine research which is needed." The NIH, he contended, should stick to basic research. "The peer-reviewed, investigator-initiated fund-allocation mechanisms of NIH are a great engine of discovery, but they are ponderously slow, ignore public health priorities, and lead to major gaps in AIDS vaccine development," thundered Weniger.

Once again, Weniger called for a "crash program," styled after the March of Dimes and NASA's Apollo program, to develop a vaccine. The semi-independent consortium he imagined could act rapidly, pursue several vaccine strategies at the same time, and contract basic and applied research with government and industry.

Weniger went so far as to spell out specific budgets for the federal government's AIDS vaccine effort. At least 25% of the $1.4 billion that went to the NIH for AIDS research should go to AIDS vaccine research—which roughly would more than triple the amount this area had been receiving. Half of that $350 million should go to NIH for basic research, while the difference should equally support targeted programs at the U.S. Army and IAVI.

Finally, Weniger reiterated his call on Clinton to declare a goal of developing an AIDS vaccine by 2005.

After other members of the research committee reviewed the draft recommendations Weniger had proposed, only two survived—and in a much watered down form. A larger portion of the federal government's AIDS research budget should go to vaccine R&D, the committee advised. Second, the U.S. government should contribute "funding, expertise, and other resources" to an "expedited" AIDS vaccine program that would target gaps in knowledge, product development, and testing. Although the committee did not tap IAVI for the job, it might as well have, offering that "a non-profit, international AIDS vaccine initiative" should run the effort.

Weniger sent out these recommendations via e-mail to many of his like-minded colleagues, and they eventually made their way to OAR director Bill Paul, who worried that the OAR review and PACHA were giving Clinton "conflicting" advice: support a stronger NIH AIDS vaccine effort versus establishing a new international effort. He voiced his concerns to PACHA, which held an emergency conference call of its research committee.[26]

PACHA's response to Paul demonstrated to Weniger, once again, that the NIH's dominance cowed scientists—and even some people on his own committee. The language "ought not to change," he wrote his fellow committee members.[27]

Tensions between Weniger and basic researchers who carried the NIH flag continued to build, erupting that June in another e-mail brawl with John Moore, a key member of the OAR's vaccine area review panel. The organizers of the 1996 international AIDS conference, slated for Vancouver in July, invited Moore to debate Uganda's Edward Mbidde about whether researchers needed to conduct more fundamental research before staging efficacy trials of AIDS vaccines. To help Mbidde prepare, Weniger e-mailed more than a dozen colleagues and solicited specific suggestions to counter Moore's anticipated arguments. When Moore learned of Weniger's behind-the-scenes efforts, he sent him an e-mail with the subject line "Running scared?" that dripped with bile and marked a new low point in the relationship between empiricists and reductionists:

> I got a copy of your memo to your cronies about the Vancouver Debate. God, you must be petrified of the truth to do such a thing. Talk about a panic reaction. Some of us don't need any help to present an intellectual debate. You have no valid argument, so getting dozens of people to help [Mbidde] defend the indefensible will do you no good at all—I will now take even more satisfaction than before in blowing you away. Previously, it was a job, now it is a cause. If you had any guts you would be on the podium yourself instead of acting by proxy. I despise you [and] Don Francis and all the evil corporate politics you stand for—trying to make money out of the dying is pretty pathetic, really.

Weniger retaliated by widely circulating Moore's screed, accompanied by a gentlemanly note. "I regret that you see these important issues regarding AIDS vaccine development in such black-or-white terms," wrote Weniger.

In truth, both John Moore and Bruce Weniger understood the nuanced complexities of the AIDS vaccine search. And both agreed on the need for a goal-oriented, targeted research program. But the contempt they felt for each other revealed that the division between the two camps they represented had grown. A fault that ran through their common ground had become a canyon.

• • •

THE VANCOUVER SHOWDOWN turned out to be a most civilized debate, and, anyway, reports about the awesome power of the new drug cocktails overshadowed most everything at the conference.[28] Several studies now showed that a protease inhibitor mixed with two reverse transcriptase inhibitors, triple combo therapy, could wallop HIV so effectively that people with AIDS rose from their beds, Lazarus-like, and began living again. Although the drugs had not rid anyone's body of the virus, they made it more and more difficult to find—"undetectable" became the buzzword—and researchers began to talk openly about the possibility of "eradication," complete clearance of an HIV infection. No one's damaged immune system returned to normal, but CD4 cell counts, the most widely used gauge, began to rebound dramatically and stay elevated.

Unlike the vaccine effort, industry had invested heavily in anti-HIV drug research, assigned top scientists to goal-oriented projects, and used the NIH largely to run the clinical trials of different compounds. The biggest pharmaceutical companies in the world—Merck, Glaxo Wellcome, Hoffmann–La Roche, Bristol-Myers Squibb, Abbott, Boehringer Ingelheim, Pharmacia & Upjohn—truly raced each other. Fifteen drug makers had even formed an Inter-Company Collaboration for AIDS Drug Development to speed the testing of different anti-HIV cocktails.[29] And a well-organized, angry, intelligent, and imaginative constituency of AIDS activists relentlessly criticized the speed at which new treatments came forward, writing detailed reports, staging widely covered public demonstrations, and running sophisticated lobbying campaigns on Capitol Hill.

The new treatment regimens had serious drawbacks. Not only did their $15,000-a-year price tag make them too costly for most HIV-infected people in the world; scant data addressed how long these new treatments could prevent disease and extend life. The drugs caused nausea and could have serious toxicities. In some people, the drugs worked only for a short time. In others, HIV became undetectable, but their immune systems were too damaged to rebound. Research had shown that HIV could take refuge in body tissues (as opposed to blood), which the drugs had a harder time reaching. The virus also routinely mutated into strains that could resist every drug proven effective against it. There were practical limits, too: Treatment required taking two dozen or more pills a day, on a schedule, a daunting task to carry out for years on end; and missed pills lowered the potency of the treatment, paving the road for drug-resistant HIV mutants.

Despite the caveats, journalists went hog-wild. The *New York Times*

Magazine ran a cover story headlined "When AIDS Ends" that pronounced "this ordeal as a whole may be over" and explored the impact that the "twilight of an epidemic" was having on gay communities. A *Newsweek* cover more coyly asked the question, "The End of AIDS?" The *Wall Street Journal* ran a front-page, first-person article by an HIV-infected editor who described how the new drugs had brought him back from death's door and made him "more likely to be hit by a truck than to die of AIDS." And *Time* magazine named David Ho, who had done some of the cutting-edge studies of new drug cocktails and promoted the possibilities of eradication, its Man of the Year for 1996.[30]

This treatment triumphalism wrongly created the sense that AIDS was a disease of the past, and drowned out two other big stories in the HIV vaccine world. One was a publication, "Industry Investment in HIV Vaccine Research," issued in December 1996 by the nascent AIDS Vaccine Advocacy Coalition (AVAC), the first activist-led organization of its kind.[31] After interviewing researchers and leaders from 23 companies that had "active or once-active" AIDS vaccine programs, AVAC concluded that "the overall picture is quite discouraging," noting that the field faced such "extraordinary hurdles" as "inadequate funding, insufficient focus on the scientific roadblocks, lagging industry investment, few candidate vaccines in the pipeline, little urgency among affected communities and a lack of leadership in the overall effort." Many of AVAC's recommended fixes sounded familiar: boost the NIH's AIDS vaccine budget, target specific scientific questions, enact liability legislation, have the President declare a goal, and create more incentives for industry. But the group promoted radical, novel ideas too, such as the call for companies with approved anti-HIV drugs to either invest in AIDS vaccine development or devote money to an independent group like IAVI.

The second piece of big AIDS vaccine news came on December 12, when the NIH announced that Nobel laureate David Baltimore had agreed to head the new AIDS vaccine panel called for in the OAR review. As leader of the AIDS Vaccine Research Committee, Baltimore theoretically would direct and oversee the new restructured, trans-NIH AIDS vaccine program, which would enjoy a 1997 budget jump of 18% to $129 million. An esteemed basic researcher who won the Nobel for his codiscovery of the reverse transcriptase enzyme, Baltimore had many admirers, and activists and scientists alike roundly hailed the choice. "The NIH has often been accused by others of not taking AIDS vaccine development seriously," OAR director Bill Paul told me. "We think this appointment, aside from scientific value, is saying to the nation and the pharmaceutical industry we are taking it seriously."

In a *New York Times* op-ed that ran on January 4, 1997, Weniger and Harvard's Essex took a heretical stance, questioning the choice of Baltimore and the very idea that the NIH should lead the AIDS vaccine search. "A newcomer to vaccine research, [Baltimore] will become part-time chairman of yet another committee of outside advisers," wrote Weniger and Essex. "This is unlikely to overcome the handicaps of the N.I.H. in promptly developing a vaccine." Weniger and Essex then decried the ARAC decision, highlighted how the whole killed approach had been "untried," and charged that the NIH culture had "impeded" the effort because it "undervalues applied research." Set up an expedited, targeted, applied research program, they urged, along the lines of a March of Dimes, that has as its single purpose the development of an AIDS vaccine. Appoint a vaccine or pharmaceutical company veteran to run it. And have Clinton declare a goal of developing a vaccine by 2005. "Reinventing the Federal AIDS vaccine effort would be essential for such an inspiring achievement," they concluded.

The *Times* ran a letter-to-the-editor from NIH Director Harold Varmus, who contended that there was no imbalance in the NIH's AIDS vaccine portfolio.[32] NIH had played "major roles" in the development of several vaccines, wrote Varmus, who found Weniger and Essex's views "cynical, misplaced, and incorrect." Baltimore also sent Weniger a sharply worded e-mail. "I do not take kindly to debates set off by missives sent to the NYT," Baltimore wrote. He called the op-ed "strident" and criticized Essex and Weniger for not notifying him of their objection to his selection, which he said was "just plain uncivil."

Essex and Weniger had made important arguments in their op-ed. But the criticisms from Varmus and Baltimore indicated that serious scientists did not take them seriously. And that was a problem.

• • •

AT PACHA'S INVITATION, David Baltimore, Bill Paul, and Tony Fauci attended an AIDS vaccine meeting held a few blocks away from the White House at the Madison Hotel on April 7, 1997. The PACHA meeting asked these panelists to consider six draft recommendations for President Clinton, which included asking him to declare a goal to develop an AIDS vaccine by an as-yet-undetermined date "in order to galvanize international collaboration, public opinion, political will, and bureaucratic priorities."

Baltimore, Fauci, and Paul all argued against setting a goal with a specific date. Baltimore maintained that the declaration could have the opposite effect of its intent, depressing the effort by making the development of a vaccine seem too distant. Bill Paul, too, warned that announcing a date could lead to disappointment. "Giving a date of that sort is, I think, not a wise tactical decision," said Paul. And Fauci roundly blasted the suggestion. "I'm probably the only person in the room that was standing there when then Secretary Heckler said that we'd have a vaccine in 2 years," said Fauci. "I certainly don't want to repeat that history, so I would say that I believe it's folly to give a date."

PACHA ignored their advice, forwarding their goal-setting recommendation to Clinton.

• • •

THE WHITE HOUSE
Office of the Press Secretary
For Immediate Release May 18, 1997

COMMENCEMENT ADDRESS BY THE PRESIDENT AT MORGAN STATE UNIVERSITY
Edward P. Hurt Gymnasium
Baltimore, Maryland
10:30 A.M. EDT

THE PRESIDENT: . . . science is about more than material wealth or the acquisition of knowledge. Fundamentally, it is about our dreams. America is a nation always becoming, always defined by the great goals we set, the great dreams we dream. We are restless, questing people. We have always believed, with President Thomas Jefferson, that "freedom is the first born daughter of science." With that belief and with willpower, resources and great national effort, we have always reached our far horizons and set out for new ones.

Thirty-six years ago, President Kennedy looked to the heavens and proclaimed that the flag of peace and democracy, not war and tyranny, must be the first to be planted on the moon. He gave us a goal of reaching the moon, and we achieved it—ahead of time.

Today, let us look within and step up to the challenge of our time, a challenge with consequences far more immediate for the life and death of millions around the world. AIDS will soon overtake tuberculosis and malaria as the leading infectious killer in the world. More than 29 million people have been infected, 3 million in the last year alone, 95 percent of them in the poorest parts of our globe.

Here at home, we are grateful that new and effective anti-HIV strategies are available and bringing longer and better lives to those who are infected, but we dare not be complacent. HIV is capable of mutating and becoming resistant to therapies, and could well become even more dangerous. Only a truly effective, preventive HIV vaccine can limit and eventually eliminate the threat of AIDS.

This year's budget contains increased funding of a third over two years ago to search for this vaccine. In the first four years, we have increased funding for AIDS research, prevention, and care by 50 percent, but it is not enough. So let us today set a new national goal for science in the age of biology. Today, let us commit ourselves to developing an AIDS vaccine within the next decade. (Applause.)

There are no guarantees. It will take energy and focus and demand great effort from our greatest minds. But with the strides of recent years it is no longer a question of whether we can develop an AIDS vaccine, it is simply a question of when. And it cannot come a day too soon.

If America commits to find an AIDS vaccine and we enlist others in our cause, we will do it. I am prepared to do all I can to make it happen. Our scientists at the National Institutes of Health and our research universities have been at the forefront of this battle.

Today, I'm pleased to announce the National Institutes of Health will establish a new AIDS vaccine research center dedicated to this crusade. And next month at the Summit of the Industrialized Nations in Denver, I will enlist other nations to join us in a worldwide effort to find a vaccine to stop one of the world's greatest killers.[33] We will challenge America's pharmaceutical industry, which leads the world in innovative research and development, to work with us and to make the successful development of an AIDS vaccine part of its basic mission.

My fellow Americans, if the 21st century is to be the century of biology, let us make an AIDS vaccine its first great triumph. (Applause.)

• • •

THAT WAS THE LONGEST, most aggressive and compassionate speech any President of the United States had ever given about the drive to find a vaccine that could thwart HIV. But when Bill Clinton, resplendent in a graduation robe, invoked JFK and announced his goal to develop an AIDS vaccine within a decade to the packed gymnasium at Morgan State, a historically black university, it also was a perfect Clinton moment: camera-friendly, politically correct, emotionally charged—and a grand compromise. Bill Clinton, the first President in the era of AIDS who unabashedly attacked the disease, had the opportunity to spearhead a truly targeted AIDS vaccine program that, with new money, would aggressively work toward finding a vaccine in a decade. But instead of "reinventing" the U.S. government's AIDS vaccine effort, as Essex and Weniger had called for, Clinton announced a new crusade that offered little of substance.

Clinton took credit for a budget increase for AIDS vaccines that would have occurred without him. He announced a new NIH research center dedicated to AIDS vaccines that an accompanying press release said would be "fully operational within the next several months";[34] the center, which actually was OAR director Bill Paul's idea, in the end would focus on all vaccines—not just AIDS—and by the end of 1998 did not even have a director, let alone a fully functional program.[35] And, most egregiously, Clinton created another illusion that a problem had been solved when, in reality, it had only been acknowledged.

The other new agenda setters who took the stage following the fateful ARAC decision of 1994 similarly were better at articulating their dreams than they were at making them come true. Although the budget for IAVI steadily grew, in 1998 it had only $7.7 million to spend, a far cry from the $50 million or more it had hoped to devote each year to a targeted research program. Burt Dorman and Acrogen had no luck finding funding to make a whole, killed HIV vaccine. The visionary Levine committee report led to many important changes in the NIH's AIDS vaccine program, but more than three years later, the institution still had not devoted $25 to $50 million a year to a targeted program, established cooperative or public ownership arrangements for riskier AIDS vaccines, noticeably changed the culture to attract more top-notch immunologists to the field, or analyzed the financial disincentives facing industry. Baltimore's AIDS Vaccine Research Committee quickly created a new program to support innovative ideas, but it did not, as the Levine report called for, set up and oversee a new "trans-NIH" AIDS vaccine effort. Nor did the White House follow the Levine committee's advice and establish a new National AIDS Vaccine Task Force to coordinate the workings of "all U.S. Government agencies."

Thailand did move ahead with its testing of the gp120 vaccine, but the projects progressed slowly. One problem was that Chiron, discouraged by the lack of enthusiasm from leading scientists and the failure in efficacy trials of their genetically engineered envelope vaccine against herpesvirus, lost nearly all interest in that approach. Another problem was that VaxGen struggled to raise the needed money to conduct trials with its product.

VaxGen, however, did end up raising $27.5 million, and by the end of 1998 had launched an efficacy trial of its gp120 vaccine in the United States, and began a similar trial in Thailand in early 1999. VaxGen's progress, though, meant little to most researchers at the front of the AIDS vaccine search. An e-mail exchange between Bruce Weniger and David Baltimore captured the essence of mainstream thought. Weniger had questioned why Baltimore had continued to criticize these trials, even though they did not involve NIH funds. "I doubt anyone [today] would consider making such preparations as vaccine candidates," Baltimore wrote Weniger in June 1997. "I think it is time that you started thinking rationally and became a help in getting decent preparations for test so that we can have a chance of getting a vaccine in 10 years. Simply repeating the old mantras will help no one."

With the guidance and support of Baltimore, the OAR, a revamped NIAID, IAVI, and even industry, many new vaccine ideas would move forward during the next few years. But the problem repeatedly highlighted over the years by Weniger and many others remained: promising leads did not move forward quickly enough.

Empiricism versus reductionism in the final analysis came down to time versus resources. The reductionists maintained that, given the limited bankroll the field had to work with, they had distributed money wisely, especially after the OAR reforms; tilting the scales more toward applied, empirical science, these pragmatists believed, likely would shortchange progress on more basic questions, potentially *slowing* progress. Weniger and other empirically minded scientists had a more romantic vision. They believed the money existed for a targeted program, which they hoped would identify and fill critical research gaps, better coordinate primate research, and effectively address nonscientific issues such as liability, providing an extra push to significantly shorten the time it took for an idea to reach an efficacy trial in humans. Had they found funding outside the U.S. government for this program—from, say, philanthropists—they might well have silenced their main critics. No one, after all, stopped VaxGen from conducting gp120 efficacy trials on the company's nickel. But for the most part, the empiricists kept appealing to different branches of the U.S. government either to reslice the AIDS research pie or serve up a new one just for their project. And these people, individually and collectively, lacked the scientific respect or the political clout necessary to shake loose those federal dollars, so their dream, worthwhile as it was, died.

CHAPTER 13

Running in Place

In early 1996, Thomas Kindt watched as his long-standing research quest, which held the promise of revolutionizing the way scientists tested AIDS vaccines, died and then came back to life.

Kindt, an immunologist at the National Institute of Allergy and Infectious Diseases, for nine years had pursued the dream of infecting rabbits with HIV. If HIV could infect rabbits and make them sick, this animal model potentially would offer extraordinary advantages over either monkeys or chimps. Not only do rabbits cost researchers a fraction of the amount they pay for primates; monkeys can only be infected with SIV, not HIV, and chimpanzees, an endangered species, are always in short supply. A small, cheap animal model would make it feasible for labs routinely to conduct comparative challenge experiments with different vaccines. If an approach looked promising, scientists feasibly could test it in dozens of animals and robustly assess its worth. Large studies with rabbits also would make it much easier to isolate the antibodies and cell-mediated immune responses that accounted for a vaccine's success.

Although Kindt reported in 1989 that he could infect some rabbits with HIV, the model had many problems.[1] During the next six years, he improved the ability of HIV to infect rabbits by engineering their white blood cells to contain human CD4 receptors, the surface molecule that the virus first docks onto during the infection process. These cells proved much easier to infect, but a few months after he published these results in 1995, scientists working on the Office of AIDS Research review of the NIH's AIDS research program began scrutinizing his work.[2] NIAID had been spending $3.5 million per year on the rabbit model, which the reviewers cited as one of the "striking examples of the disparity between the quality, productivity and importance of the research and the resources devoted to it."[3] *Newsday* ran an article about the

OAR review, headlined "Mi$$pent," that quoted an unnamed reviewer mocking the rabbit work too. "It was like a standard joke in the review meetings," the reviewer said. "Whenever something smelled fishy we'd all yell, 'Bunny rabbits! More bunny rabbits!' "[4] Kindt was furious. "They dumped all over it without really looking at the data," he said.

Then, coincident with the completion of the OAR review, other researchers at NIAID made a stupendous discovery that rocked the entire AIDS field—and resurrected the rabbit model. Scientists had for a decade struggled to understand how HIV infected a cell. They knew that the virus relied on CD4 receptors, but experiments had shown that some other cofactor also was required. Over the years, several putative cofactors had been dragged into the interrogation room and put under the harsh light, only to reveal that they could not be guilty. In May 1996, however, NIAID's Edward Berger and coworkers reported in *Science* that they had collared a suspect which other labs soon confirmed was indeed the elusive cofactor.[5]

Stoking the excitement about Berger's discovery, two other laboratories recently had published reports that beautifully dovetailed with his. The cofactor Berger identified normally served as a docking slot, or receptor, for chemokines, immune system messengers that play a key role in the inflammatory response.

Most AIDS researchers first learned about chemokines six months earlier when a team led by Robert Gallo announced that CD8 cells from HIV-infected people secreted high levels of these little-known natural chemicals. These chemokines, in turn, could, for some unknown reason, block HIV's ability to infect cells. Berger's discovery unraveled the mystery: the chemokines had attached to their receptors, which blocked the ability of HIV to use them, locking the virus out of the cell.

The second study that meshed with Berger's, and which appeared the month before in *Nature Medicine*, focused on 25 exposed, uninfected people (EUs). Working with researchers from seven other institutions, Richard Koup and coworkers at the Aaron Diamond AIDS Research Center found that the CD4 cells in these EUs resisted infection by HIV. The researchers associated this resistance with higher levels of chemokines.[6]

By the summer's end, chemokines and their receptors had become the hottest basic research topic in AIDS, a bona fide breakthrough. Soon it became clear that several different chemokine receptors could open the cellular doors for HIV. But, curiously, HIV relied on one particular receptor—dubbed CCR5—when it first entered the body, and, as a person progressed to disease, the virus started to favor other chemokine receptors. Aaron Diamond's Koup and Nate Landau showed that two of the EUs who had high chemokine levels actually had defective CCR5s, which presumably prevented HIV from ever entering the cell and establishing an infection.[7] Although it turned out that defective CCR5s occurred in only about 5% of the EUs studied and thus did not explain why most of them had resisted HIV, the finding underscored that this receptor, by an unknown mechanism, played an essential role in the initial infection process, making it a sudden star in the AIDS vaccine arena. CCR5 also greatly intrigued researchers like Kindt who wanted to develop a small animal model in which to test potential vaccines.

Had a targeted, goal-oriented AIDS vaccine program existed, a forward-thinking group of scientific advisers could have decided during the fall of 1996 to fast-track the engineering of a small animal model that had the human receptors for both CD4 and CCR5. As it happened, a handful of research groups began attempting to put these human genes into mice, mostly by borrowing money from other grants or with funds supplied outside of the NIH. These "transgenic" mice, though, from the outset had a distinct disadvantage to rabbits: HIV did not replicate in their cells, which meant that it could not cause disease. Kindt had stopped most of his lab work by then, taking on a new job as an NIAID administrator, but fortunately a researcher at the Gladstone Institute of Virology and Immunology—a privately endowed affiliate of the University of California, San Francisco—called and asked him to collaborate on the rabbit model. The researcher, Mark Goldsmith, had some seed money from the Gladstone Institute, and with that, the effort to stitch human CCR5 and CD4 into rabbits began.

So the great chemokine breakthrough of 1996 led one research team in the world to investigate the possibility of making a CD4/CCR5 transgenic rabbit, an animal model that potentially could hasten the entire AIDS vaccine search. To its credit, the newly established AIDS Vaccine Research Committee headed by David Baltimore recognized at its first meeting on February 17, 1997, that the NIH could markedly help the field by instituting a new grant program that targeted key areas and encouraged risky, innovative work. NIAID a month later announced its new Innovation grant program, which from the start targeted animal models. Goldsmith applied for one of these grants and, after his proposal went through peer review, that October was awarded the first of two $150,000 annual installments.

Not surprisingly, the award thrilled Goldsmith: $300,000 is a handsome sum for an academic researcher. But view this from the perspective of the NIH guiding a field that badly needs a small animal model. The NIH invests $3.5 million a year on the rabbit/HIV model, scientists run into a roadblock, the NIH cuts the project, scientists find a promising way around the roadblock, and a year later the NIH funds a new project at $150,000 a year.

Something is wrong with this picture. And it is especially dispiriting because Goldsmith's small group made steady progress, showing in test-tube experiments that HIV readily could infect rabbit cells engineered to express human CD4 and CCR5 receptors. Encouragingly, the HIV replicated nearly as well in the rabbit cells as it did in human cells, and these new viruses could then infect human cells, proving that they were viable.[8] While many scientific obstacles stood in the way of translating this advance into a living rabbit, Goldsmith readily acknowledged that he could have done more with more. "There's no question in my mind, if we had more resources, we could make progress at a faster pace," he told me in January 1999. Imagine, too, what the rate of progress would have been had other groups simultaneously attempted to make HIV-infectable rabbits.

Not only was the rabbit model advancing slowly; Goldsmith was having trouble finding funds for what he believed was an even more exciting possibility: transgenic rats that could be infected with HIV. Goldsmith's lab found, to his astonishment, that CD4/CCR5 transgenic rat cells support an HIV infection as readily as the rabbit cells do. A rat model holds many advantages over the rabbit; in addition to being cheaper and easier to breed, rats are a much more common lab animal, which means that scientists have a better understanding of their immune systems and have developed many rat-specific reagents needed to study them. When Goldsmith told NIH officials about his advance, he was grateful that they pledged to find $75,000 to help him. But that amounted to a fraction of what he needed. "I have to figure out a way to launch *the* most promising avenue by borrowing money from the rabbit program," said Goldsmith.

Many exciting, alluring leads surfaced in the late 1990s, but few moved forward with a detectable sense of urgency. Researchers discovered better ways to make whole, killed vaccines. They found antibodies that could readily neutralize primary isolates. They devised wondrous new vectors—harmless bacteria and viruses that give HIV genes a piggyback ride into cells. They showed that naked DNA vaccines containing little more than HIV's genetic material could protect monkeys and chimpanzees. They designed powerful new methods to evaluate the relative merit of antibodies and killer cells. They uncovered mind-stretching immunologic mechanisms of protection in exposed, uninfected people.

Just as the effort to make a rabbit model received support, all of these promising leads did too. But once again, the money—whether from governments, industry, or foundations—invariably was meted out slowly and in small amounts. And because no central leadership directed the enterprise, few groups typically went after each lead. Time simply was not of the essence, which left many excellent researchers running in place.

• • •

WHEN RESEARCHERS WITH SHODDY REPUTATIONS and wild theories have trouble advancing the leads that they find promising, a field does not suffer, and may well even benefit. But when the likes of Dennis Burton do not receive the support they need, red flags should pop up.

Burton was the type of HIV immunologist that the OAR report said the field desperately needed to attract. Trained at Oxford University, Burton ran a much-respected lab at the Scripps Research Institute in La Jolla, California, and had made a name for himself as one of the foremost experts on HIV antibodies. Indeed, when the members of the NIH's AIDS Vaccine Research Committee wanted a comprehensive update on neutralizing antibodies, they tapped Burton and three other leading investigators in the field to bring them up to speed.[9] Burton also consistently received generous support from the NIH—he had five separate grants from the institution, including one from the Innovation program, funding him in 1999. Yet one of his most exciting findings, which had direct relevance to a critical question about AIDS vaccines, languished for years.

Burton believed that immunologists like himself could help guide vaccine developers by elucidating the relationship between HIV's gp120 and antibodies. In the language of immunology, gp120 is an antigen—a molecule that prods the immune system into action—and different parts of the protein, called epitopes, elicit different antibodies. Further complicating the picture, HIV, like a pitcher hurling a baseball at the batter, presents gp120 to the immune system in many different ways, which also leads to different antibody responses. This explains why Burton found more than 200 different kinds of gp120 antibody in one HIV-infected person.

Along with his friend John Moore, Burton in scientific journals and talks clobbered the gp120 vaccines made by Chiron and Genentech, which he contended did not take into account fundamental discoveries about the infection process.[10] In particular, studies of the surface of HIV have shown that gp120 clusters into groups of threes. These "flowers," as Burton called them, play an essential, if little understood, role in the presentation of the molecule to a cell. But the gp120 vaccines contain single molecules of gp120, not the flowers, a shortcoming in Burton's eyes that strongly predicted failure.

Most of the gp120 antibodies that Burton harvested from patients had little ability to stop various strains of HIV, whether primary isolates or laboratory adapted, and did not bind to the flowers. This led him to conclude that the immune system mainly sees "viral debris"—pieces of gp120 that stud HIV's surface—rather than the properly formed molecules that have clustered into flowers. But after sifting through thousands of human antibodies, Burton found one that could effectively trip up most any strain of HIV thrown at it. Only two other such "monoclonal" antibodies had previously been discovered.[11]

Burton wanted to make this monoclonal antibody in large quantities, inject the antibodies into monkeys, and see whether they could prevent infections from highly virulent strains of SHIV—the chimeric SIV engineered to contain HIV's surface protein. If this passive immunization strategy worked, it would both bolster the argument that antibodies could, by themselves, stop HIV, and provide a benchmark for the type of antibodies a vaccine should elicit. Sleuthing with the tools of molecular biology might also lead to the epitope on gp120 that triggered production of the antibody. A gp120 vaccine, then, could prominently feature this epitope.

But Burton did not have the setup to make the large quantities of antibody—which means 100 grams—needed to do the passive immunization study. NIAID's Cliff Lane, who wanted large quantities of the antibody himself to test its ability to help HIV-infected people, in 1995 began exploring the possibility of having the NIH use its facilities at Frederick, Maryland, to do the manufacturing job. In 1999, the NIH still had not made the antibody. "It's very ad hoc–ish, either because nobody believes antibodies can do much and we're a sideshow, or there's a lack of decisiveness," a frustrated Burton told me. And, in his judgment, the passive immunity experiment had much to offer the field at large. "It's just such an important principle to establish," said Burton. "If antibodies don't work, then why bother?" Alternatively, he suggested, potent antibodies might buy time for cell-mediated immunity to kick in and clean up any cells that did become infected. The experiment might help elucidate that too.

Burton, in the end, devoted three people in his lab to making as much of the monoclonal antibody as they could. "We just sort of grind it out," he said. "It's kind of infuriating: we're doing it in 100-milligram quantities."

If the NIH made large quantities of the antibody for Burton and quickly provided him with monkeys, the world would not have an AIDS vaccine within weeks, months, or even a year. Basic research, when everything runs optimally, moves forward slowly. But when you start to add up the number of *respected* AIDS vaccine researchers who, like Burton, remain confounded by the obstacles that eat up their days without moving their research forward, well, a year here, a year there, and pretty soon you are talking about real delays.

• • •

WHEN JACK NUNBERG HEARD about NIAID's new Innovation grants program, first announced in March 1997, he hurried to send in a proposal. NIAID's novel program specifically targeted the structure and function of the HIV envelope protein, which meshed beautifully with Nunberg's research aims. Unlike the traditional investigator-initiated grants—the so-called R01s—the Innovation program also encouraged proposals from researchers who had little, if any, preliminary data to back their ideas. And Nunberg had solid credentials: before moving in 1996 to his current job as director of the biotechnology center at the University of Montana, he had worked on the gp120 vaccine at Genentech for four years and had spent three years at Merck working on anti-HIV drugs.

Nunberg's idea centered around the recent discoveries about chemokine receptors and how they help HIV fuse with a cell to establish an infection. The virus enters a cell through a three-step process, with HIV's gp120 first attaching to a CD4 receptor. Next, the gp120/CD4 complex leans over and binds to a chemokine receptor. After completing these handshakes, the virus slips into the cell. Nunberg wanted to derail the infection process in between steps of this infection process.

Picture a man wearing a T-shirt whose navel becomes visible when he lifts his arms; Nunberg similarly knew that when HIV's gp120 bound to a CD4 receptor, the shape of the viral protein changed, exposing new epitopes for the immune system to see. Similarly, in the next stage of the infection process, in which the gp120/CD4 complex bound to a chemokine receptor like CCR5 that resides on the same CD4 cell's surface, more new epitopes presumably became visible. Nunberg postulated that by chemically fixing the infection process at different stages, he might be able to trigger production of novel antibodies.

To evaluate his idea, Nunberg proposed that he would inject mice with what he called "fusion-competent" vaccines, bleed them, and then assess whether the resultant antibodies could neutralize a variety of primary isolates and lab-adapted strains.[12] His proposal and 132 others went to a specially selected study section for expedited review. A whopping 58 of the applicants—43.6%, about double the success rate of investigators who submit traditional R01 grants—learned in the fall of 1997 that they had won funding. Nunberg was not one of them.

Grants are given priority scores, with low numbers indicating a better proposal. The cut-off for these grants was a priority score of about 200. Nunberg's score: 360. "I got all these cockamamie reviews," said Nunberg.

The NIH gives comments from reviewers to applicants. At least three reviewers noted that Nunberg had impressive credentials, but they deemed his proposal only "moderately innovative." One complained that "the likelihood of success at each of the required steps appears low." A second reviewer allowed that if the vaccines did lead to antibodies that could neutralize primary isolates, "this will be an important result which would trigger a number of subsequent studies." But this reviewer also doubted that the experiment would work, concluding that "there is a small likelihood that this work will contribute broadly to the search for an effective HIV vaccine."

The response did not make sense. This program specifically offered "support for preliminary studies of a highly speculative nature which are expected to yield, within this time frame, sufficient information upon which to base a well-planned and rigorous series of further investigations."[13] Nunberg was beside himself. "There's a lot vested in the status quo," he concluded.

Nunberg turned to the American Foundation for AIDS Research, which had just announced its first-ever program that explicitly aimed to fund AIDS vaccine work. Modeled after the NIH's Innovation grants program, AmFAR devoted $1.1 million to creative ideas that would stimulate the field. In February 1998, AmFAR announced that its peer-review committee decided to fund Nunberg and 10 other vaccine projects. They concluded that Nunberg had "a very novel approach" and awarded him $142,456. "AmFAR was a godsend," said Nunberg.

In January 1999, less than one year later, Nunberg had a report in *Science*—which rejects 90% of the original research papers it receives. The report showed how, in test-tube experiments, antibodies made in response to his fusion-competent vaccines could neutralize 23 of 24 primary HIV isolates from many parts of the world, representing five different subtypes of the virus. The impressive results even warranted an accompanying perspective by two leading HIV-antibody experts.[14]

The NIH's rejection of Nunberg's proposal did not, thanks to AmFAR, significantly slow his progress. Nor will it be easy, as the *Science* perspective stressed, to translate Nunberg's findings into a human vaccine. But the fate of Nunberg's idea provides an important window into the serious limitations of the NIH's granting system. Nunberg submitted his grant to a program that asked for high-risk, innovative AIDS vaccine ideas. A special study section that had vaccine experience reviewed the applications. NIAID liberally made awards, doubling the typical success rate of grant applicants. Yet Nunberg slipped through the cracks.

• • •

IRVIN CHEN, head of the NIH-funded Center for AIDS Research at the University of California, Los Angeles, also wrongly thought the Innovation grants program would help him launch a novel HIV vaccine project that, besides his reputation, had little suggesting that it would work.

In 1996, Chen became intrigued by Acrogen's Burt Dorman and his campaign to make a whole, killed HIV vaccine. Dorman had come to Chen's UCLA office and explained that no one had yet tested a properly made inactivated HIV preparation. "As he was talking, it really struck me that he had a good point," remembered Chen. "There hadn't been that much done on inactivated HIV vaccines." While the Immune Response Corporation had made a whole, killed HIV vaccine and even tested it in a chimpanzee challenge study, the inactivation process destroyed the envelope protein, creating what some called a "nude" HIV. The few other researchers who tested whole, killed HIV vaccines in chimpanzees also did not optimize the inactivation procedure to retain envelope.[15] And without an intact envelope, an HIV vaccine had little chance of triggering a potent antibody response against the virus.[16] Monkey tests of whole, killed SIV vaccines had prematurely taken a nosedive, Chen concluded, because of the 1991 finding that researchers had contaminated their experiments with human proteins.

Although Dorman had made a dozen whole, killed veterinary vaccines, Chen offered three critical attributes: wide scientific credibility, HIV experience, and nonprofit status. Chen, the son of a prominent plant biologist, earned his Ph.D. working with Howard Temin, the University of Wisconsin retrovirologist who shared the Nobel Prize with David Baltimore and Renato Dulbecco for the discovery of reverse transcriptase. Chen enjoyed a solid reputation in HIV basic research circles for his work illuminating the role of various viral genes, and he regularly published in top-flight journals.

Dorman's interest in nonprofit status arose because he already had struck out everywhere he looked for support as a for-profit company. The NIH repeatedly had turned down his grant proposals. The Walter Reed Army Institute of Research showed some interest, but ultimately no money came from that route either. Private investors liked the idea until they learned about the daunting liability issues. Now Dorman had another idea: with Chen's help, Dorman thought the two of them might be able to win a grant from the California Endowment/California Healthcare Foundation. These two new sister foundations, which California forced Blue Cross to start because it changed to for-profit status, had a combined $3.3 billion in assets, at least 5% of which the law required them to spend each year.[17] And the Healthcare Foundation's incoming president, Mark Smith, was a doctor who specialized in AIDS at the Henry J. Kaiser Family Foundation.

On November 11, 1996, Chen, Dorman, and Donald Kennedy (the former FDA commissioner and Stanford University president) met with Smith to see whether they could interest the foundation in an ambitious multimillion-dollar program to develop a whole, killed HIV vaccine. Smith, though, did not think the proposal would fly, and the foundation ultimately did turn them down. "So I gave up, pretty much," Dorman told me. "But Irvin had gotten infected by the idea."

When Chen learned about NIAID's Innovation grants and the focus on envelope structure and function, he thought he and Dorman would have a good shot. He applied and, though he received a better priority score than Nunberg had—280—he did not make the cut. From the reviews, Chen deduced that the reviewers did not believe it was truly innovative. "They have the perception that it has been thoroughly tried and it hasn't worked," said Chen. Dorman, once again, felt dejected. "At that point, I said, This is one of those worthwhile programs that isn't going to happen until after my lifetime," said Dorman.

Chen decided to ask the NIH for a $100,000 supplement to the grant UCLA received for its Center for AIDS Research. The supplement, which would simply support his lab's attempt to make an inactivated HIV that retained its envelope and then to study its structure and properties, did not have to go through peer review. He received the funding. It was a far cry from millions, but it was a start.

Chen and his coworkers soon overturned the dogma, showing that, by using heat, they could kill HIV without having the virus lose its envelope proteins. "It wasn't that hard," says Chen. "What it really depends on is the strain of the virus you use."

In 1998, Larry Arthur and coworkers at the National Cancer Institute—whose work did not have to go through peer review—published that they, too, had devised a way to kill HIV and not destroy the envelope proteins.[18] Using a chemical, these researchers crippled a core protein inside of the virus that it needs to copy itself. "With R01 grants, this is just not an attractive thing to do," acknowledged Arthur. "[Whole, killed vaccines] really haven't been given a good chance. These types of experiments are not intellectually exciting, but they're probably more important than ones that are."

Chen shared this sentiment. "I'm not a martyr—Burt definitely is," said Chen. "This is one part of my program. I do a lot of stuff that's sexier. But this, if it works, will have far greater impact than anything else I've been working on." So Chen again applied for an Innovation grant in 1998. His priority score: 228.

By 1999, Chen wanted to test a well-characterized, envelope-containing whole, killed-virus vaccine in monkeys. "I'd love to put it into monkeys, but I can't get anyone interested," he said. So even though the OAR report smartly suggested that the NIH should make monkeys more accessible to academic researchers, Chen's experiences once more spotlighted how good recommendations, if unheeded, can foster a sense that a problem has been solved when, in truth, it only has been put into bold type.

• • •

BRUCE WALKER, an immunologist at Harvard Medical School's Massachusetts General Hospital, in 1997 applied for an Innovation grant to test a vaccine in both mice and HIV-infected humans to see whether it could boost a component of the cell-mediated immune system that he deemed critically important. "I thought this was sort of a shoo-in when we submitted it," said Walker.

Walker's proposal grew out of his examinations of Bob Massie's immune system. Massie, an HIV-infected man, first came to see Walker in July 1995 because he had a pressing question that he hoped this renowned AIDS cellular immunologist could help him answer. Massie had hemophilia, and he likely became infected by HIV in 1978 from a tainted batch of factor VIII, the clotting agent in blood that his body could not properly produce. The year before, Massie ran for lieutenant governor of Massachusetts and lost, his marriage of 12 years had broken up, and he had started to date an old college friend. The woman "was seriously wondering where this was going to go, and what it would mean to be in a physical relationship with someone with HIV," Massie explained to me. "And her mother was not thrilled about the idea, as you can imagine." He also intensely wanted to understand his fate. "I thought my God, here it is 1995, and I was exposed in 1978, and I am still fine."

Bob Massie had been infected for 17 years and had never taken any anti-HIV medication, which, statistically speaking, meant Massie should not only have developed AIDS but also died from the disease. Yet in all of those 17 years, Massie never once had had an abnormal CD4 count. "Bruce jumped out of his seat," recalled Massie. "He just thought this was the most exciting thing. It made me feel really good. Frankly, I thought maybe we'll get to the bottom of this."

Shortly after that visit, Walker phoned Massie to discuss the results of a test that quantified his viral load—the amount of HIV in his blood. Typically, infected people who do not take anti-HIV drugs have tens of thousands of copies of the virus per milliliter of blood. (The body has 5 liters of blood in all, so a person who has a viral load of 10,000 actually has at least 50 million copies of the virus floating around.) Massie's viral load was "undetectable." This did not mean that Massie was uninfected: he made antibodies against the virus, clear evidence that HIV was still there, albeit at levels below the minimum of what the then-most-sensitive test could detect (500 copies of HIV per milliliter, or 2.5 million HIVs bodywide). Still, the news thrilled Massie—whom scientists classify as a long-term nonprogressor—and Walker.

Although Bob Massie might simply have been infected with a harmless, defective form of HIV, it seemed to Walker more likely that the immune system somehow had managed to control the virus, and he set out to find the specific immunologic warriors responsible for this success. Antibodies, either in quality or quantity, could not explain Massie's odds-defying health. Once a person is infected with HIV, the virus can slink directly from one cell to another, a covert move that antibodies don't detect. People also routinely die from AIDS with robust anti-HIV antibodies patrolling the blood. And if you take antibodies from the blood of a person like Massie and repeatedly infuse them into a sicker infected person, they don't slow the virus.

Walker had made his name in AIDS research with his studies of more sophisticated immunologic warriors, the killer cells, technically known as cytotoxic T lymphocytes, or CTLs. In contrast to antibodies, killer cells obliterate cells that already have been infected. In a test-tube experiment, Walker prodded Massie's blood cells with pieces of HIV. "He had the strongest CTLs of anyone we'd ever seen," said Walker.

If Walker simply had demonstrated that Massie's blood was teeming with CTLs trained to annihilate HIV-infected cells, the AIDS research community would not have paid much attention. The power that CTLs have against the virus long has been revered, almost mystically so. But Walker looked beyond these immunologic Green Berets, because he had analyzed the blood of other unusual patients who had steadily declining CD4 counts despite having lots of CTLS. In an exquisite twist of plot, that strange response led him to investigate the CD4 cells themselves.

When it comes to ridding the body of an invader, CD4s play one-on-one, with individual cells devoting themselves to distinct invaders. Specifically, when HIV confronts a CD4 cell, the cell takes note of what the virus looks like and then copies itself furiously. These new battalions of HIV-specific CD4s then can mount an attack on the virus. Immunologists call these trained CD4s "helper" cells, because rather than assaulting invaders, they serve as what Walker calls "Command Central," coordinating the entire immune response by secreting chemicals that in turn marshal antibodies, CTLs, chemokines, and other immunologic forces.

Scientists have called HIV insidious, diabolical, evil, and worse because the virus exploits the helper cells for its own destructive ends. HIV turns the meaning of helper cells on its head first by leading the immune system to expand its populations of CD4s, which creates many more targets for the virus. Once HIV has infected large numbers of these newly minted CD4 helper cells, the virus has thoroughly infiltrated Command Central. Over time, the immune system runs out of new recruits—healthy CD4s—and no longer can manufacture more HIV-specific CD4s. "When command central gets knocked out, you have a bunch of armies wandering around the desert," said Walker.

Bob Massie, it turned out, had what Walker called "unbelievable" HIV-specific CD4 helper responses. Walker's lab soon began studying another untreated, long-term nonprogressor who similarly had a whopping CD4 helper response. Walker applied for an NIH R01 grant to expand his work. "They said it wasn't generalizable," guffawed Walker. "Give me a break. That's the whole point here."

Walker resubmitted the R01 grant application and won funding. This allowed him to analyze HIV-specific helper responses in 25 other untreated, HIV-infected people. "It looked like most people didn't have these [helper responses]," said Walker. "And if you had help, you had a low viral load, and if you didn't have help, you had a high viral load. There was a strict correlation." Another odd case helped hammer in this point: This man, infected for 10 years, had a screaming high viral load—175,000—yet a normal CD4 count. As Walker suspected, this man had no HIV-specific helper cell response to speak of.

In November 1997, Walker's lab published these findings in *Science*.[19] The paper received much attention and received high praise from AIDS researchers. "The observation is really quite exciting and opens up whole new areas," NIAID's director Anthony Fauci told me.

Walker, logically enough, submitted a proposal to the Innovation grant program that outlined a plan to turn these findings into a vaccine. Walker's strategy aimed to give the immune system a head start against HIV. From the moment HIV enters the body, it races the immune system. If the immune system responds quickly and fiercely enough, the virus never has a chance to establish a chronic infection. A slow and weak immune response, in contrast, resembles the invasion of a fort at dawn: the CD4 soldiers come running out of their bunkers unarmed and half-dressed, providing easy targets for the virus. As Walker explained in his proposal, he and his coworkers had found pieces of HIV's internal proteins that, when injected into animals, trained the immune system to produce a stronger, faster CD4 helper cell response.

To make the vaccine, Walker hooked up with Peptimmune, a Cambridge, Massachusetts, biotech company that had a technology for stringing together these key protein pieces. Walker and Peptimmune first hoped to test the vaccine in HIV-infected people, who, ideally, would build stronger troops of HIV-specific CD4 helper cells and begin to more closely resemble Bob Massie. A preventive vaccine would come next, based on the theory that priming this arm of the immune system would give it a chance to shuck off an infection. And even if vaccinated people became infected, the head start their immune systems had against the virus might protect their HIV-specific CD4 helper cells, turning vaccinees into long-term nonprogressors.[20]

To Walker's dismay, his Innovation grant proposal did not much impress the reviewers, who complained that he did not have enough preliminary data and had not carefully thought out the way he planned to measure different immune responses. "Do they think we're stupid? Why don't they say, 'That's a great idea, why don't you add this on'?" asked Walker.

Next, Walker turned to AmFAR's AIDS vaccine innovation grant program. AmFAR, too, turned him down.

Study sections, Walker concluded, no matter what their charter, prefer middle-of-the-road ideas over innovative ones. "I'm the chair of a study section, and I know the dynamic that goes on," he said. "You are looking for what *not* to fund." So instead of evaluating an idea, Walker explained to me, they look for something to criticize, and then encourage you to submit the proposal again. "If time is no problem, that's fine," said Walker. "But we have an incredible urgency here."

• • •

THE 100 ACADEMICS AND ACTIVISTS who wrote the landmark 1996 review of the NIH's entire AIDS program had many excellent ideas about how to speed the search for a vaccine. They called for a targeted research program, run by a team of extramural and intramural experts, to coordinate, among other things, the "rapid evaluation of promising new vaccine candidates or animal models" and to provide the "missing link" between basic research and clinical trials. They recommended that the NIH establish the AIDS Vaccine Research Committee to "provide leadership, direction and oversight to a comprehensive AIDS vaccine effort" to "rapidly exploit new advances." These were solid ideas, and, to the degree that the NIH took the advice, they made a difference. But the experiences of Mark Goldsmith, Irvin Chen, Jack Nunberg, and Bruce Walker illustrate that the NIH, even at its best, still left serious gaps that other funding bodies needed to fill.

There of course is no guarantee that any of these research projects would have paid off. But then, high-risk research, by definition, is research that likely will fail. Without question, a fine line often separates high-risk research ideas worthy of funding from those that have no hope of succeeding. The trick is to gamble wisely, which often means betting on people and not just the idea itself.

The NIH, try to change as it might, does not have much of a stomach for high-risk ideas. Its egalitarian review process—smartly—offers few rewards based on a person's reputation. But when the research involves a public health crisis, the rules must change. Goldsmith has a strong reputation and works for an excellent institution. Why didn't the NIH provide him with more money to explore transgenic rabbit and rat models and, better yet, seek out other top-notch researchers to compete with him? And what of those well-respected researchers who failed to win grants from the Innovation program, a mechanism tailor-made for funding creative AIDS vaccine research? Because the Innovation program came about at the suggestion of the AIDS Vaccine Research Committee, I asked its chair, David Baltimore, about its failure to fund Chen and Nunberg.

I chose these two names for a reason: I knew that Baltimore particularly liked the idea of a whole, killed vaccine with virus that retained its envelope, as well as the notion of making stronger antibodies by exposing new epitopes. I suggested to Baltimore that if he were the scientific director of an organization that had the power to fund whatever he chose, he would have decided, unilaterally, to give money to these two researchers. "I would have," he agreed.

Yet Baltimore stressed that the fact that the NIH Innovation program failed to award them grants hardly proved that the system did not work. "Look, you can give hundreds of examples in the whole history of science of important things that weren't recognized as important, didn't get funded, that people had to do on their own," he said. "But in some ways the story is the story about how the system works, because one of the things that is very important in all funding is pluralism, is having other places to go. So when the first guy gets it wrong the second guy can have a shot to get it right."

As Walker said, if time is no problem, that's fine. But if time does matter, then a granting system that does not rely on peer review potentially could have much impact.

More troubling still, research supported by the other funders of AIDS vaccine research, all of which had less money and influence than the NIH, ran into the same roadblocks.

• • •

BY THE START OF 1999, only a handful of other countries had brought serious money to the AIDS vaccine table, with a little more coming from philanthropic organizations like AmFAR. According to an analysis of annual AIDS vaccine R&D funding published in 1994 by the Rockefeller Foundation, after the $111 million spent by the U.S. government—mostly the NIH, but also including the U.S. military—the next biggest contributors were France ($6.9 million), the United Kingdom ($2.5 million), and Germany ($2.1 million).[21]

Work funded by non–U.S. government sources often contributed cutting-edge ideas to the AIDS vaccine search: Consider the intriguing leads uncovered in the late 1990s by England's Thomas Lehner, Italy's Mario Clerici, and Canada's Kelly MacDonald. Lehner demonstrated in monkey experiments that *where* you inject a vaccine can determine whether it succeeds or fails. Clerici revealed that exposed, uninfected people have novel antibodies in their genital secretions. And MacDonald found that the reason most infected mothers do not pass HIV on to their infants overlaps with the reason people often reject transplanted organs. Promising as these three leads were, they, too, did not speedily move forward, because industry showed no interest in translating them into human vaccine studies. And, as academics, none of these researchers had the facilities to take that crucial manufacturing step themselves.

Lehner's work, first published in the July 1996 issue of *Nature Medicine*, took a novel approach to stimulating immune responses at mucosal surfaces, like the walls of the vagina and the rectum.[22] Vaccines traditionally boost immunity in the bloodstream, which flows through the entire body and so creates a "systemic" barrier to infection. Lehner, an immunologist at London's United Medical & Dental Schools of Guy's and St Thomas' Hospitals, instead wanted to see whether he could protect monkeys from rectal infection by teaching the immune system how to secrete special "mucosal" antibodies at the rectal surface. With funding from the U.K.'s Medical Research Council, the European Community Concerted Action against AIDS, and NIAID, Lehner conducted a most unusual experiment.

To stimulate this mucosal immunity, Lehner injected his vaccine, which contained genetically engineered versions of SIV's surface and internal proteins, into a lymph node located near the rectum.[23] He then challenged these seven vaccinated monkeys by putting SIV into their rectums. Four of the animals completely resisted infection and the other three had transient infections or extremely low viral loads. In contrast, seven of eight unimmunized controls became readily infected upon challenge. When six other monkeys were given the same vaccine via different routes—which ranged from traditional injections under the skin to placing the vaccine in the nose and the rectum—all of them became infected, too.

Lehner and his coworkers analyzed the immune responses in the monkeys and discovered a few surprising correlates of protection. While he did find correlations between protection and the number of cells in the lymph nodes that secreted mucosal antibodies, the most intriguing, clear-cut correlates were the immune messengers secreted by CD8 cells. In particular, protected animals had higher levels of the specific chemokines identified by Gallo's lab as potent inhibitors of HIV. The researchers further found that these protected monkeys also secreted higher levels of the mysterious CD8 antiviral factor, or CAF, earlier identified by Jay Levy's lab.[24]

In an organized pursuit to find effective HIV vaccines as quickly as possible, Lehner's approach—which clearly differed from every other one under study—quickly would have been moved into humans to see whether it could trigger higher levels of these chemokines and the mucosal antibody-producing cells in the lymph nodes, as it had in monkey experiments. By the winter of 1999, three years after Lehner first submitted his paper to *Nature Medicine*, no such development had occurred. "We are planning trials both in the United States and England," Lehner explained to me, "but, as you know, funds are not readily available, particularly if one is not following the orthodox strategy of vaccination."

• • •

THE UNIVERSITY OF MILAN'S MARIO CLERICI, an immunologist who spearheaded the study of exposed, uninfected people with Gene Shearer of the National Cancer Institute, made a discovery in 1997 that also advanced with little sense of urgency.

Clerici and his colleagues, with funding from the Italian government and the European Community, investigated the immune systems of 16 heterosexual couples in which only one person was infected despite their having had repeated, unprotected sex. As expected, none of the uninfected partners had HIV antibodies in their blood, and nine of them (56%) showed evidence of cell-mediated immunity when the researchers prodded their CD4 cells with HIV. More remarkably, as they detailed in a November 1997 *Nature Medicine* paper, eight out of these nine (89%) had HIV specific antibodies in their urine or in swabs taken from their vaginas. HIV mucosal antibodies were found in the urine of only 4 out of the 50 (8%) low-risk controls analyzed in the study; only 1 out of 50 (2%) low-risk controls had HIV mucosal antibodies in vaginal swabs.[25]

This finding put a new spin on Shearer and Clerici's thesis about cell-mediated versus antibody immunity, the Th1/Th2 theory they had popularized in the early 1990s. Initially, they hypothesized that low doses of HIV created the state of cell-mediated immunity—without antibody—that protected exposed, uninfected people; indeed, stimulating the antibody, or Th2, arm of the immune system could dampen the Th1 response. Now Clerici's data suggested that these apparently protected people had Th1 immunity in their blood and Th2 immunity in their genital secretions.

Clerici teamed up with a group led by Frank Plummer, the University of Manitoba epidemiologist who had been following a cohort of exposed, uninfected prostitutes in Kenya since 1985. That summer, at the 12th World AIDS Conference[26] in Geneva, Switzerland, Plummer's group showed that they, too, had found mucosal antibodies in the genital tracts of most of these women. In all, 16 of 21 (76%) women studied had HIV-specific mucosal antibodies in their genital tracts. The researchers further noted that 3 of the 28 (11%) controls who had the antibodies reported higher-risk behaviors.

More confirmation of this phenomenon came in January 1999, when a multinational research group reported results from their study of female "commercial sex workers" in Northern Thailand who repeatedly had been exposed to HIV but remained uninfected.[27] The researchers collected secretions from the genital tracts of 13 of these women and found that six (46%) had evidence of HIV-specific mucosal antibodies.

A coauthor of this study, Thomas VanCott, a researcher with the Henry M. Jackson Foundation for the Advancement of Military Medicine, had studies underway to design a vaccine that mimicked the responses observed in EUs. VanCott and his coworkers wanted to explore in monkeys which vaccines and routes of administration could create systemic cell-mediated immunity, no systemic antibodies, and mucosal antibodies at the genital surfaces.[28] Once this state had been achieved, he planned to challenge the animals. If they resisted the challenge, then the vaccine theoretically would be moved into humans with confidence that it was worthy of an efficacy trial.

Whether the Jackson Foundation—a nonprofit affiliate of the Walter Reed Army Institute of Research that had an intensive, but limited, AIDS vaccine program—could follow through on these ambitious goals remained to be seen, but the approach had a satisfying logic to it. And it was a logic that was absent everywhere else. Many research groups were developing vaccines that aimed to stimulate both cell-mediated and mucosal immunity, and indeed some even had entered human trials. The most advanced such study, launched by NIAID's AIDS Vaccine Evaluation Group in 1998, injected volunteers with Pasteur Mérieux Connaught's recombinant canarypox virus to stimulate cell-mediated immunity and then, to build mucosal immunity, boosted them with the vaccine through the nose, the mouth, the vagina, or the rectum. Finally, the researchers planned to inject the volunteers with the VaxGen or Chiron gp120 vaccine. This groundbreaking study promised to chart the largely unexplored terrain of using HIV vaccines to trigger mucosal immune responses in humans. The problem, though, is that the intellectual underpinning of the study was exploratory rather than goal oriented.

Basically, the researchers said, let's try to turn on every arm of the immune system and see what happens. The only putative advantage of the canarypox vector was what most researchers saw as its flaw: it hardly triggered any systemic antibodies against HIV. As for stimulating secretory mucosal antibodies, other agents—such as cholera toxin, modified salmonella, and Venezuelan equine encephalitis—had a track record. And canarypox also had a decidedly mixed performance when it came to triggering cell-mediated immunity against HIV. So researchers chose the canarypox-HIV vaccine for this study simply for a practical reason: it existed. If the vaccine worked, it would be due to chance, not foresight.

• • •

KELLY MACDONALD, a University of Toronto researcher who worked with Plummer's group in Kenya, looked at yet another group of humans who resisted HIV-infection despite repeated exposures: babies.

With financing from Canada's Medical Research Council, MacDonald investigated 141 HIV-infected pregnant women there and asked why some passed the virus on to their babies while most did not. Her findings showcased a novel potential immune correlate in exposed, uninfected people that, again, vaccine developers largely ignored.

Special proteins that stud the surface of the body's cells mark them as belonging to a specific person. Babies and their mothers, however, often share the same markers on some cells. As MacDonald reported at the 1998 World AIDS Conference in Geneva, she found that babies who had *different* markers on their immune cells from their mothers were significantly less likely to become infected.[29]

MacDonald and others contended that this difference was protective because when HIV buds from a cell, it carries proteins from the host cell membrane on its surface. So babies who build an immune response against their mothers' discordant proteins, theorized MacDonald, also are building a response against HIV from their mother. This is the same mechanism that leads people's immune systems to reject a kidney or a heart from a donor who has a dissimilar—or "allogenic"—genetic background.

These findings had an obvious parallel to the Stott experiment, which in 1991 called into question the relevance of data from monkey tests of whole, killed SIV vaccines. In that famous experiment, Stott showed a link between protection from these vaccines and anticell—rather than antiviral—antibodies. MacDonald similarly found that immune responses to cellular, rather than viral, material could lead to protection.[30] But, curiously, MacDonald and coworkers revealed that allogenic antibodies did not explain the protection seen in the babies. More likely, she suggested, the babies had allogenic cell-mediated immune responses to these cellular proteins.

Further support for the idea that these allogenic proteins could trigger effective anti-HIV cell-mediated immune responses came from Gene Shearer's lab. In November 1998, Shearer and coworkers published a study in the journal *Blood* that described a clever test-tube experiment.[31] The researchers showed that when they added HIV to white blood cells taken from HIV-*un*infected people, the virus, as expected, easily established an infection and replicated to high levels. But when they put HIV into a mixture of white blood cells from two allogenic, uninfected people, the virus could hardly copy itself at all. Something about the allogenic mixture, apparently, inhibited HIV.

Through a process of elimination, the researchers tracked this inhibition to chemicals secreted by CD8 cells—the main actors in cell-mediated immunity—in response to the allogenic intrusion. The chemicals, they proposed, were not chemokines, but, rather, the CD8 antiviral factor identified by Jay Levy. (Shearer, incidentally, with an academic collaborator, [32] applied for an Innovation grant to pursue an allogenic HIV vaccine and was turned down.)

In conjunction with a U.S. primate center, MacDonald won an NIAID grant that she believed would allow her, on the side, to test vaccines in monkeys based on cellular proteins. But she had strong doubts that industry would pick up the ball, and already had had her hopes dashed. Pasteur Mérieux had shown some interest but wanted support from the Canadian government. In February 1997, the company and several government agencies held a meeting with AIDS researchers, including MacDonald, that faulted Canada for not targeting any money for AIDS vaccine research and urged the country "to develop a planned and coordinated HIV vaccine development strategy." The group suggested that the government annually invest 10 million Canadian dollars toward the effort.[33] That June, Canadian Prime

Minister Jean Chrétien joined Bill Clinton and leaders of the world's six other wealthiest countries for a "G-8" summit in Denver, which led to a communiqué that declared, "We will work to provide the resources necessary to accelerate AIDS vaccine research, and together will enhance international scientific cooperation and collaboration."[34] MacDonald felt certain that Canada would soon follow through with this pledge. By the start of 2000, the Canadian government still had not targeted any money for AIDS vaccine research.

MacDonald well understood Pasteur Mérieux's reluctance to pay for these studies by itself. "This is not the most attractive vaccine to industry," said MacDonald. For one thing, she said, it will never be used in developed countries, because people who become immunized against cellular proteins will severely limit their ability to tolerate transplanted organs. "For the developing world, it may well be appropriate, but that's not where the money is," said MacDonald. The second problem, she said, is that the approach has never been used to make a vaccine. "There's no model, and every time you bring it up with industry, they have a hard time getting their heads around it."

MacDonald had no plans to ask the Canadian government for money again. "When you've written your grant and it's fallen flat, you need to lick your wounds," she said. And she stressed that the approach still had a long way to go before proving itself. "I don't say that if you gave me money tomorrow, I'd be there." But here, once again, a promising idea had, from its inception, such rickety support that it stayed alive the way a vagrant might, picking through the remains of others, perpetually searching for handouts, and shuffling from place to place instead of aggressively moving forward with direction and purpose.

• • •

A FEW DAYS BEFORE JONAS SALK DIED in June 1995, he told me that he wanted his biography to be titled, "There Must Be a Better Way." There must.

CHAPTER 14

Better Ways

In May 1998, the AIDS Vaccine Advocacy Coalition issued the type of kick-ass-and-take-names report that, until then, had only come from AIDS activists enraged about the slow pace of drug development. Entitled *9 Years and Counting*, the 52-page report came out on the first anniversary of Bill Clinton's declaration of a goal to develop an AIDS vaccine within a decade. "At the current level of effort, we will not have an HIV vaccine in 9 years," the report asserted. "Pretending to fill the leadership gap, marginally increasing public funds, and improving part of the grant evaluation and awards process does not add up to the full mobilization we need to develop an AIDS vaccine."[1]

The AVAC members slammed the National Institutes of Health for not yet naming a director to the Vaccine Research Center that Clinton announced in his May 1997 AIDS vaccine speech. The institution, they noted, also had failed to follow the Office of AIDS Research report's recommendation that it establish a coordinated, trans-NIH AIDS vaccine program. "Although there are many chiefs at the NIH, none seem to have the clear authority and responsibility for making sure something will come out of the pipeline," the report warned. They also urged the directors at the National Institute of Allergy and Infectious Diseases and the National Cancer Institute to organize a large, comparative monkey study of different vaccines.

Outside of the NIH, they assailed the G-8 nations for failing to take any "significant coordinated action." Clinton took it on the chin for not offering a coordinated plan to achieve his goal. Clinton and NIH director Harold Varmus, they suggested, should invite the CEO of SmithKline Beecham—which, with $1.2 billion a year in vaccine sales, was the world's largest vaccine maker—to the White House to learn why this behemoth had "watched the epidemic from the sidelines" with only a "minuscule" AIDS vaccine program. The Walter Reed Army Institute of Research, they contended, should receive more support for its AIDS vaccine effort, which "more closely resembles a [*sic*] objective-driven private company than any other government program currently involved in HIV vaccine research."[2] Foundations like Ford, Kaiser, and Rockefeller should pony up more for AIDS vaccine R&D, they declared, as should philanthropists like Microsoft's Bill Gates, Dreamworks' David Geffen, and investment mogul George Soros. Finally, they insisted that government, the public, and industry all should do their part to see that more vaccines moved into human trials.

Although the report did not directly lead to any obvious actions, it did succeed at spreading AVAC's message. Most notably, the *Wall Street Journal*, which had run few stories about industry's lack of commitment to making an AIDS vaccine, devoted an article to the report that was headlined "Clinton Is Failing to Fulfill a Pledge on AIDS Vaccine, Activist Group Says."[3]

The report had another, more meaningful, impact too: It gave voice to the impolitic argument that, despite all the cheerleading from the highest levels of power, despite the U.S. government's increases in the AIDS vaccine budget, despite the NIH's sincere attempts to improve the way it does business, despite the industrialized nations' vowing to do more, everything was not fine. And AVAC, in a most radical move, presented a thoughtful, nonscientific view of not just the field's many problems, but possible solutions to them.

Other problem solvers soon would step forward with their own prescriptions for the field, and some would actually see their ideas transformed into actions. But the most important and daring suggestions for speeding the AIDS vaccine search would go nowhere fast.

• • •

THE AVAC REPORT threw what amounted to a right-hand punch and, the next month, at the World AIDS Conference in Geneva, the International AIDS Vaccine Initiative followed with a left hook. In its manifesto, *Scientific Blueprint for AIDS Vaccine Development*, IAVI overlapped in many ways with AVAC, agreeing that "the world is not on track to meet the bold goal set by the U.S. President, Bill Clinton, to identify a safe and effective vaccine by the year 2007."[4] But IAVI, whose scientific director, Peggy Johnston, had previously worked at NIAID's Division of AIDS, offered a much more penetrating analysis of the NIH. And IAVI's blueprint had a more tangible aim: to set its own agenda.

IAVI welcomed NIAID's Innovation grants and another new, ambitious program designed to encourage academics to move their vaccine ideas into human trials.[5] But IAVI cautioned that the programs ultimately would foster basic research and new vaccine designs without speeding the time it took to move a preparation into efficacy trials. And the report had even less enthusiasm for a new NIH program that promised to contract out the development of vaccines. "A previous attempt by NIH to fund HIV vaccine manufacturing attracted only one offerer, precluding a competition among offerers," the report said. "It is questionable to what extent private companies will respond to this new contract solicitation, given their concerns about government control of their in-house processes and decision making."[6]

IAVI particularly chastised the NIH's parochial practice of developing vaccine test sites in poor countries "with no discernible effort" to coordinate, at an early stage of product development, with industry. "As a result, companies are left with tremendous uncertainty about whether they can evaluate their candidate HIV vaccines in developing countries, and where," the report complained. "It also becomes more difficult for companies to make decisions about which HIV subtypes to use."

IAVI offered a solution, proposing to create international product development teams that would link vaccine designers with companies and clinical researchers to focus on testing a specific vaccine at a specific site. For $350 million to $500 million, IAVI said it could fund the manufacture and testing of six vaccines through phase II trials, and then move three of those through efficacy trials—all by 2007. The selection of which vaccines to test would rely on safety, immune responses in animals, probable efficacy based on animal challenges, and the cooperation of whoever held the patent rights to the approach. A process of "thoughtful empiricism," which evaluated safety and immune responses in small human studies, would determine which candidates to move into efficacy trials.

IAVI won much attention at the Geneva conference, in part because the group announced that the William H. Gates Foundation had just given the organization a $1.5 million grant. With new money coming from the World Bank, the Levi Strauss Foundation, and the U.K. government, IAVI said its cash on hand and commitments totaled $15 million—enough to create at least one international development team.[7] At long last, IAVI had money to put where its mouth was. And IAVI, which until then had attracted little media coverage, now had such popular appeal that the group made it on CNN, the Reuters wire, and even in the pages of *USA Today*.[8] As *Newsweek*'s online magazine noted, IAVI "has had a good week at the conference."[9]

• • •

COINCIDENT WITH THE PUBLICATION of IAVI's scientific blueprint, the NIH hired a new director of its Office of AIDS Research: Neal Nathanson, a 70-year-old viral epidemiologist from the University of Pennsylvania who flung into the scene in Geneva with more energy than scientists half his age. Running from one scientific session to the next, Nathanson had the zeal of a postdoctoral student, feverishly taking notes and intensely questioning many of the results presented. Where OAR director Bill Paul had a calming, smooth diplomacy—which served him well during the turbulent years he led the monumental review of the AIDS research portfolio at the institution where he had worked for more than 30 years—Nathanson was frenetic, blunt, and decidedly not part of the NIH establishment.

Nathanson's career began with a definitive analysis of the Cutter incident, the tragic mishap that forever haunted Salk's killed poliovirus vaccine, but he had gone on to study the sheep retrovirus called visna, hepatitis B, rabies, mad cow disease, and such exotic viruses as Bunyanwera, Tamiami, West Nile, and Langat. He had done limited HIV research on how the virus infects cells, and had even coauthored an obtuse, little-noticed 1989 rumination about the search for an AIDS vaccine entitled "Human Immunodeficiency Virus: An Agent That Defies Vaccination."[10] Now Nathanson, who would oversee a $1.8 billion AIDS research budget in 1999, would have a chance to apply his eclectic experience. At the top of his list: AIDS vaccines, which would receive an estimated 200 million of those dollars, a 100% jump from four years before.

Nathanson came to the job with a strong vision of how to speed the search, and while he shared IAVI's sense that a more targeted approach would help, he took issue with a key feature of the group's scientific blueprint. IAVI, in his eyes, wanted to move a few vaccine approaches forward without rigorously comparing the attributes and weaknesses of all the different possibilities. "There must be dozens of potential vector systems, a number of possible adjuvants, a whole variety of immunization protocols using different routes," Nathanson told me during an early morning breakfast meeting in Geneva. "A lot of those have been stimulated by NIH money, which watered the basic science field—we've let 100 flowers bloom. Now we have to figure out some way of harvesting those and getting them eventually into phase I trials."

To sort through the myriad of different AIDS vaccine strategies, Nathanson proposed a concept that had been suggested as early as 1987 by Merck's Maurice Hilleman: a massive comparative study in monkeys that used standardized protocols and reagents.[11] "I'd like to take an orderly, logical approach," Nathanson said. "We'd like to bracket the system from the very effective to the ineffective." Human versions of the most promising vaccines then would move forward. Proceeding the way IAVI proposed, Nathanson argued, would "arbitrarily exclude 90% of the candidates." He readily conceded that, as IAVI stressed, the monkey model might not reflect what happens in humans. "If you really don't believe [in the monkey model], then this all falls apart—you could be discarding the best things and testing the worst things," he said. But here, at long last, was an NIH official who had a large say in the AIDS vaccine research budget, declaring his willingness to gamble that the monkey model, while not perfect, offered the best guide at hand and should be exploited fully. For an NIH official, he had a refreshing single-mindedness about solving the problem, whatever it took.

I asked him whether he worried, as some investigators did, that the surge of new money might support second-rate grants, decreasing the average quality of AIDS vaccine research. "I wouldn't lose a moment's sleep over that," he shot back. "That's exactly how *not* to look at this problem. The exploding global epidemic has produced an urgency, and the response to the urgency is to accept the idea that speed and cost efficiency are totally different parameters, and you can't have both at the same time. So, in fact, we are deliberately, if anything, overfunding to do many different things in parallel, with the understanding that this is a very cost-inefficient process and we have only one priority: getting things into the field as fast as possible. The way this epidemic is going, any other approach would be intellectually absurd and ethically unconscionable."

A few weeks after the Geneva conference, the NIH made another bold hire: Peggy Johnston, the scientific director at IAVI. Johnston, an iron-willed biochemist with fully baked ideas of her own about how to improve the field, before IAVI had worked at NIAID's Division of AIDS for nine years, rising to the deputy director position. Now she would return as NIAID's point person for AIDS vaccine research. "I wouldn't go back unless I was absolutely convinced that the NIH was taking AIDS vaccine development more seriously," Johnston told me.

At IAVI, Johnston had learned much about industry and international issues, knowledge she wanted to put into action at the NIH. She wanted to see more academics run human tests of their vaccines, which she thought the new Vaccine Research Center could manufacture. "In the past, NIH almost exclusively relied on companies to bring products into clinical trials," she explained. She also thought the NIH now could establish international product development teams similar to the ones outlined in IAVI's blueprint. "I'm good at putting a little chaos into some structure," Johnston said. "An organization needs chaos in structure for creativity to flourish."

With the hiring of Johnston and Nathanson, both of whom had the strong backing of the highly influential David Baltimore, the NIH entered a new era of AIDS vaccine R&D. Coupled with the new blood, the NIH formed a study section, as the OAR report had called for, that for the first time would specialize in vaccine proposals, which often received short shrift in the standard peer-review system. "A lot of vaccine research does tend to be more empirical," explained Johnston. "It's not asking a key fundamental virology or immunological question."

In a surprising move that symbolized just how powerfully the winds were shifting, NIAID announced on August 18, 1998, that it would collaborate with VaxGen on the efficacy trials of its gp120 vaccine. VaxGen, which had raised $27.5 million, that June received FDA approval to stage efficacy trials in 5,000 people in the United States and 2,500 people in Thailand. NIAID's renewed interest in the approach it had abandoned four years earlier had nothing to do with the company's reformulation of the vaccine.[12] Rather, NIAID, recognizing that the company had the wherewithal to stage these trials on its own, wanted to fund substudies that would analyze immune mechanisms behind the vaccine's successes and failures. "Potentially valuable science will be captured because VaxGen itself will do limited studies with licensure as the goal, not scientific understanding," David Baltimore, an outspoken critic of the vaccine, explained to me.[13]

Nathanson, who had been a prime player in the backroom negotiations with VaxGen, worked hard to spread this pragmatic point of view. "Obviously, the VaxGen trial can make important contributions in a variety of different ways to developing a vaccine," he said. "That's all I really care about."

• • •

MORE WINDS OF CHANGE blew that November, when IAVI announced that it had formed two international product development teams, one of which would develop a vaccine for trials in Kenya and the other for tests in South Africa. The details of the plan simultaneously highlighted IAVI's ingenuity—and limitations.

After reviewing six proposals, IAVI's scientific advisory committee awarded a total of $9.1 million to two vaccine approaches, both of which originated in academic labs and used state-of-the-art technologies that had shown promise in early animal studies. Up front, both teams agreed to base their vaccines on strains of the virus circulating in the region where the preparation would be tested. IAVI simultaneously negotiated creative agreements with each team to provide developing countries with the vaccines, if they proved effective, "at very reasonable prices." IAVI additionally would receive a 1% royalty on sales of the vaccines in developed countries.

Andrew McMichael, an Oxford University immunologist whose lab intensively studied cell-mediated immunity, would head the team developing the vaccine for Kenya. In collaboration with a Kenyan researcher and a German pharmaceutical company, McMichael planned to test an unusual combination of two vaccines.[14] The first of these vaccines would rely on naked DNA.

The much-heralded approach had, at its core, a circular piece of bacterial DNA known as a plasmid. Like an engineered bacterium or virus, the plasmid could vector foreign genes into a cell. But the plasmid—and this was its great advantage—was not, unlike *E. coli* or vaccinia, a pathogen itself.

McMichael had little interest in antibodies, favoring instead a vaccine that triggered the strongest possible killer cell response. McMichael and his coworkers had spent years exploring how the immune system responded to different HIV peptides, the name given to pieces of the viral proteins. They found peptides that they believed most powerfully spurred the production of killer cells against the virus. The plasmid, then, contained the genetic instructions to these peptides.

McMichael had shown that mice given this vaccine developed strong killer cell responses, as he had hoped. But McMichael found he could elicit an even stronger response by subsequently injecting the mice with HIV genes stitched into a variant of the smallpox vaccine, a virus known as modified vaccinia Ankara, or MVA.

Both naked DNA and MVA had shown great promise as HIV vectors for years, but their development had, for different reasons, stumbled forward. Naked DNA, which first wowed scientists as a vaccine strategy in 1993 mouse experiments with influenza, potentially offered a cheap, safe vector that scientists could easily modify.

Traditional bacterial and viral vectors had another built-in drawback besides being pathogens: the immune system would learn to recognize them, destroying the ability to use them repeatedly in the same person. The plasmid vector in DNA vaccines, in contrast, evaded immune detection, so scientists could, theoretically, boost a person repeatedly. Many companies recognized the potential of this approach, and the field of DNA vaccines exploded.[15] But the fate of naked DNA and HIV provides yet another object lesson in industry's lackluster interest in developing preventive AIDS vaccines.

Apollon, a biotech firm in Malvern, Pennsylvania, that opened its doors in 1992, built its vaccine around a technology developed by David Weiner at the University of Pennsylvania. With help from the NIH, the company in 1995 began human tests of its vaccine in HIV-infected people and, the next year, in uninfected volunteers.[16] In keeping with the field's illogic regarding the link between challenge experiments and human trials, in May 1997, Weiner and coworkers published a study showing that the approach could protect two chimpanzees, albeit against the "wimpy" HIV-SF2 challenge strain.[17] One year later, Wyeth-Ayerst, a major vaccine maker, bought out Apollon, but excitement about this specific approach, as IAVI noted in its July–September 1998 newsletter, remained muted. Early results from the human studies showed mediocre killer cell responses, and, as the IAVI newsletter reported, "a number of leading researchers" doubted that the current formulation of the vaccine would have much success.[18]

Most AIDS researchers put their chips on Merck, which played a central role in the 1993 findings about DNA vaccines and influenza. Merck's team quickly began working on HIV DNA vaccines, but ran into scientific and not-so-scientific impediments. Margaret Liu, the researcher who headed the effort, told me that their first vaccines did not produce HIV proteins at as high a level as they had anticipated, and that the company for many years had a bias toward antibodies, which was the vaccine's weak suit. Although she had high praise for the amount of resources the company devoted to her work, she also acknowledged that more resources would have been a great help. "I saw a lot of promising things that needed to be tried."

Merck, working with Harvard's Norman Letvin, in 1997 showed impressive protection in a monkey study with a DNA vaccine followed by a boost with recombinant protein.[19] Merck, though, in early 1999, still had not moved an AIDS vaccine into clinical trials, which IAVI's newsletter asserted was a "key indicator" as to whether a company was "serious" about its HIV DNA project. Chiron, which hired Liu in 1997, and Pasteur Mérieux Connaught both had HIV DNA vaccine projects, but had no immediate plans for human tests.

The other vector in McMichael's vaccine, MVA, had shown in an SIV experiment published in 1996 that it could powerfully immunize animals, but had yet to be used in an HIV vaccine for humans.[20] MVA was a weaker version of its close cousin vaccinia virus, the vector used in the ill-fated Oncogen AIDS vaccine. Vaccinia, a powerful vector, had a critical shortcoming: the virus could cause a fatal disease if given to people who had damaged immune systems, and just such accidents occurred when Daniel Zagury used a vaccinia/HIV construct as a therapeutic AIDS vaccine.[21] MVA, on the other hand, which had been tested in 120,000 people before smallpox was eradicated, did not appear to cause problems in immunocompromised people. In a monkey experiment in 1998, MVA also had shown a remarkable ability to stimulate killer cells.[22] Yet despite MVA's promise, no researchers had yet done clean, comparative studies of MVA with other vectors and no company had announced its intent to test an MVA HIV vaccine in humans.

From industry's vantage, MVA had a big drawback, too. At least three parties already had patent claims to its use: Hoffmann–La Roche, Pasteur Mérieux Connaught, and the NIH. While Hoffmann–La Roche may have licensed MVA, Pasteur Mérieux, which had a broad patent for all engineered vaccinia viruses, had only licensed its technology to one company, Therion Biologics. And the NIH, which also had a broad patent for vaccinia vectors, was engaged in a protracted dispute with Pasteur Mérieux. So any company that entered this mess, did so at its peril.

McMichael's team planned to first test its DNA/MVA vaccine in small human trials in Oxford, which the researchers hoped would start in late 1999. If the vaccine appeared safe, they would start small trials in Kenya about six months later.

The second approach IAVI funded had sexy science too—and potential patent mayhem in its future. Developed by Robert Johnston and Nancy Davis at the University of North Carolina, this strategy exploited a vector known as Venezuelan equine encephalitis virus, or VEE.

Johnston and Davis, who formed a company called AlphaVax to promote this technology, repeatedly had taken a beating by NIH study sections when they asked for money to develop a VEE-based HIV vaccine. "The NIAID program officers liked it, but the study section nailed it—too early, not enough preliminary data," said Johnston. Encouraged by NIAID's staffers, they submitted a second proposal with more preliminary data. Again, the study section gave it the boot. "I threw up my hands and I said, 'I'm going to call the people at the Army.' " The Army had just decided to award grants to novel AIDS vaccine ideas with the $20 million originally appropriated for a trial of the MicroGeneSys therapeutic vaccine, and VEE was one of the 10 lucky winners. "All of the basic data we have now that has generated a little bit of excitement came from that," said Johnston.

Johnston and Davis jiggled VEE, a horse virus, such that it could hold foreign genes and then produce a sham virus they called a "replicon." This replicon could then infect cells and produce high levels of the proteins coded for by the foreign genes. But the replicon could not copy itself, a key safety feature as it could not spread and cause disease. The replicon also targeted lymphoid tissue, which itself led to robust and broad immune responses, including high levels of mucosal antibodies.

AlphaVax later won a small, one-year Innovation grant from the NIH, but had another, more ambitious one turned down. "It was very frustrating to think NIH was charged with making this vaccine," said Johnston.

At Robert Gallo's annual lab meeting in August 1998, Johnston presented data from a monkey challenge experiment with this vaccine. Using a highly virulent strain of SIV, Johnston showed that the vaccine failed in one monkey, but completely protected two others, and contained the virus in a fourth. In contrast, two of four control animals quickly developed AIDS, and two others had climbing viral loads.[23]

The data much impressed IAVI. "In IAVI's competition, the goal was to make a vaccine, and they loved our stuff," said Johnston. With IAVI's help, AlphaVax believed it would take three years to ready this VEE replicon vaccine, which would contain genes for both HIV's surface and core proteins, for human trials.

By 1999, AlphaVax had six patents issued and a deal with Wyeth-Ayerst to develop VEE-based vaccines against five diseases other than HIV. But Chiron's Margaret Liu cautioned that her company had a broad patent issued about using the entire family of alphaviruses, which VEE belongs to, as vectors for vaccines. "IAVI, frankly, didn't look at it," cautioned Liu.

IAVI actually looked carefully at patent positions for both VEE and MVA. "They are real issues," IAVI president Seth Berkley told me. "We've struggled with them and spent a lot of money on lawyers about them." Berkley said that Chiron may not have as strong a patent position as it thought, but acknowledged that litigation could badly hurt a small company like AlphaVax. He hoped that IAVI's earnest mission would allow it to sidestep the problems. "One thing IAVI would bring to the table as a not-for-profit is some reality to the process," he said.

But the potential patent pitfalls raised troubling questions about IAVI's pledge to offer the vaccines to poor countries at reasonable prices. IAVI gave away its money with the condition that vaccine manufacturers could not charge more than 10% above their production costs—and if IAVI deemed production costs too high, it could solicit bids from a third-party manufacturer. Yet licensing rights could have a substantial impact on the actual costs—and that is assuming the patent holders would be willing to issue licenses. So IAVI—although it might relish the fight—could find itself with an exorbitantly expensive vaccine or a patent fight. And as the price tag goes up, then the question that the NIH's Nathanson raised becomes even more acute: How strong was the rationale to back these two vaccine approaches over 90% of the others?

• • •

IN 1999, THE AIDS VACCINE SEARCH was in better shape than ever. The NIH finally had a healthy AIDS vaccine budget, a strong leadership team, and smart grant programs. IAVI, which that January hired NIAID's former AIDS vaccine chief, Wayne Koff, finally had something of a war chest to work with, and had launched two vaccine projects that aimed to move academic-originated ideas into human trials of vaccines designed specifically for poor countries.[24] Researchers at universities all over the world, at private foundations like the Henry M. Jackson, and even in industry had interesting AIDS vaccine projects underway. And a group of AIDS activists, AVAC, finally devoted themselves to bird-dogging industry and the government.

But chaos still ruled the enterprise as a whole, which gave it the look of a powerful storm front that energetically moves from place to place, impervious to concepts like coordination, guidance, and purpose.

The solution to this dilemma is to combine all of the good ideas coming from all of these sources into one well-funded, goal-oriented, targeted research program that exists solely to speed every potential promising lead. An adjunct to the NIH and every other government-funded program, this fantasy organization—call it the March of Dollars—would have its own money, in the neighborhood of $1 billion, donated by philanthropists and foundations, to solve the problem.[25] Let the NIH, AmFAR, IAVI, and every other funder do what they do. If their efforts lead to an effective vaccine, individually or collectively, clang the church bells and dance in the streets. The premise of this organization, however, is that the effort can be speeded up, that not everything is being done that can be done to take advantage of scientific opportunities and remove obstacles. If this sounds a good deal like the March of Dimes, that is no mistake.

The March of Dollars would serve two main purposes: filling research gaps and staging clinical trials. At its helm would be someone with stature, clout, and connections to the highest levels of power. Preferably, this person would share many characteristics with the March of Dimes's Basil O'Connor: a lawyer with excellent negotiation skills whom the disease had personally impacted. Although it would take a careful search to find the right person, I am thinking of someone like Randall Robinson, a Harvard-trained lawyer who runs TransAfrica, an organization that successfully pressured the United States to enact economic sanctions against the South African government because of its apartheid policies. Randall's brother Max, a former *ABC News* anchor, died from AIDS in 1989.

Under this leader, a scientific director would oversee two preeminent scientific advisory boards. One of these boards, which would meet, say, every three months, would consist primarily of basic researchers who would determine the gaps that needed to be filled. A financial officer, checkbook in hand, would sit at the table when the gap-filling advisers met. If the scientists decided that the rat model needed help, they'd send Mark Goldsmith funds and simultaneously contact two or three other researchers who had experience in transgenic rodents to see whether they would like money to study the problem, too. Everyone who received these funds would sign an agreement to share data freely. Any patent rights that came from the work would be assigned to the organization, with the inventors, like U.S. government-employed researchers, receiving limited royalty payments.

The second group of scientific advisers would mix epidemiologists, primate researchers, molecular biologists, clinicians, and others who had expertise in making and testing vaccine candidates. This group's first order of business would be to stage a massive monkey trial, such as the one that Neal Nathanson described and AVAC advocated, to compare every sound vaccine idea in a standardized way.

In the year 2000, the master monkey trial would mean a comparison of at least 12 different vectors that already have been engineered to contain SIV genes: canarypox, MVA, vaccinia, the attenuated vaccinia strain called NYVAC, naked DNA, salmonella, VEE, a VEE relative called Semliki Forest virus, herpesvirus, adenovirus, BCG, poliovirus, and *Listeria monocytogenes*. Scientists could construct these vectors, in turn, to contain different combinations of SIV genes. Included in this comparison would be SIV vaccines made from whole virus that had been attenuated or killed, recombinant proteins alone, cellular proteins, and strings of peptides.

Aside from using different parts of the virus in different vectors, the comparative test could analyze the impact of different adjuvants—immune potentiators that range from the commonly used alum to more exotic concoctions of cytokines. Some vaccines could be tested in combinations. Additionally, specific concepts could be tried with different vaccines, such as low-dose immunization, route (injecting into a lymph node or swabbing onto a mucosal surface), and potential correlates of immunity (antibody only, antibody plus cell-mediated, cell-mediated only). Challenge viruses of different strengths could be used, and they could be given intravenously, rectally, and vaginally. When new vaccines came along, they could be plugged into the same challenge protocol.

The purpose of this grand experiment, which would require enormous resources and possibly more monkeys than are currently available for research, would be to ascertain which vaccine approaches appeared safe and demonstrated the most robust protection.[26] A point system could define the quality of a protection. If a challenge with a hot strain of SIV showed that animals vaccinated one year earlier completely resisted infection, that might be worth 10 points. A wimpy challenge virus with the same outcome might be worth only 9. A vaccine that did not prevent infection from a challenge but did significantly delay disease or slow the process of immune destruction might yield 8 points. Other endpoints that scientists could assign values to include infection but clearance of the virus, protection from a strain that differs widely from the one used to make the vaccine, protection from a higher dose of virus, protection from challenge given via different routes (intravenous or rectal), and better containment of virus (a lower viral load, say, six months after becoming infected).

The grading system itself is not of course the point: It's simply a tool to help researchers select the half dozen vaccines that they believe work best in monkeys. These vaccines then logically would move into human trials with the idea, from the outset, of advancing them to efficacy trials unless they appeared too dangerous in phase I and II tests. Assuming the monkey trials did not reveal a specific correlate of immunity with a chosen vaccine, the immune responses measured in the early human trials would not be used as criteria for moving the vaccine into efficacy trials. In other words, the level of neutralizing antibodies or killer cells in a vaccinated person—the central gauges used to evaluate phase I and phase II vaccine studies conducted by the NIH and others—would take a back seat to the empirical finding that the chosen vaccine had, for whatever reason, worked best in monkey experiments.

Obviously, the scientific advisory board would not want to test every possibility and would have to make some arbitrary decisions. A vast scientific literature already exists about SIV vaccine studies that could guide them in this process. Another limiting factor would be the willingness of companies and academic researchers to share different vaccine constructs. This might be overcome by the intent of the organization—which the NIH never stated, as NIAID's Alan Schultz explained during the famous ARAC meeting—to select vaccines for human trials based on how they performed in these animal experiments. And if a vaccine maker did not want to subject a preparation to this comparative trial, then the leader of the March of Dollars could play the shame card: Imagine the *Wall Street Journal* headline that says "March of Dollars Charges Vaccine Maker with Slowing Search for AIDS Vaccine." Finally, it would take time and lots of money to make SIV analogs of some promising HIV vaccine strategies. But the organization could contract out much of that work.

I am not arguing that researchers should ignore the interactions between gp120 and CCR5, abandon efforts to stimulate higher production of CTLs, or stop analyzing questions about genetic variation, neutralizing antibodies to primary isolates, and the role of dendritic cells in initial infection. Those types of mechanistic studies surely complement empirical vaccine testing. But it is worth remembering that it is possible to solve the problem without having a deep understanding of such parameters. Vaccine history amply proves that case. Will reductionist studies help here? Most likely. Are they a requirement? No.

I ran these admittedly grandiose and technically challenging ideas by David Baltimore on a balmy Pasadena day in January 1999. We met in his office at the California Institute of Technology, where he recently had taken over as president. His office walls were dressed with two photographs of the late Richard Feynman, the randy and brilliant Caltech physicist who worked on the Manhattan Project. Baltimore, a bearded, Solomonic man—who himself, in the early years of the epidemic, called for a Manhattan Project for AIDS—passionately but politely dismissed the whole scheme.

When I raised the idea of the jumbo monkey experiment, Baltimore countered that such comparative studies already were taking place. In reality, however, Neal Nathanson's idea had led Harvard's Letvin and NIAID's

Schultz to organize a relatively modest experiment that compared canarypox to MVA, vaccinia, NYVAC, and maybe a DNA vaccine. Ostensibly, this experiment would address whether the NIH should fund efficacy trials of the experiment using canarypox followed by recombinant gp120. This prime-boost experiment had become the front-runner in NIAID's clinical trials, and now Letvin and others had preliminary data suggesting that canarypox led to underwhelming killer cell responses.

This experiment well illustrated the limits of the NIH-led effort. As both Letvin and Schultz acknowledged, much haggling occurred before researchers would agree on which challenge viruses to use. Then a company Letvin and Schultz thought would supply a DNA vaccine backed out. Finally, and most damningly, the NIH did not declare that the data would definitively determine the fate of canarypox as the vector of choice. Once again, the NIH wanted to just do the experiment and let the chips fall where they might. The writing on the wall said NIAID was about to relive the gp120 brouhaha of 1994. Pasteur Mérieux, the maker of the canarypox-based vaccine, would argue that the monkey model wasn't relevant. Clinicians would trot out killer cell data from human studies and debate the merit of the various assays used to measure them. ARAC or some other group of advisers would be gathered to make the call.

Baltimore said he could "not agree more" that the NIH could have moved the monkey experiments forward more quickly and should have organized a standardized system earlier. But he contended that it made little sense to organize a giant monkey experiment now to select the best candidates for efficacy trials. "New ideas are coming forward, immunologists are contributing, all the things that at least I wanted to see happening are happening so that we have the pot boiling sufficiently to bubble up ideas that will work effectively," said Baltimore. "I don't think we were in that situation before. Now we're on a cusp. You might want to do it now, but I think it's too early, and you might not need ever to do it because these things are going to make themselves known to us just by their effectiveness."

If Baltimore believed that the world needed a more targeted AIDS vaccine program, he said, he would advocate for it. "We could get the federal government to fund this," he predicted. "The federal government is just made of people." But he placed great hopes in both the recent changes that had occurred and the ones on the horizon, especially the NIH's nascent Vaccine Research Center. The director of this center, he assured, would provide the real leadership that the AIDS vaccine effort had been missing. "There has been no one at NIH with the clout to see that something gets done," he said. And he anticipated that the center would run the $25 million to $50 million targeted AIDS vaccine program that the OAR report envisioned.

I do not share Baltimore's optimism. For one thing, the NIH did not name a director for this center, which first was announced by Clinton in May 1997, until March 1999. "I can't tell you and I won't tell you how many people were cajoled, invited, and inveigled to think about taking that job who turned it down," Baltimore allowed. The researcher chosen, Gary Nabel, did not, as the NIH had hoped, have industry experience—indeed, Nabel, a gene therapy expert, had little vaccine experience. Still, he had excellent credentials as a basic researcher, including having done postdoctoral work in Baltimore's lab.

Whatever Nabel's qualifications were for the job, the delay in finding the right person clearly broadcasts the main inadequacy with the status quo: Everything takes too long. And as became ever more apparent in 1999, as the clock ticks, it becomes increasingly difficult, for ethical reasons, to even stage AIDS vaccine efficacy trials.

CHAPTER 15

Disparate Measures

Remarkable success stories with new cocktails of anti-HIV drugs had become commonplace by March 1997, but Barry Bloom, a researcher at New York's Albert Einstein College of Medicine, realized that this triumph came with a painful irony: it raised staggering new ethical quandaries for AIDS vaccine developers.

As the recently appointed head of the Vaccine Advisory Committee for UNAIDS, the Joint United Nations Programme on HIV/AIDS, Bloom that March began organizing a meeting to address what he saw as a mind-boggling dilemma. If everyone who became infected during an AIDS vaccine trial quickly began taking powerful anti-HIV drug cocktails, Bloom reasoned, it might become impossible to distinguish whether the vaccine or the drugs had delayed or prevented disease. In such a setting, an efficacy trial could assess only whether a vaccine prevented an infection, which researchers already had agreed was far too high a standard. Maybe, then, researchers could only stage meaningful AIDS vaccine efficacy trials outside of wealthy countries, where populations had little access to anti-HIV drugs. But that strategy created its own quagmire, as ethics, to prevent exploitation of the poor, mandated that vaccines first be tested in the country that makes the product.

Bloom was an ideal scientist to direct the confusing traffic of opinion that surely would clutter the intersection of AIDS drugs and AIDS vaccines. A fireplug of a man whose intense bulging eyes hid behind lenses as thick as headlight glass, Bloom had a scholarly, judicious manner—he was a *mensch*—that he mixed with a ferocious commitment to public health and an overt disdain for anyone who stood in the way. He enjoyed an impeccable scientific reputation as an expert in immunology and mycobacterial diseases like tuberculosis and leishmaniasis, a distinction that had earned him a fellowship with the Howard Hughes Medical Institute, an elite organization that bankrolled 300 or so of the top scientists in the United States. Leading scientists and health-care policy makers from around the world routinely sought his advice. So it was that Bloom wanted to take the lead and organize a robust, constructive discussion of these vexing ethical issues before they even had occurred to most AIDS researchers.

But this vision fell apart the next month when the Public Citizen Health Research Group, a biomedical watchdog organization founded by consumer advocate Ralph Nader, blitzed the media, Congress, and public health officials about what they called "blatantly unethical" drug trials that aimed to prevent transmission of HIV from infected, pregnant women to their babies. Although Public Citizen's outrage focused on human tests of drugs, not vaccines, the principles at stake were identical. "Once it hit the newspapers, I realized it was a bigger problem than we had anticipated," said Bloom.

Public Citizen's frontal assault began with a scathing letter to Donna Shalala, secretary of Health and Human Services, that assailed her agency for supporting trials in poor countries of the anti-HIV drug AZT in pregnant, infected women. The April 22, 1997, letter recalled that three years earlier, a trial in the United States and France had shown that AZT could cut the transmission of HIV from an infected mother to her baby by nearly 70%. But this treatment regimen had little value to HIV-infected women in poor countries, as it required treating the mother for an average of 11 weeks before delivery, giving her an intravenous drip of the drug during labor, and then treating the baby for six weeks after delivery. So the National Institutes of Health and the Centers for Disease Control and Prevention—both of which fell under Shalala's purview at HHS—and, separately, UNAIDS, sponsored drug trials in developing countries that compared simpler, more affordable treatment regimens against no treatment at all.

The researchers conducting these "placebo-controlled" studies believed they had debated the ethical issues publicly before launching them—a point that Public Citizen ignored—and decided to go forward because the trials were a fast way to evaluate practical strategies for reducing maternal-infant transmission in poor countries.[1] To Public Citizen, these "exploitative" trials violated the Nuremberg Code set up in the wake of Nazi human experiments and had "echoes of the notorious Tuskegee syphilis study. . . . We are confident that you would not wish the reputation of your department to be stained with the blood of foreign infants," Public Citizen's Sidney Wolfe and Peter Lurie, as well as four cosigners, wrote Shalala.[2] The group distributed this letter and a press release, headlined, "1000 FOREIGN INFANTS TO DIE

Unnecessarily in US–Funded HIV Studies: Human Experiments Are Tuskegee Part Two."

Public Citizen's argument revolved around an ethical precept spelled out in the World Medical Association's Declaration of Helsinki.[3] These famous ethical guidelines explicitly stated that everyone who participates in a medical study, including those in the control group, "should be assured of the best proven diagnostic and therapeutic method."

Two weeks later, on May 8, a congressional subcommittee held a hearing on bioethics at which Peter Lurie aired Public Citizen's views.[4] NIH director Harold Varmus, one of several officials who defended the ethics of the U.S.-government-sponsored trials, came armed with letters of support that he had received. Edward Mbidde, chair of the AIDS Research Committee in Uganda, wrote Varmus that he found Public Citizen's campaign "patronizing" and confused. "It is not the NIH conducting the studies in Uganda but Ugandans conducting their study on their people for the good of their people," declared Mbidde, who worked at the Uganda Cancer Institute in Kampala. "If this is not acceptable and the only way to do it is that which has been suggested in the news release, then this is tantamount to ethical imperialism." Researchers from the University of Malawi warned that "misplaced zeal and misinformed advocacy can jeopardize the progress that has already been achieved" and that it was "an insult" to suggest that "ethics are the monopoly of the United States and that all of us have to wait upon those citizens to give us ethical standards."[5] Similar letters came from researchers in Tanzania, South Africa, and the United States.

During the next several months, the din from this debate would grow louder and louder, providing the backdrop for the AIDS vaccine world's own heated exchange about whether to provide known, effective drugs to people who became infected during vaccine trials. There was, in the end, no satisfying solution to the dilemmas raised, no consensus of opinion about how to proceed, no resolution. But the process exposed a fundamental truth about the testing of new medicines on humans that "ethicists" often failed to embrace: desperate times call for disparate measures. This perspective does not mean, as Public Citizen charged, that a "research double standard" exists, with ethics shifting at a country's borders. Rather, it simply recognizes the unfortunate reality that the poor have fewer health-care options than do the rich. Or, as a coalition of researchers from Johns Hopkins University put it in a letter to Varmus, "The guidelines call for universal principles of ethical procedures, not universal standards of medical care."[6] Put simply, poor countries that participate in medical research have their own scale for measuring potential risks versus potential benefits.

From the very first human AIDS vaccine test, finding the proper balance between risks and benefits has confronted researchers with dizzying, deeply emotional ethical quandaries. Who decides whether a trial meets ethical standards? Should children be included in tests of new vaccines? When is a placebo control justified? What if one country wants to test a vaccine that has been rejected by another? Where is the line between a fully informed person volunteering for a trial and being coerced to join? Is it an ethical violation to delay trials of vaccines that appear safe and might work? How much effort must researchers put into teaching volunteers how to avoid becoming infected, which works at cross-purposes with the vaccine trial? What happens when people volunteer to test a vaccine that most scientists think carries too much risk to warrant human trials? What if a poor country tests a vaccine that proves effective, but then cannot afford to buy it? These are but some of the unruly ethical questions that by 1999 had troubled—and slowed—AIDS vaccine researchers as profoundly as the many scientific uncertainties they confronted.

• • •

ETHICAL QUESTIONS circled France's Daniel Zagury from the moment the *New York Times* on December 17, 1986, revealed sketchy details about his Zairian AIDS vaccine trials, the first ever conducted anywhere. But after the initial uproar died down, including a *Le Monde* article about the danger of "wildcat" trials, scientists, lawyers, and health officials who pontificated about the ethics of AIDS vaccine trials never delved into the details of Zagury's experiments. Instead, they explored at great length such general ethical principles as individual autonomy (chiefs should not be allowed to offer their tribes for a trial), informed consent (researchers should fully explain risks and benefits to volunteers), beneficence (researchers should do everything possible to maximize benefits and decrease risks), and distributive justice (populations that participate in a trial should have access to any resultant product). And so the discourse about the ethics of AIDS vaccine trials for many years remained constrained to august, little-read journals like the *Hastings Center Report* and *Public Health and the Law*, or to small, invitation-only conferences held by the World Health Organization or the Institute of Medicine.[7] Then, on March 10, 1991, the ethics of AIDS vaccine trials—and Daniel Zagury—became headline news.

John Crewdson, the *Chicago Tribune* investigative reporter who had been scrutinizing Zagury's friend Robert Gallo for years, broke the story. Headlined "AIDS lab may have ignored ethics rules," the front-page jaw-dropper revealed that the NIH, in response to questions from Crewdson submitted the previous July about the role Gallo played in Zagury's AIDS vaccine trials, suspended their collaboration. The NIH also barred any of its scientists from working with the controversial French immunologist until an investigation led by the NIH's Office for Protection from Research Risks, OPRR, could determine "the adequacy of the current system for assuring the protection of human research subjects in NIH collaborations with foreign scientists."[8]

With input from a team of outside consultants, OPRR delved deeply into Zagury's human tests of AIDS vaccines that he had made with the help of technical advice, reagents, and a vaccinia vector supplied—unwittingly, as it turned out—by Gallo and other NIH scientists. OPRR kicked over every stone it could find, even interviewing Zagury for two days, and on July 3, 1991, issued a 338-page interim report that revealed "a general failure" by the NIH to protect the humans who participated in the vaccine trials. According to OPRR, the NIH scientists wrongly "assumed that they had no responsibility in this area as long as they did not directly inject human beings with experimental materials." Zagury and his coworkers, wrote the OPRR, did not understand that because they collaborated with NIH scientists, they, too, had to abide by NIH rules. "These scientists assumed incorrectly that adherence to the legal and ethical requirements of the countries in which they worked was sufficient," the report stated.

Although much of the OPRR's criticism focused on the failure of Zagury and his NIH collaborators to receive the proper approvals before sharing material and expertise, reports submitted by the seven outside consultants ultimately had a more damning tone. One consultant[9] hammered on Zagury for using nine uninfected, Zairian children, aged 2 to 12, in the first tests of his vaccine. "[N]othing that has been presented to the panel or in other forums on vaccine development to date seems to me to justify the inclusion of seronegative children in Phase I vaccine trials," the consultant wrote. WHO's AIDS vaccine guidelines explicitly said that people who participate in phase I trials "should be fully informed volunteers," the critique continued. "It is hard to see how they could qualify as 'fully informed volunteers,"' this consultant concluded, emphasizing that "young, developmentally immature children would not be able to understand all the relevant issues."

Another consultant[10] spotlighted the overlap between Zairian investigators who worked with Zagury on the trials and simultaneously sat on ethical review boards that evaluated their merit. As this consultant gingerly put it, "care needs to be taken to avoid the potential for conflict of interest."

OPRR included in its report a rebuttal from Zagury, who dispensed with all the dispassionate verbiage favored by the NIH investigators and, with an operatic brio, declared his innocence.[11] Obviously put out by OPRR's investigation, Zagury first unloaded on the *Tribune*'s Crewdson, accusing him of "conducting a campaign of denigration and slander" against his studies. "I am surprised, if not beside myself, at these accusations," wrote Zagury, who complained that OPRR had been "particularly kind" to Crewdson and had made itself his "accomplice." He further allowed that he was doing OPRR a favor by cooperating, as no foreign organization "should be entitled, without being authorized by my government, to interrogate me concerning my professional activity."

Zagury strongly attacked most every charge levied against him. "I have always respected the ethical and deontological regulation of my country and that of Zaire," wrote Zagury. French law, Zagury emphasized, allowed children to participate in vaccine trials. "Did not the first vaccine produced in the world by Pasteur concern a child?" he fumed. And he included these children, Zagury contended, out of compassion. Their fathers all had died from AIDS, and their surviving mothers, who all had AIDS themselves, "begged us to do something for their child." The vaccination, in the end, "was a source of comfort and hope" for these families, Zagury insisted, stressing that the vaccinations had not harmed any of the children.

Three weeks later, Zagury sent OPRR a more detailed rebuttal to the charges that dripped with both nationalism and sarcasm. "We deplore the entire action of the O.P.R.R.," thundered Zagury, who accused the NIH of using him "as a pawn" in its attack on Gallo. He especially recoiled at OPRR's request to have its own experts review his files, which, he queried, "is this by international law, imperial law or divine law?"

The OPRR's investigation of Zagury's vaccine trials ultimately fell apart. Not only did France tell the NIH to take a hike; the government of Zaire never even responded to the OPRR's request to do an on-site review.[12] France did agree to send its own expert to Zaire to review the vaccine studies there, but in September 1991, Zairian soldiers, angry that they had not been paid, began rioting in Kinshasa, leading to widespread looting and mayhem—and the mass evacuation of foreign researchers.[13] On March 26, 1993 the OPRR issued a final report on the whole mess. Aside from the interesting detail that

Zagury's initial studies actually involved twice as many children as originally reported—10 who ranged from two to nine years old and eight more between 10 and 18—the final report offered few new insights. It also, much to Zagury's delight, lifted all restrictions on his collaborations with Gallo and other NIH researchers.

In the end, the Zagury investigation led the NIH to revise the oversight of its scientists who either conduct research on humans themselves or who collaborate with outside researchers. But the controversy transcended the mere reorganization of a bureaucracy: it established a floor for the ethics of AIDS vaccine research, and said, in effect, this is not the way to do business. "These experiments represent an unfortunate beginning for the HIV vaccine effort, perhaps especially in the conduct of collaborative international research," wrote Christine Grady in her 1995 book, *The Search for an AIDS Vaccine: Ethical Issues in the Development and Testing of a Preventive HIV Vaccine.*[14] A bioethicist, NIH AIDS nurse, and wife of NIH AIDS point person Anthony Fauci, Grady condemned the experiments not so much for violating NIH regulations, but for using children, which she bluntly said "was not justified." And Grady noted that in part because of the Zagury controversy, when other investigators from developed countries ventured to Africa or Asia to test their AIDS vaccines, "distrust and suspicion about the motives and methods of foreign scientists and governments are common."

From a more distant perspective, Zagury's Zairian trials, although no one would dare say it, had a positive impact on the field, too: without, apparently, hurting anyone, they offered a lesson to an inexperienced world about how to identify and prevent unethical AIDS vaccine trials, much in the way that an inexperienced immune system learns how to defeat invaders after being exposed to vaccines.

• • •

ON APRIL 18, 1996, the New England Regional Primate Research Center received a package, wrapped in plain brown paper, addressed to Ron Desrosiers with no return address and a postage stamp that said Fresno, California. The next day would mark the one-year anniversary of the bombing of a federal government building in Oklahoma City and the two-year anniversary of the fiery end to the Branch Davidians in Waco, Texas, and Desrosiers had just read an article in the newspaper warning people to be on the lookout for strange packages. The Unabomber, who targeted molecular biologists like Desrosiers and often mailed packages from northern California, also was much in the news because Ted Kaczynski, then only a suspect, had been arrested two weeks earlier. Added to these concerns, the primate center always worried about violent animal rights activists, and Desrosiers himself, who recently had been the feature of a PBS documentary, certainly had stirred much high-profile controversy by becoming the world's most vocal advocate for the development of an AIDS vaccine made from the live, weakened virus. Still, Desrosiers, a leading vaccine researcher who routinely infected monkeys with SIV, did not believe anyone would actually try to harm him. But, to be safe, he ran the small box through his lab's X-ray machine.

Two of his technicians who had military training stood by. They both agreed: it looked like a cluster bomb. Desrosiers phoned the police.

Shortly after the Massachusetts State Police rushed over and evacuated the buildings, camera crews from two Boston TV stations raced to the scene, just in time to observe the arrival of a bomb squad, which had its own X-ray machine. The bomb experts deemed the package harmless and opened it. Inside, they found a small ball studded with 40 metal spikes that had suction cups on their ends. It was a model of HIV. The package also contained a note from a high school class: Desrosiers had spoken to the students, who contacted him after seeing the PBS show, and they wanted to thank him.[15]

The bomb incident richly symbolized the deeply polarized reaction AIDS researchers had to Desrosiers's push to develop an attenuated HIV vaccine for humans. No one, of course, physically threatened him, but his detractors maintained that a live, weakened HIV vaccine, no matter how safe it appeared, was no less than a Pandora's box that held a viral bomb waiting to explode. To Desrosiers, fear of the approach made sense, but he believed that by moving forward cautiously, scientists could reduce the risk to an acceptable level so that they could, in effect, open the box without detonating the viral bomb.

From the outset, Desrosiers met strong resistance when he spoke of developing a live, attenuated vaccine; indeed, the NIH funded his initial studies with weakened SIV in monkeys, which began in 1989, with the understanding that he wanted to examine how the virus causes disease. When he first demonstrated in 1992 that his weakened virus had a powerful ability to thwart a subsequent SIV infection, he received many plaudits from leading researchers who hoped his work might reveal the immune correlates of protection—but they insisted that the idea would never move into humans. In 1994, Ruth Ruprecht of the Dana-Farber Cancer Institute, validated, at least in the minds of the detractors, the contention that an attenuated HIV vaccine had no future.

At NIAID's annual AIDS vaccine meeting in November 1994, Ruprecht showed sobering data that moved the debate out of the theoretical realm—and she emerged as Desrosiers's most ardent critic.[16] From the beginning, Desrosiers had planned to make as many genetic deletions as he could in SIV without losing the ability of this attenuated "vaccine" to properly teach the immune system how to ward off a subsequent infection by a fully competent virus. He had moved well beyond the initial vaccine, which, as he described in his landmark 1992 *Science* paper, simply contained SIV minus its *nef* gene. His latest attenuated vaccine had three key deletions. As with his *nef*-deleted SIV, the so-called delta-3 vaccine soundly protected monkeys in challenge experiments. But then Ruprecht, a native of Switzerland whose perfect posture captured her no-nonsense manner, asked what would happen if mothers given this delta-3 virus passed it to their infants, which are born with immature immune systems. "We were in for surprises," Ruprecht told the audience.

About eight months after putting the vaccine virus down the throat of an infant, the animal—which had never been challenged—suffered a steep drop in its CD4 cells. Unlike adults, the newborn also could not contain the replication of the delta-3 SIV, and Ruprecht's group consistently found high levels of the virus in the baby's blood. A subsequent analysis revealed that the delta-3 had *not* mutated back into a known virulent strain, emphasizing that the virus in its original form was causing the disease.[17] Ruprecht then infected two more newborns with the virus. At the time of the meeting, the SIV had replicated to high levels in the blood of all four infants, one had died, and two had developed signs of AIDS.[18] "Based on our data, I truly believe that the live, attenuated vaccines are dangerous," said Ruprecht. "I don't see [attenuated strains of HIVs] as a viable approach for human vaccines."

Desrosiers acknowledged that Ruprecht's findings were "damaging to the concept" but urged his colleagues not to write off the live, weakened virus vaccine. "We're continuing to search for that right balance of attenuation and potency, and it's not going to be easy to establish where that line is," he said. Desrosiers also offered intriguing data from a hemophiliac who had been infected in 1983 by the contaminated clotting factor he used to stop his bleeding. The 44-year-old man, who then had been infected for 11 years, had perfectly stable CD4s and was in fine health. And an analysis of HIV taken from his body revealed an extraordinary detail: the virus was missing a large section of its *nef* gene. "What we like to think is that this is an individual who already got the vaccine," Desrosiers said. "[H]e is our first safety study."[19]

This "vaccine" not only appeared safe—it may have been effective as well, said Desrosiers, noting that after this "long-term nonprogressor" became infected with the attenuated virus, he likely received transfusions of more HIV-infected lots of clotting factor and yet still resisted infection.

One year later, an Australian group of researchers reported in *Science* that a *nef* deletion in HIV appeared to answer a mystery that had been dogging them about eight long-term nonprogressors who shared an odd feature.[20] Jennifer Learmont, a nurse at the New South Wales Red Cross Blood Transfusion Service, first linked these people to each other. Learmont ran the "Lookback" unit, which notified people who had either donated blood or received a transfusion that they might be infected. Twice in 1987, the same doctor came to Learmont with infected patients who had received transfusions and asked her to find the donor. Both patients were perfectly healthy—and both traced back to the same donor, who also was healthy. Learmont phoned the doctor and noted the oddity of this, but it wasn't until 1989, when she learned that both patients and the donor still had normal CD4 counts, that she decided to investigate whether the man's donated blood might have infected other people. "I guess it sounds funny, but I really did seem to have a flash," Learmont said in a documentary about the case made by the Australian Broadcasting Corporation.[21] "It just came to me."

Learmont's search ultimately linked the donor to seven recipients. As detailed in the November 1995 *Science* paper, one recipient had died from what the researchers concluded was a non-AIDS-related disease, and everyone else, including the donor, had stable and normal CD4 counts. The donor, a gay man, had been infected since at least April 1981; the recipients had received their transfusions between that month and July 1984. So the people in this group had been infected between 10 and 14 years, a duration that statistics said should have caused AIDS, if not death, in at least half of them. And now, as the *Science* paper explained, the researchers finally had an explanation for the phenomenon: HIV isolated from four of the so-called Sydney cohort all shared a large *nef* deletion. They concluded that this strain of HIV "could perhaps"—an interesting double conditional that indicates how touchy the suggestion was—"be the basis for a live attenuated vaccine."

The agenda setters acknowledged the significance of the finding. "It's a very important experiment of nature," NIAID's Fauci told me. But he still was holding out for something safer. "We may ultimately have to go for a live, attenuated vaccine," said Fauci, "but there are many reasons for concern about the approach."

The finding much buoyed Desrosiers. Despite Ruprecht's data pushing the field further away from the approach than ever, Desrosiers had continued testing a variety of mutant SIVs in monkeys and HIVs in chimps, discovering that four deletions appeared to offer the maximum safety without sacrificing too much efficacy. The Australian data, he predicted, are "going to refocus attention on the live, attenuated vaccine."

Ruprecht and her coworkers wrote a sharp letter to the editor, strongly questioning the conclusion that the one patient who died did not have AIDS and chastising the Australian group for not mentioning her lab's monkey data.[22] They did offer what, in the eyes of Desrosiers, amounted to the standard sop: animal experiments with attenuated vaccines "could play a major role" in revealing the correlates of protection.

This letter starkly framed Ruprecht's message: data would never make a compelling case that an attenuated HIV vaccine was safe enough to test in humans. And her negative data about attenuated vaccines simply reinforced that point of view. Yet that line of reasoning dodged Desrosiers's main argument, which was a risk/benefit equation. He did not claim that he could make an attenuated vaccine that carried no risks. "Is it going to be absolutely, 100% safe?" he rhetorically asked me. "Forget it. It never will be. If you put it into enough people, there will be problems. That's true of every live, attenuated vaccine." But the real question, Desrosiers said, is what is the likelihood of a person becoming naturally infected by HIV versus being injured by the vaccine? "We're never going to know until we put that into humans, and that's why people have different best guesses."

Ruprecht remained a formidable Desrosiers critic, because unlike other detractors who argued on purely theoretical grounds, she continued to produce new monkey data that highlighted the dangers of the approach. Desrosiers countered with data showing that neonates only became sick from the attenuated vaccine when scientists gave them high doses of the altered virus; indeed, no neonate became infected directly from the vaccine virus given to a mother, which would be the main mode of transmission in the real world.[23] And Desrosiers continued his efforts to make a safer attenuated vaccine, for which he received generous NIH funding, deleting more and more genetic elements of the virus in a search for the most crippled virus that could still protect animals from a subsequent challenge.

The debate among Ruprecht, Desrosiers, and their proponents turned into a stark ethical dilemma in 1997 when the little-known International Association of Physicians in AIDS Care, IAPAC, launched a campaign to recruit a few hundred volunteers for safety tests of an attenuated HIV vaccine. AIDS clinician Charles Farthing, medical director of the AIDS Healthcare Foundation in Los Angeles, headed the drive. In "a call to physicians" that appeared in the August 1997 issue of IAPAC's journal, Farthing, who had the fastidious look of a watchmaker, urged his colleagues "to respond to the moral imperative of doing everything possible to bring low-cost effective HIV vaccines to market." Farthing noted that physicians long have volunteered themselves to test new medicines. "It seems to me that the only way human trials will begin is if some volunteers step forward," explained Farthing, who accused the WHO and NIH of being "too timid to even propose this."[24]

Farthing, who told me he decided to make this move because he had become "progressively irritated" about the lack of movement toward human trials of an attenuated vaccine, hoped a safety trial would show after a few years that humans, like monkeys, can control replication of the weakened virus and not suffer any immunologic damage. "We're never going to be able to tell the Africans to do a trial unless we do it ourselves," said Farthing.

Fauci cautioned that such a safety study would not address many of his worries about the approach, such as the possibility that people harboring this supposedly benign infection might develop cancers at a higher rate after, say, 10 years. Farthing recognized the risks and acknowledged that the Food and Drug Administration might never approve his proposed test. But, he said, "if you just assume everybody's going to say no, you don't do anything." His aim, he said, was to "create the debate."

With great media savvy, Farthing and his IAPAC supporters the next month parlayed an informal meeting with researchers at NIAID into worldwide television and newspaper coverage about their campaign. NIAID officials were beside themselves about the ruckus. "A number of people had to be sedated around here that the meeting was even being held," NIAID's Alan Schultz, then head of its AIDS vaccine branch, told me. But in concert with the publicity, Ruprecht and several other researchers—including Desrosiers—revealed that adult monkeys "protected" by the delta-3 SIV "vaccine" had, after several years, begun to experience high viral loads and disease. "The triple-deleted viruses we've worked with are pathogenic," Ruprecht told me. "If they're called a vaccine, I'm highly concerned."

Desrosiers countered that he had moved on to a delta-4 vaccine, which was a "considerably" weaker version of SIV than the delta-3. "This is a vaccine for targeted, high-risk groups," said Desrosiers. "This is not for babies."

The new data did not change NIAID's official position about the strategy, because the institution already had determined that the approach carried too much risk. But Schultz, while wearing his unofficial hat, captured the feeling that many scientists inside and outside the NIH had. "Anyone who would sign on to this trial after reading an informed consent form is certifiably nuts," he told me.

The next summer, a packed scientific session at the world AIDS conference in Geneva featured a remarkable showdown between Farthing and Ruprecht over the pros and cons of the attenuated vaccine approach.[25] Farthing opened, noting that more than 300 people volunteered to test an attenuated AIDS vaccine and that this approach had worked successfully against many other viruses. He then referred to the Sydney cohort, and recounted Desrosiers's monkey data, explaining that the most recent version of the vaccine was a delta-5.

Farthing explained that John Sullivan at the University of Massachusetts in Worcester had proposed a way to add a measure of safety before staging the IAPAC-sponsored safety trial. Sullivan, who, with Desrosiers, had first identified the healthy hemophiliac who had a *nef*-deleted HIV, proposed testing a delta-5 HIV on five volunteers who were terminally ill with cancer. Ideally, tests after six months would show that these patients could contain the infection with this virus such that they maintained low viral loads. If they could not contain the infection, however, Farthing stressed that they could be given the powerful antidrug cocktails now available. "The safety concerns of a live attenuated trial in human volunteers are being way overplayed," he said. "There seems to be a zeitgeist in the late 20th century that a clinical trial can have no risk whatsoever. I disagree with this. This is what informed consent is all about. I think this is really an ethical question, not a scientific one. And we at IAPAC would like to take this ethical debate out of the vaccinologists' domain, and so we asked the physicians, and obviously many think that this degree of risk is acceptable to see if this most promising area of vaccine research is safe [in order] to pursue the goal of achieving an HIV vaccine for mankind. The urgency could not be greater."

Farthing made a compelling case, but strong data argued against it. The disease in "vaccinated" monkeys was sobering, and could not be dismissed, as Farthing attempted to do, by noting that SIV and HIV differed: if they were so different, than how could he build his case for the superiority of the approach, which relied largely on Desrosiers's monkey data? Three people in the Sydney cohort, which Farthing did acknowledge, also recently had seen CD4 drops. "It would have to change your thinking that it's not causing damage," Learmont, the nurse who discovered the cohort, told me. "Obviously this now needs to be examined very hard." Farthing, however, countered that these people only had one crippled gene—*nef*—in their HIVs, and he emphasized that none of them had developed illness or immune suppression, despite being infected for between 13 and 17 years. "This is remarkable," he said.

Strong scientists also took Farthing to task—first Ruprecht and then, more trenchantly, Robert Gallo. Following Farthing's presentation, Ruprecht reviewed the negative data, allowing only that, theoretically, researchers might one day determine the genetic factors that make HIV dangerous. But until then, she said, the concept of an attenuated AIDS vaccine was dead. When the microphones opened to the audience, Gallo came out swinging. Unlike attenuated vaccines made against viruses like polio or smallpox, "a retrovirus is forever," Gallo said. "How long would you be prepared in a pilot study to wait before you thought it was safe?"

Before Farthing could answer the question, Gallo, being Gallo, went on the offensive. Every single retrovirus in every single species, from chicken to man, causes disease, Gallo declared. "Every. Single. One." This is not strictly true, as SIV does not cause disease in African monkeys, but Farthing failed to point this out. Instead, he countered that Ruprecht's monkey data seemed inconsistent with data from humans who had weakened strains of HIV. "We'll never know unless we do an experiment in humans," said Farthing.

Gallo allowed that Farthing made some good points, but he appealed to his common sense. "Let's just talk facts like we were in a barroom, instead of posturing in front of a big audience," Gallo said. "I don't know when you'd be able to tell in a pilot study that it was safe. It's going to be after your lifetime. That's the problem. And it's going to take a lot of people, too." Retroviruses, said Gallo, can take 30 years to cause disease.

The firefight stopped when Ruprecht asked Learmont to come to the mike and explain her new data. But right after she finished, Farthing took another shot. "Now that Dr. Gallo's away from the microphone, I can maybe attempt to answer his question," Farthing chided. Farthing gave a slight, exasperated laugh, and he looked thoroughly defeated. He reiterated his hope that the small trial in terminally ill people would reveal how well the virus copied itself, but, indeed, right then and there, you could see the wind leave the sails of his campaign for a live, weakened, attenuated HIV vaccine trial. It did not help when in January 1999 Desrosiers and Sullivan reported in the *New England Journal of Medicine* that the hemophiliac they had been following who had a *nef*-deleted HIV also had begun to suffer a CD4 decline.[26]

Regardless of whether Farthing underplayed the dangers of the approach, he was spot on about the essence of the debate: it was an ethical predicament, not a scientific one. Farthing had recruited physicians who, presumably for altruistic reasons, knowingly volunteered to test a vaccine that many scientists deemed too dangerous for human use. Sullivan similarly proposed injecting a weakened HIV vaccine into informed terminal patients who volunteered for a trial. Maybe these trials would prove meaningless, as Gallo and other critics asserted, and that is a debate that took place repeatedly. But why was there no debate about the ethical question that Farthing raised, namely, Should volunteers be stopped from staging such a test if they so choose?

• • •

PRAPHAN PHANUPHAK, director of the AIDS program at Thailand's Red Cross Society, on April 18, 1997, sparked a similar ethical debate when he sent a missive to the deputy governor of Bangkok that criticized the planned tests of another HIV vaccine strategy. Praphan, who often played the gadfly, wanted to derail the efficacy trials in Thailand of the gp120 vaccines made by VaxGen and Chiron.

Praphan became involved after meeting John Moore at an AIDS conference in France. Following their talks, Praphan solicited data from Moore's boss, David Ho, and Steven Wolinsky of Northwestern University, who were leading the analysis of gp120-vaccinated people in U.S. trials who became infected—the infamous breakthrough cases that had helped convince the ARAC to scuttle the efficacy trials in 1994, and that Moore had described in detail at the Cent Gardes meeting in October 1995. Ho and Wolinsky described how the vaccines did not stimulate "meaningful" neutralizing antibody or killer cell responses and had no impact on the subsequent viral load. "It is not the fact of the breakthrough infections that particularly concerns us, it is what we have learned from studying these cases in detail," they concluded. Yet they did not explicitly offer advice as to whether Thailand should proceed with the efficacy trials. "We'd be remiss if we didn't provide the Thais with that information," Wolinsky told me a few weeks later. "But it never was intended to stop or start a trial. Our colleagues in Thailand are very intelligent, and they don't need David Ho or Steven Wolinsky telling them what to do."

Praphan's letter to the deputy governor included Ho and Wolinksy's data, which he said led him to conclude that these vaccines were "not useful in preventing HIV infection." Therefore, wrote Praphan, it was "not appropriate for Thailand to allow (approve) an efficacy trial" of them.

VaxGen's Don Francis and Phillip Berman received Praphan's letter, and fired off a letter of their own to the deputy governor that said "a large segment of the scientific community feels strongly that this vaccine deserves to be tested for efficacy."[27] They challenged Ho and Wolinsky's data, point by point, turning negatives into positives and stressing their success in chimp experiments and the safety of the vaccines in more than 2,000 volunteers in the United States. "If there is no safety downside, we see no ethical argument not to move ahead as fast as possible towards answering the essential question of efficacy," they wrote. "We recognize that in each advancement of science, there are and will be nay-sayers who want to keep things as they are—to do more research or just wait and see. We feel this is especially inappropriate at this time with this vaccine and this disease."

A few weeks later, David Ho visited officials from the Ministry of Public Health in Bangkok, and, at a press conference afterward, he took off the gloves and, without naming names, pummeled VaxGen and Chiron. "For me as part of the Asian minority in the U.S., I feel it's important for the Thai people to be aware of the possibility of exploitation," he said, according to an article that appeared in the *Bangkok Post* on May 17—the day before Clinton's historic announcement of a 10-year goal.[28] The article, headlined "Researcher Warns over Vaccine Trials," said he had strong feelings about the matter. "If a product is rejected elsewhere, why should you take it?" he asked. "It's wrong for some U.S., European, and other researchers to look at this only as an opportunity to develop a product."

E-mails began to zip around the world. "Thanks for letting me know what the *Time* Magazine Cover Laureate is saying in Thailand," snickered the CDC's Bruce Weniger to a Thai colleague who sent him the article. "The press ought to be informed of the fact that many accomplished vaccine scientists (of which Dr. Ho is not one) feel [ARAC] was a bad decision, reflective more of the politics of money for scientific research, and the ignorance of many basic science researchers of how vaccines are developed." William Heyward, the head of AIDS vaccines at the CDC, similarly found Ho's comments "misleading and inappropriate," he wrote me. (A year earlier, Heyward had coauthored an article that said "it would not be ethical" for hard-hit developing countries to wait for more basic research before staging efficacy trials.[29]) "Depressing to see how low it can go."

VaxGen's Francis wagged his finger at Ho in a letter that blended impassioned ethical rhetoric with dispassionate scientific arguments. "For you to travel to Thailand and make statements without such knowledge is, in my opinion, irresponsible, and not typical of the David Ho I know," wrote Francis. "There has never been nor will there ever be an attempt on my part to exploit anyone. Our studies in Thailand have been based on truth, honesty and a sincere attempt to answer scientific questions which will lead to the interruption of transmission of this virus in Thailand and the rest of the world." Francis reiterated the chimp protection data and said their current formulation of the vaccine for Thailand "was even more exciting." Come to VaxGen, said Francis, "and allow us to update you on our work."

Ho never took Francis up on the offer. And the mudslinging intensified the next month when John Moore received calls from journalists who told him of Weniger's viewpoints. "In many respects, whatever you say is of no concern to me, since you are not a person I take seriously," Moore wrote Weniger on June 23, stressing that this was a "personal and confidential" letter.[30] "I regard you as a fool who is messing in an area of science about which you know precious little; your 'recipe' for vaccine research would more or less guarantee that we fail to deliver, for you propose simply to repeat the failures of the past." Moore explained that his objections to the gp120 tests in Thailand centered on the VaxGen trials only, and had nothing to do with his own need for more basic research money. "It must be obvious even to an individual whose thought processes are as muddled as yours so plainly are, that what goes on in Thailand affects in no way the scientific programs of researchers such as myself. It therefore follows that my opposition must be based on scientific factors—nothing else is remotely relevant."

Although the VaxGen and Chiron gp120s were largely similar vaccines, Moore did not object to the Chiron trial because the company had linked with the U.S. Army's research group "for whom I have considerable personal and professional respect." He "strongly opposed" the VaxGen trials, however, because he asserted that the company was "abusing the Thai people for selfish reasons," and he "thoroughly disapprove[d]" of the way in which the scientists interpreted the scientific data.[31]

U.S.-based scientists like David Ho, John Moore, Don Francis, William

Heyward, Steven Wolinsky, and Bruce Weniger of course have every right to slug it out over the ethics of a U.S.-based company conducting efficacy trials in Thailand. But ultimately, as all of them realized, the decision fell to the Thais, who, rattled by this din, asked UNAIDS for advice. UNAIDS convened a panel of 13 experts to review the scientific arguments on both sides of the fence. A final report noted that of the 13, 10 encouraged Thailand to stage efficacy trials with these vaccines. "The document contains gems of wisdom and experience of high intellectual content which are most valuable," wrote Natth Bhamarapravati, chair of Thailand's Subcommittee for HIV Vaccine Trials.

Natth, an experienced vaccine maker himself, in truth could not have learned much from the expert advice he received, because it offered precious little new information. But Natth, like Fauci before him, needed the blessing of independent outsiders before he could advocate staging a trial. Unlike Fauci, however, Natth was not asking advisers whether his government could justify spending money on trials of vaccines that many scientists had little faith in. Natth was asking whether they could ethically justify the trial that outsiders offered to fund—a more subtle, yet every bit as critical, a question.

• • •

PICTURE YOURSELF at the edge of a river, a river that is roaring down, and someone is being swept away. You do not know how to swim. There is no ethical violation for failing to throw yourself into the water, because no one likely will benefit from your action and the risk is very high. But if you are a lifeguard, a trained lifeguard, and you simply stand and look and do nothing, then you clearly have violated your ethical duty to help.

These are nearly the exact words that Jonathan Mann used on March 15, 1998, when he spoke to a subcommittee of the Presidential Advisory Council on HIV/AIDS.[32] As former head of WHO's Global Programme on AIDS and, before that, Projet SIDA in Zaire, Mann well knew the power of sermon. "So the failure to proceed with vaccine field trials, given the history of vaccine development for every other vaccine that's been successful . . . is patently unethical," said Mann, who recently had left the Harvard School of Public Health to become dean of the Allegheny University School of Public Health.

Mann then upped the rhetorical ante. Not only was it an ethical violation; the U.S government's "failure" to proceed with the gp120 efficacy trials "represents a clear violation of the human rights of American citizens." Mann then intimated to the PACHA subcommittee that the failure to proceed with efficacy trials had a social, if not racial, bias. He said it in code, but he said it. HIV-infected people had been "marginalized" and "discriminated against," he said. "Lyme disease, as you probably know, is a disease that occurs in places like Cape Cod and Martha's Vineyard. The progress toward developing this vaccine has been rapid. If 40,000 college students in this country were infected with HIV each year, then I assure you there would have already been field [efficacy] trials of AIDS vaccines."

A member of the PACHA subcommittee asked Mann why he thought the delays had occurred. "I think it's actually very clear, and I'm actually quite fortunate to have had you ask that question of me because I don't have any NIH grants," he said. The room laughed. "And I think it's time to be as clear as possible." His questioner interrupted to clarify that they had never spoken before. "Correct," said Mann. "The people who have been placed in charge of AIDS vaccine development in this country, number one, have no experience developing vaccines. Number two, they come from a stream of science which deserves great respect, but it is only one view. It is the stream of science that would like to have all the answers before it proceeds."

Mann was in full throat now. "We eradicated smallpox from the world without ever knowing—ever—exactly how [the vaccine] worked," he intoned. "So what we have here is a scientific world that is basically holding a monopoly lock on the process, and basically saying, 'Until we, the scientists, get answers to the questions which are probably unanswerable, but until we have all those answers, we will not allow this to proceed.' And the names of David Baltimore and Harold Varmus are right at the top of the list." He went on. "I think there's an abdication of responsibility at the highest levels of Health and Human Services, where the responsibility for public health resides. The government has ethical and human rights responsibilities. There, the NIH cannot be allowed to monopolize or paralyze the process." Proceeding with trials no longer is the ethical problem, Mann said—*not* proceeding is. "Something is desperately wrong."

Like a man who had held his tongue too long, there was no stopping Mann now, and he stepped up his attack on David Baltimore, first obliquely, then by name. At the 1987 AIDS conference in Stockholm, Mann recalled, a Nobel laureate said the problem with the Nobel Prize is that its recipients think they're experts in everything. The crowd laughed again. "I would never take anything away—I mean, David Baltimore made tremendous contributions to science—but David Baltimore is not equipped to make tremendous contributions to vaccine development." Baltimore, said Mann, maybe should be on the NIH's advisory committee, but with his lack of vaccine experience, he should not be the chair of it.

PACHA's Bruce Weniger concurred, invoking the name of Basil O'Connor and the March of Dimes. "He would solve every problem as they arose," said Weniger. Yes, said Mann, emphasizing that the key was having an accountable person in charge. "I imagine that both Donna Shalala and the President have so many other things to do, that what's necessary is their appointing of a person who is given the authority."

Jonathan Mann made several valid points during this diatribe. Clinton and Shalala had not appointed a true AIDS vaccine champion, à la Basil O'Connor, to head the search. NIH director Harold Varmus and David Baltimore, although they had done outstanding science that won them Nobel Prizes, had no vaccine experience. The NIH had monopolized the clinical trials of AIDS vaccines, despite the vast experience of the CDC in staging such tests. But Mann's claim that these shortcomings violated human rights and ethics devalued those words. And what Mann refused to consider is that the federal government, as was the case during the polio years, might not be equipped to solve this problem.

Seeing as how Mann aimed to change the federal government, he made one other miscalculation: His bold, frontal attack on the powers that be severely crippled whatever hopes he had of influencing change within that system. First, 57 researchers and activists banded together, writing a letter to the editor critical of Mann that appeared in the May 8th *Science*.[33] An e-mail from Neal Nathanson to Mann more bluntly demonstrated how ostracized he had become. Mann had phoned Nathanson to congratulate him on his new job as head of NIH's Office of AIDS Research. On June 5, Nathanson sent an e-mail note to Mann, copied to both Baltimore and Varmus, that read, in part:

> in light of your recent statements regarding dr. baltimore and dr. varmus, i am very loath to talk with you. frankly, i am not comfortable how anything i say might be quoted by you in some public context which could be inimical to the programs for which i am responsible.

• • •

NEAR THE END OF THE SUMMER OF 1997, Barry Bloom got wind of the *New England Journal of Medicine*'s plan to run an editorial by Public Citizen's Peter Lurie and Sidney Wolfe about the "unethical" placebo-controlled trials in poor,

HIV-infected pregnant women. Worse still, from Bloom's perspective, Marcia Angell, the executive editor of the journal, planned to write an accompanying editorial, siding with their argument that these trials should compare experimental treatments to the "best proven" ones. "I had the chutzpah to try to get in touch with Marcia Angell," said Bloom. "I gave her my best 15 minutes."

Bloom had just planned a meeting at UNAIDS in Geneva to debate whether ethics demanded that people in AIDS vaccine trials who became infected should receive the "best proven" treatment, as Public Citizen and Angell maintained about the pregnant women, or a treatment that was the "highest attainable" in that region. Bloom urged Angell to consider that the existing guidelines needed refinement and were not handed down by Moses. Angell would not budge. "My strongest wish is somebody should give her an airplane ticket to see a hospital in Uganda," said Bloom.[34]

The editorials, which ran in the September 18, 1997,[35] issue, received the choicest media coverage that the *NEJM* and Public Citizen could have hoped for: A page 1, above-the-fold *New York Times* story with this headline deck:

RESEARCH ON AIDS IN POOR NATIONS RAISES AN OUTCRY
SOME CALL IT UNETHICAL
COMPARISONS WITH A SYPHILIS STUDY ARISE BECAUSE SOME DO NOT GET THERAPY

David Ho and Catherine Wilfert, scientific director of the Pediatric AIDS Foundation, resigned from their positions on the *NEJM*'s advisory board. Neither had been asked to review the editorials beforehand, and Angell's comparison to the infamous Tuskegee syphilis experiments had deeply offended them both. "Here they have something as important as this," a disgusted Ho told me, "and they don't even run it by us."[36]

At the UNAIDS meeting held five days later, the participants decided that they needed to stage a full airing of the best-proven-versus-highest-attainable debate, so they organized workshops in Brazil, Uganda, Thailand, and Washington, D.C. Come June 1998, representatives from each of the workshops would gather to cobble together a set of recommended guidelines, which UNAIDS would forward to the Council for International Organizations of Medical Sciences (CIOMS), the group that had cowritten with the WHO the most current guidelines in use.[37] At least that is the way it was supposed to work.

The well-attended workshops thoroughly explored the issues, and by the time of the June meeting, held at WHO headquarters in Geneva, it was clear that different regions of the world had different opinions. Brazilians fervently believed in the best-proven treatment standard, and any country that did not want to offer state-of-the-art cocktails to people who became infected should not be allowed to conduct an AIDS vaccine trial. In the United States, Uganda, and Thailand, the majority contended that it was ethically sound to provide the highest attainable treatment.

The Geneva meeting, held June 25 and 26, mixed 85 AIDS vaccine developers, ethicists, public health officials, lawyers, and activists from more than two dozen countries.[38] They reached agreements on some important points, such as the recommendation to end the requirement that a vaccine be tested first in the country where it is made, and they said trials should be more closely monitored to make sure that participants truly give their consent. These recommendations could lead to "major changes in the way trials are done," said Bloom. But the central controversy over how to treat those who become infected—the question that led to calling the meeting in the first place—remained unresolved. The debate indeed became so absurd that at one point the group refused to vote on the issue—and then refused to vote on whether they should vote on it. "This is a monstrous responsibility," said a clinician from Morocco. "There is nothing wrong in saying the topic was so difficult that we did not reach a decision," said the health minister from Zambia.

The issue boiled down to whether a trial that offered potent anti-HIV drugs to everyone who became infected could detect efficacy. The widespread use of anti-HIV drugs could "have tremendous scientific impact," warned Mary Lou Clements-Mann, head of AIDS vaccine testing at Johns Hopkins University (and, as of 1996, Jonathan Mann's wife), "making it impossible to design a scientifically valid [vaccine] trial."

Don Francis contended that the whole point was moot. Francis said not everyone would start on treatment immediately, and since the researchers would be taking blood from participants every 24 weeks or so, they should be able to catch at least one viral load measurement in many untreated people who had become infected. If the vaccine had an effect, said Francis, it should be relatively easy to see. "Statistically, the power we have to determine even a 30% decrease in viral load is immense," said Francis. David Ho, on the other hand, remained skeptical. "I think it's tough in a country like the United States," said Ho. "Patients are going to be treated very quickly."

Many participants contended that offering people in developing countries expensive anti-HIV drugs raises its own set of ethical concerns. Dwip

Kitayoporn of Thailand's Mahidol University worried about people not being able to receive the drugs after the trial ends. "It's like leaving a Cadillac or Rolls Royce in our country but no one can afford to drive it or even repair it," said Dwip. Major Rubaramira Ruranga, an HIV-infected Ugandan who works at a research center in Kampala, warned that people may also sign up for vaccine trials just to get access to drugs. "We're going to create a safe haven now for people who are going to be put on the trial," Ruranga said. This, others noted, would violate the ethical principle that says researchers must not "unduly influence" people to join trials.

But an impassioned, ardent minority, including Public Citizen's Peter Lurie, rejected the idea that trial volunteers in poor countries should be treated any differently from those in wealthier ones. Dirceu Gerco, coordinator of an AIDS vaccine center in Brazil, worried that setting a lower standard for poor countries was a slippery slope. "When you put the level of ethics below the maximum, it's very easy to lower it more," said Gerco, whose sentiments were shared by several other Brazilians at the meeting.

At the meeting's end, there was no consensus about whether ethics demanded that people in poor countries who participate in a vaccine trial and subsequently become infected should be offered state-of-the-art drug cocktails. But Ruth Macklin from the Albert Einstein College of Medicine and chair of the AIDS vaccine ethics subcommittee at UNAIDS, tried mightily to craft a consensus statement anyway. "When people disagree or are morally ambivalent, a solution is to turn it to a procedural solution, that is to say, who should decide this question," explained Macklin. "We in this room are unable to decide it. Again, I don't know whether a smaller subset locked in this room for 10 days will be able to decide it. But we do know that there is a procedural solution, and it's precisely the one you just named: this should be left to each country in which the trials are to be conducted. That answers the question of who should decide, but it doesn't come close to answering what should be available."

At that point, Christine Grady, a professional AIDS vaccine ethicist of sorts, spoke up. "How is that different?" asked Grady. Her point was that some people in the room did not think each country should be allowed to decide this question. Rather, they insisted that there be one standard for the world.

UNAIDS issued a press release on June 29 that further confused the issue. "Is there an ethical obligation to provide expensive retroviral drugs which are not available to the rest of the population or that may not be available once the trial ends?" rhetorically asked the release. "The consensus reached was that this decision should be left to the participating country to decide."[39] The August issue of *Nature Medicine* compounded this error with an article headlined "Hard-Won Consensus on AIDS Vaccine Trial Guidelines."[40]

On September 2, Mary Lou Clements-Mann boarded Swiss Air Flight 111 to attend a small UNAIDS meeting in Geneva on AIDS vaccines. With her was her husband, Jonathan Mann, who had business of his own at WHO headquarters. Shortly after takeoff, the plane crashed into the frigid waters off Nova Scotia. No one survived. Two of the strongest voices in the debates about the ethics of AIDS vaccine trials were silenced.

• • •

TOMORROW NEVER DIES, as the James Bond movie put it—and neither, it seems, do ethical debates. Should someone have stopped Zagury from conducting his Zairian trials? Should volunteers for a live, attenuated vaccine be allowed to test it? Is it ethically wrong to test a vaccine in one country that another country has rejected? Should volunteers who become infected receive the best-proven treatment?

No organization can solve these ethical dilemmas. UNAIDS certainly can offer guidance, but countries are free to take it or leave it. The NIH can steer with its fat checkbook and unparalleled intellectual and technical resources, but Zagury conducted his controversial trials with that institution's help. IAVI can set examples of how best to proceed, the AVAC can issue critical manifestos, and the media can watchdog the enterprise, but none of these forces solve the problems, either.

A March of Dimes–like organization could help calm these waters some, if it took a page from the NIH branch that runs the Human Genome Project and established a formal program to address ethical issues. The National Human Genome Research Institute sponsors the Ethical, Legal and Social Implications Program, which, in turn, funds research into these issues, arrives at policy guidelines, and sponsors workshops.[41] Although the ethical dilemmas surrounding genomics remain every bit as acute as those facing AIDS vaccine researchers, the overt attempt to anticipate issues, organize discussions, and arrive at policies helps reduce some of the ambiguity. The program also furthers the sense that researchers are doing everything possible to address these uncomfortable questions.

Ethical dilemmas are messy—and will remain so. But the lesson of the AIDS vaccine search, again and again, is that as HIV continues to infect ever more people around the world, the ethical questions will continue to become ever more complex. The only solution, ultimately, is to find a working AIDS vaccine, which surely will unleash another rash of ethical dilemmas of its own about access and distribution. Yet that is a problem that the world, at this bleak point, can only hope it will one day face.

Epilogue

The world learned today that its hope of finding an effective weapon against paralytic polio had been realized.

—WILLIAM LAWRENCE, *New York Times,* April 13, 1955

For several years, I have kept this sentence from the *New York Times* pinned to the bulletin board beside my desk. I once was certain that I would write a similar sentence about AIDS. As a Bob Dylan refrain goes, "I was so much older then, I'm younger than that now."

There are no guarantees when it comes to scientific research, and making predictions, as Robert Gallo noted at the famous Heckler press conference, is a mug's game. A research community, however, can improve its odds of success if it does everything possible to remove obstacles, organizes itself, logically uses animal models, fills research gaps, takes appropriate risks, and fully exploits promising leads. The March of Dimes directed just such an effort in the search for a polio vaccine. Despite many concerned, earnest scientists attacking the problem, that has not yet happened with AIDS vaccine research.

Vaccines are not drugs. After clean water, vaccines are the most powerful way to improve public health. They also can completely rid a country, even the entire planet, of a disease—which means nothing need be spent on it ever again. Governments realize this, but even the wealthiest ones cannot organize an efficient vaccine R&D program. Governments simply do not have the freedom to distribute money quickly. They tend to punish bold leadership too, and remain vulnerable to political shenanigans like the MicroGeneSys affair. And the U.S. government, which in 2000 spent $200 million on AIDS vaccine research—far more than the rest of the world combined—has the added handicap of having created a scientific culture that looks askance at targeted research programs.

Striking the balance between basic and applied science, between reductionism and empiricism, between investigator-initiated and targeted research, requires a delicate touch. Basic research has, without question, led to an untold number of serendipitous discoveries that have solved staggering mysteries about biology. But when a car is making a ticking sound, you may not need to take out the engine and ratchet off every bolt, crack every weld, and spread every part on the garage floor—you may just need to adjust the valves.

There is a great danger that directing science will lead to shoddy, unimaginative approaches and embarrassing, even scandalous excesses. AIDS vaccine researchers, however, have put too much faith in serendipity and not enough in organization.

Industry can, and sometimes does, mount aggressive, targeted vaccine research programs. But companies exist to make money, not to make the world a better place, and the AIDS vaccine equation, especially given the scientific unknowns and the liability issues, has yet to attract much interest from major pharmaceutical houses.

Philanthropic organizations are free of the constraints of both government and industry, with no need to answer to constituents or stockholders. Bill Gates Jr., the Microsoft co-founder, indeed amply demonstrated this truth in 1999 when he and his wife, Melinda, donated $100 million to various vaccine research efforts. A whopping $25 million went to the International AIDS Vaccine Initiative, with another $50 million going toward malaria vaccine research and $25 million to a group pushing for development of a better vaccine against tuberculosis. The Bill and Melinda Gates Foundation (formerly the William H. Gates Foundation) pledged an additional $750 million over five years to a new fund that will purchase existing vaccines for children in poor countries.

The Gates Foundation, now the largest philanthropy in the United States, had assets in 1999 of $17.1 billion—5% of which it must spend each year or it will face substantial tax penalties. Single-handedly, this foundation could bankroll the type of program I am advocating. Thanks to the stock market boom of the 1990s, many other philanthropies that fund scientific research now have much fatter endowments than they ever imagined and similarly could support such an enterprise; in 1999, for example, assets at the U.K.'s Wellcome Trust (then the *world*'s largest charity) totaled $19 billion, the Packard Foundation had grown to $13.5 billion, and the Howard Hughes Medical Institute enjoyed an endowment of $12 billion.

I spoke with top officials at each of these foundations in 1999 and asked them what they thought of targeted research programs. Unlike the other giant foundations, Bill Gates Sr., who coruns the Gates philanthropy for his son and daughter-in-law, explicitly had little interest in funding basic research. And while the Gates Foundation's giving pattern makes clear that it has no bias against targeted research programs, Gates Sr. also stressed that the philanthropy has no intention of launching a scientific program of any sort. "We're substantially reactive," Gates explained to me, sitting in a conference room of his law firm that had a commanding view of Seattle. "We're favorably impressed by the possibilities that exist for doing things without ourselves becoming a monstrous bureaucracy."

Wellcome, Howard Hughes, and Packard, which have pronounced differences, do launch scientific programs, but they share a philosophy that they can do the most good by providing an alternative source of funding for basic research. As Hughes's Max Cowan explained, "Our sense has been that most of the targeted programs have not been very successful."

Money for a March of Dollars need not come from the mega foundations. Several smaller philanthropies could band together with individual donors who have shown an interest in the world's health, such as CNN founder Ted Turner, investment banker George Soros, Microsoft cofounder Paul Allen, and C-SPAN cofounder John Evans. Or a wealthy individual could decide, single-handedly, to devote a billion dollars to the effort, just as Turner did with the United Nations in 1997. The public might contribute substantial amounts too, especially if the March of Dollars stoked the sense of urgency as successfully as the March of Dimes did.

This prescription for the AIDS vaccine search, which builds off the ideas of many others, may well not lead to the day where I write a sentence that says the world has realized its hopes and found an effective vaccine against HIV. But the world at least could declare that it did everything in its collective power to develop an AIDS vaccine as quickly as possible, which is not something it can now say. And if the effort succeeded, it would provide compelling evidence that, as Jonas Salk suggested, there is a better way for scientists to research and develop vaccines. The impact of that message would reach far beyond AIDS. It would apply to most every other infectious disease that destroys human lives but has remained invulnerable to a vaccine, as well as to future plagues that surely will appear from nowhere, the way polio and AIDS once did.

POSTSCRIPT

Breaking the Silence

On a clear and cheery rain-washed morning in Kampala, Uganda, in March 2000, Rose Busingye, an AIDS social worker and nurse, led four visiting Italian AIDS researchers on a dark and depressing tour that began at an orphanage. "They have no one," Busingye explained as she walked through the well-kept house that served as the new home for a dozen children whose parents had recently died from AIDS. She introduced a boy whose mother, a prostitute, had hanged herself when she learned that her blood test for HIV was positive. A girl walked by whom Busingye had found wrapped in her dead mother's arms. Busingye pointed to siblings the orphanage had found eating from a garbage pile.

The tour proceeded to Namuongo, a nearby village, where Meeting Point, the group Busingye works with, provides care and support for several people near death from AIDS. The researchers walked through fields of sugarcane, jackfruit, and yams, arriving at a cinder-block home with a corrugated tin roof. Inside, a 36-year-old woman, who had tuberculosis and the disorientation typical of AIDS dementia, was bedridden in a dark, two-room space she shared with her brother, his wife, and four children. She had been in this bed, a thin mattress on the floor, for three weeks now. "I'm not OK," said the woman, who had a large red Bible by her side.

The last stop on the three-hour tour was Kireka, a poor village that abuts a rock quarry. Packs of children roamed the small slum, which has rows of huts made from mud, sticks, and straw. Busingye said 40 of these children had lost both parents—because of either AIDS or the civil wars here and in neighboring Rwanda—and now were being cared for by extended family. One orphan with withered limbs, who himself had late-stage AIDS and was wearing nothing but a long-sleeved pink turtleneck, took the hand of an Italian AIDS researcher and walked with her. Meeting Point helps 120 AIDS patients in Kireka, including a 32-year-old mother of two who lay on the ground in the village's marketplace selling vegetables. Her husband had died from AIDS in December 1998. "He had another woman," the vegetable seller explained. Now she had AIDS, as did her 15-year-old son.

Mario Clerici, the University of Milan immunologist and AIDS vaccine researcher, pursed his lips and shook his head. "All AIDS researchers should come here to see what AIDS is," Clerici said. "Most AIDS researchers have barely seen AIDS patients. Here, you see how horrible it is." Clerici held up a hand and rubbed his fingers together. "Come here and you can touch it."

The start of the new millenium marked the point in time when many European and North American AIDS researchers—and indeed the world at large—finally began to recognize just how horrible the epidemic truly is in Uganda and other countries in sub-Saharan Africa. The surge of interest mirrored the surge in HIV itself. Only one African country had an HIV prevalence rate in adults higher than 2% in 1982. By 1998, UNAIDS estimated that HIV had infected more than 7% of the adults living in 21 African countries. An unusually impassioned World Bank report, "Intensifying Action Against HIV/AIDS in Africa," noted that AIDS soon would kill more people than the 14th-century plague that devastated Europe. "The most disturbing long-term feature of the HIV/AIDS epidemic is its impact on life expectancy, making HIV an unprecedented catastrophe in the world's history," the report concluded. "HIV/AIDS has already reversed 30 years of hard-won social progress in some countries. Now is the time for Africa—and the world—to fight back."[1]

On many fronts, the world did begin to fight back more aggressively. Attention to Africa's AIDS problems gained momentum in May 1999 when the South African government, in collaboration with the electrical company ESKOM, pledged $8 million to support the newly formed South African AIDS Vaccine Initiative to fund R&D. Later that month, Bristol-Myers Squibb, maker of two anti-HIV drugs, launched "Secure the Future," a $100-million program to help five countries in southern Africa combat AIDS. A few weeks later, the Bill and Melinda Gates Foundation made its $25 million donation to the International AIDS Vaccine Initiative (IAVI). In October, the European Union announced that it had formed a $9.2 million initiative, EuroVac, to design vaccines based on strains of the virus circulating both in developed and developing countries. The international media also began paying attention: In November and December 1999, the *Village Voice*, a free weekly paper based in New York City, ran a remarkable eight-part series by reporter Mark Schoofs on AIDS in Africa, which later won the Pulitzer Prize.

The U.S. budget in fiscal year 2000 included a $100 million program, the Leadership and Investment in Fighting an Epidemic (LIFE) initiative, that hoped to slow the transmission of HIV in sub-Saharan Africa, improve treatment for infected people, and provide care for AIDS orphans. In a January 10, 2000, speech to the United Nations Security Council session on AIDS in Africa, Vice President Al Gore declared that "we are putting the AIDS crisis at the top of the world's security agenda," and announced that the Clinton administration would ask for an additional $150 million in the 2001 budget to fight AIDS. In April, World Bank president James Wolfensohn declared that the bank would place "no limit" on the amount it spends on AIDS. "I have said to our African clients, if you have programs, we'll fund them," Wolfensohn declared.[2] On May 10, Clinton issued an executive order that reaffirmed the government's commitment to relaxing U.S. laws regarding intellectual property rights connected to anti-HIV drugs in sub-Saharan Africa, effectively allowing countries to manufacture or import drugs without fear of violating patent law. Five pharmaceutical companies the next day announced they would slash prices of their anti-HIV drugs sold in the region.[3] On top of these commitments, several groups in 2000 donated money to help Africans take advantage of arguably the most meaningful advance yet made by AIDS researchers: the finding that relatively cheap, short courses of anti-HIV drugs can slow the transmission of HIV from mother to child.[4]

As sorely as Africa needed this assistance, many recognized that it would only have a modest impact on the continent's AIDS crisis in the immediate future. "Every war needs a war chest, but that provided by the international community is woefully empty," the World Bank's Wolfensohn told the UN Security Council in January. Wolfensohn pointed out that Africa receives a mere $160 million a year for HIV prevention; the bank estimated that Africa needs between $1 billion and $2.3 billion.[5]

All told, the Joint United Nations Programme on HIV/AIDS (UNAIDS) estimated in June 2000 that sub-Saharan Africa had 15 million deaths from AIDS, 12 million AIDS orphans, and 24.5 million people living with the virus—a whopping 70% of the global AIDS total, even though this region had a mere 10% of the world's population. "Despite millennia of epidemics, war and famine," stated UNAIDS, "never before in history have death rates of this magnitude been seen among young adults of both sexes and from all walks of life." A vaccine, the report concluded, represented "our best hope" of stemming the epidemic, "but it will take time and a concerted international effort before we have one."[6]

The World Bank, UNAIDS, the National Institutes of Health, the European Union, and several other organizations, both public and private, made dramatic moves in 2000 that specifically bolstered the search for an AIDS vaccine. But these efforts did not coalesce into a concerted international campaign to find a vaccine as quickly as possible. Rather, these groups filled niches willy-nilly. From a distance, they resembled the familiar sight of young children on the same soccer team who, instead of working together to score, form a beehive around the ball, with everyone kicking madly, hoping for contact and a lucky roll. Sometimes, of course, the ball, improbable as it seems, shoots out and makes it past the goalie. But these individual attempts, earnest as they may be, lose every time to the team that strategically spreads itself across the field, passes the ball, and then, when it sees an opportunity, strikes on goal with its best shot.

• • •

ONE WEEK AFTER MARIO CLERICI toured hard-hit AIDS communities in Kampala, Uganda, Seth Berkley visited Hlabisa, South Africa, which would soon begin a trial of a vaccine sponsored by IAVI that Berkley headed. During the four-hour drive from the bustling city of Durban to Hlabisa (pronounced SHLA-BEE-SA), a village in KwaZulu-Natal Province, contrasts abounded. A man holding a cell phone directed traffic on a lightly trafficked dirt road. A pack of impalas stood next to a yellow "impala crossing" sign. A chain-link fence topped with menacing razor wire surrounded the Hlabisa Research Clinic, Berkley's rural destination. The conversation in the car fit perfectly. Berkley told of his recent trip to the World Economic Forum, held in the ski town of Davos, Switzerland. "Clinton gave a speech on AIDS vaccines," recalled Berkley. "I just walked up and started talking to him. Where else can you do that?"

The most startling contrast of all was that Hlabisa, located in a lush region of gentle hills dotted with round mud huts on the edge of the Hluhluwe Game Reserve, did not look like an epicenter of an AIDS epidemic. But this poor rural area had one of the worst outbreaks in the world. The figures were staggering: Almost 40% of the pregnant women—a so-called sentinel population widely used to track the epidemic—in Hlabisa had become infected by HIV. A decade earlier, almost none tested positive. In the sad calculus of AIDS, this made Hlabisa an ideal place to test HIV vaccines.

By March 2000, the nonprofit IAVI had launched innovative "product development teams" in South Africa, Kenya, and Uganda. Here, IAVI teamed leading South African AIDS researchers with a North Carolina biotech company; together they hoped to tailor an AIDS vaccine by 2001 for trials in Hlabisa and other parts of the country. As part of the project, to which IAVI committed $4.5 million over three years, researchers worked closely with participating communities.

On this day, eight "community educators," whose job was to lay the groundwork needed to conduct an ethical trial, packed into a trailer that served as the headquarters of the Hlabisa Research Clinic to meet this boyish and gangly physician from New York. These community educators, college-age men and women from the area, traveled from hut to hut to teach people about vaccines and AIDS, distribute condoms, and answer questions. Berkley began by asking the educators, "What's the community saying?"

"They feel positive here," said a university student, who wore a stylish beret. "They think they're going to help and stop the epidemic."

"Do they ask any questions that stump you?" Berkley queried.

The educators squirmed a bit. Berkley asked the question again.

"The one is, 'When are we going to have a vaccine that works?' " a woman answered.

"That's a good question," Berkley replied. The educators laughed. Berkley, who well understood the scientific and ethical challenges ahead, did not offer an answer.

Halfway into the meeting, two nurses who work at the local hospital walked into the trailer, taking seats that flanked Berkley. "If we had a vaccine that worked and was safe, would people take it?" Berkley asked the nurses, Catholic sisters who wore military style shirts with epaulets.

"People would flock to it," said one sister.

"We're all praying for it," added the second sister. "Our children are dying. Our friends are dying. People 18 years old are lying in hospital beds that in the old days we filled with grannies who had tuberculosis and cardiac failure. The vaccine would be wonderful. [The situation] is very difficult for young people. This is the time when they should be falling in love, and they say, 'Now that we've come to that stage, you want us to do what you did not do.'"

Berkley thanked everyone for coming to meet with him and for helping to organize an AIDS vaccine trial in this remote corner of the earth. "This is not just a problem for KwaZulu-Natal," said Berkley. "This is not just a problem in Hlabisa or South Africa. It's a problem for the world."

• • •

FROM JULY 9 TO14, Durban hosted the 13th international AIDS conference, which drove home the message as never before that AIDS is a problem of the world. The gathering, held for the first time in Africa—or in any developing region—had the slogan BREAK THE SILENCE, a reference both to the inequities separating rich and poor and to the indifference with which the world had responded to this crisis. By the end of the meeting, the silence had not simply been broken. It had been shattered.

At the opening session of the meeting, an HIV-infected judge who sits on the country's highest court set the tone. "Amidst the poverty of Africa, I stand here before you because I am able to purchase health," said high court justice Edwin Cameron, who said he spends $400 a month on anti-HIV treatments. "I am able to purchase vigor and joy. I am here because I can pay for life itself. To me this seems a shocking and monstrous inequity." Cameron, whose talk was interrupted repeatedly by rousing ovations, concluded with this harsh verdict: "No more than Germans in the Nazi era, no more than white South Africans during apartheid, can we at this conference today say that we bear no responsibility for more than 30 million people in resource-poor countries who face death from AIDS unless medical care and treatment is made accessible to them."

Neal Nathanson, head of the Office of AIDS Research at the NIH, told me that Cameron's talk was "truly transforming," and he predicted that "it's going to be a landmark in the history of the AIDS epidemic."

Many of the 12,000 people from 180 countries who attended the meeting felt that historians will look back on the gathering itself as a turning point in the way the haves of the world view their responsibilities to the have-nots—at least as far as AIDS is concerned. "Something amazing and something profound and something unforgettable has happened here, unless I have a totally biased sample," said the University of Natal's Hoosen "Jerry" Coovadia, a Durban pediatrician who chaired the conference. UNAIDS head Peter Piot concurred. "Never before has the world's attention been so focused—finally—on the problems of AIDS in Africa," Piot said.

Although basic science and clinical research typically dominate at this event, which happens every two years, by far the most riveting sessions at this meeting revolved around the thicket of issues that prevent poor people in sub-Saharan Africa from receiving the 14 anti-HIV treatments then widely used in wealthy countries. Similarly, the call for stepped-up prevention efforts, including a more vigorous search for a vaccine, reached an unprecedented decibel level.

The meeting prompted donors and Big Pharma—which had begun to look as cornered by shame as Big Tobacco did in the 1990s—to loose a flood of promises. First Boehringer Ingelheim, the manufacturer of the drug nevirapine, announced that for the next five years, it would provide the drug *for free* to pregnant, HIV-infected women in developing countries, allowing them to take advantage of the 1999 finding that one dose to mother and baby could dramatically cut HIV transmission. Merck, the maker of two anti-HIV drugs, and the Bill and Melinda Gates Foundation pledged to split the costs of a $100 million program to help Botswana launch a new comprehensive HIV/AIDS prevention and treatment program. The Gates Foundation separately announced a $15 million gift to the Pediatric AIDS Foundation to bolster its program to provide nevirapine to pregnant, infected women and another $25 million to the Contraceptive Research and Development Program to further its work on vaginal microbicides. The World Bank announced that it had devoted $500 million to a new HIV/AIDS program that aims to help African countries that have national AIDS strategies. The European Union also revealed that it would launch a major new initiative for HIV/AIDS, malaria, and tuberculosis.

On the vaccine front, IAVI flexed all of its muscles, issuing two new reports, and announcing that an IAVI-sponsored vaccine—the first ever "designed for Africa"—had received approval for testing in the United Kingdom.[7] One of the new IAVI reports highlighted the decade or more that poor countries typically wait to receive new vaccines developed in wealthy countries. To avoid this "colossal public health failure," IAVI laid out a blueprint that strove to ensure simultaneous worldwide access to any working AIDS vaccine. At a press conference, IAVI aired a videotape of its highest profile backer, Bill Gates Jr., who decried this "trickle down" tradition. "The message to policy makers and pharmaceutical companies from IAVI's new blueprint is clear and unequivocal," said Gates. "If we do not take action now, the quest for an AIDS vaccine could eventually yield a scientific triumph and a humanitarian failure."

IAVI's second new report called for beefing up the vaccine search by introducing 25 new products into human trials, and comparing them in fast-tracked, head-to-head evaluations. By 2007, IAVI said it could have six vaccines ready for efficacy tests. IAVI estimated that this ambitious agenda would require $900 million to $1.1 billion of new funds.

Outside of IAVI, the National Institute of Allergy and Infectious Diseases announced that it had reorganized its AIDS vaccine programs to create product development teams, similar to IAVI's. Researchers from all over Africa separately released "The Nairobi Declaration: An African Appeal for an HIV Vaccine" that urged the world to mount a "coordinated effort."

The new battalions and decrees in the AIDS battle surely would help, but many at the conference stressed that much more would be needed. In a session entitled "HIV Non-Intervention: A Costly Option," health economist Jeffrey Sachs blasted the wealthier countries for their "shocking disregard" for the pandemic and for doing so little so late. "When the World Bank comes forward and says $500 million . . . that's all fine and good, but the real question is where has the World Bank been for the last 15 years?" said Sachs, director of Harvard University's Center for International Development. The packed room cheered.

Sachs estimated that sub-Saharan Africa alone needed $10 billion a year in donor support to effectively respond to HIV/AIDS, malaria, tuberculosis, and the continent's other major diseases. He noted that the 1 billion people living in wealthy countries could raise the $10 billion by each person donating $10. "It is a movie ticket and a box of popcorn," said Sachs. "The extent of underfunding of AIDS control is perhaps the most shocking feature of our generation. This will be what people ponder for decades to come. How could the world have stood by for the first 20 years of this pandemic, letting it reach 35 to 40 million people before any real funding started?"

Former South African President Nelson Mandela closed the conference with an impassioned speech that brought down the house. Mandela, after entering the conference hall to thunderous applause, with thousands of South Africans singing his name, whistling, and chanting, attempted to salve the wounds created by his protégé, President Thabo Mbeki. For months, Mbeki had roiled scientists the world over by flirting with the idea that AIDS was not caused by HIV. Mandela began by praising Mbeki, who disappointed many in the audience earlier in the week when he opened the conference with a speech that seemed to blame AIDS on poverty rather than HIV. Mandela also had kind words for his country's scientists. But he stressed that the poor, "if anybody cared to ask their opinions, wish that the dispute about the primacy of politics or science be put on the back burner and that we proceed to address the needs and concerns of those suffering and dying."

To the crowd's delight, Mandela, in a pointed departure from Mbeki, urged South Africa to adopt "large-scale actions" to thwart transmission of the virus from mother to child. Mandela also explicitly urged scientists to "continue the international effort of search for appropriate vaccines."

Echoing Justice Cameron's speech, Mandela implored the audience to attack the problems before them with an intensified sense of urgency. "In the face of the grave threat posed by HIV/AIDS, we have to rise above our differences and combine our efforts to save our people," Mandela said. "History will judge us harshly if we fail to do so now, and right now."

In 1990, shortly after I began researching this book, I walked in the light rain with hundreds of AIDS activists as they staged a demonstration at the NIH campus in Bethesda, Maryland. Dubbed "Storm the NIH," the rally shut down the institutes for the day, as the activists occupied laboratory buildings, blocked the campus roads, staged macabre street theater with mock coffins and skeleton costumes—and made a lot of noise. They honked air horns. They blew whistles. And most strikingly, they joined their voices into one hellacious choir that loudly sang angry chants, punctuated by clapping. "We die! They do nothing!" went one. "People with AIDS under attack!" another favorite complained. "What do we do? Act Up! Fight back!"

Powerful as this demonstration was, one refrain resonated in my head, because it struck me as so overinflated: "The whole world is watching!"

In 1990, the whole world decidedly was not watching. That had dramatically changed by the Durban meeting in 2000. But the question still remained: when would the whole world organize itself, with all of its brains and brawn, to wage a coordinated assault against HIV, which had by 2000 threatened to kill nearly 1% of the human population?

Notes

Many of the quotations in this book come from interviews I conducted or meetings I attended from 1989 to the present. I have made more than 200 tape recordings of these interviews and meetings, although I frequently have relied on handwritten notes alone. At this point, these notes fill more than 150 6- by 9-inch stenographer notebooks; I have numbered each one and dated each interview, delineating both questions and answers. Since the advent of the Internet, I also have saved hundreds of e-mail interviews.

I liberally used the Freedom of Information Act to obtain many documents referred to in this book. (In response to my Freedom of Information Act request, the Ronald Reagan Library processed and put into open files more than 10,000 pages of AIDS-related presidential records.) I also received dozens of verbatim transcripts from the National Institutes of Health, the U.S. Department of Defense, the National Academy of Sciences, and Food and Drug Administration meetings, as well as transcripts from NIH and White House press conferences.

For information about companies, I reviewed records at the Securities and Exchange Commission, Washington, D.C., which they now make available online via the Edgar database. I obtained lobbying reports at the U.S. Congress, Office of the Clerk, Washington, D.C. For information about campaign contributions, I did computer searches at the Federal Election Commission, Washington, D.C. To review patent information, I relied on the library and the database (now online) at the U.S. Patent and Trademark Office, Arlington, Virginia. I found some historical documents relating to the polio vaccine trials in State Department files at the National Archives, Washington, D.C.; the National Library of Medicine, NIH, Bethesda, Maryland; and the Jonas Salk Papers, Mandeville Special Collections Library, University of California, San Diego. Finally, many companies, government, and academic institutions routinely sent me their press releases, which I have kept in my own files.

The NIH's web page (www.nih.gov) provided a tremendous resource that I tapped repeatedly. To learn about specific grants, I frequently searched the CRISP system and the "Awards by State" (which have the dollar amounts of grants that the NIH, in its wisdom, no longer makes available via the CRISP system). Additionally, the MEDLINE database catalogs and abstracts articles from 3,800 biomedical journals. The AIDSLINE database also offers useful abstracts from AIDS-related meetings.

CHAPTER 1 FAST PREDICTIONS

1. The reporter, Lester Kinsolving, a well-known gadfly at the White House, then was editor of his own paper, the *Washington Weekly*.

2. Margaret Heckler's prepared text explicity acknowledged the advances made by the Pasteur Institute researchers. It read: "They have previously identified a virus which they have linked to AIDS patients, and within the next few weeks we will know with certainty whether that virus is the same one identified through the NCI's work. We believe it will prove to be the same."

3. By 1999, there still was not a single documented case of HIV being transmitted through saliva.

4. The Ronald Reagan Library, in Simi Valley, California, has copies of these three press briefings on file (October 15, 1982; June 13, 1983; and December 11, 1984).

CHAPTER 2 THE NEXT BREAKTHROUGH

1. Gallo has described his sister's death in many press interviews, though the most moving account is in his own book, *Virus Hunting: AIDS, Cancer, & the Human Retrovirus—A Story of Scientific Discovery* (Basic Books, 1991), pp. 16–18.

2. In Lawrence K. Altman's *Who Goes First? The Story of Self-Experimentation in Medicine* (University of California Press, 1998), the *New York Times* reporter explains how Salk, through a public relations officer, told Altman in 1970 that he had not experimented on himself with his polio vaccine. In 1976, Altman sat across from Salk at a symposium on vaccine development at the NIH and posed the question again. "Salk then insisted that he had actually taken it," writes Altman in a footnote on pp. 358–359. "When I asked why he told [the PR man] a different version, Salk said it was because he did not know why the question was being asked."

3. "Recovering the etiologic agent of a disease immediately conjures up dreams of developing a vaccine to prevent the infection," wrote polio researcher Dorothy M. Hortsmann in "The Poliomyelitis Story: A Scientific Hegira," *Yale Journal of Biology and Medicine*, 58:79–90 (1985). "This was as true in 1908 when Landsteinter reported the isolation of poliovirus as it is today, when the identification of HTLV3 [the old name of HIV] as the probable cause of AIDS burst on the horizon. In 1911 Flexner, who had been hard at work on poliovirus for a year and a half, was reported in the *New York Times* to have said that within six months a

specific remedy would be announced, for the way to prevent the disease had already been discovered in his laboratory." (p. 84)

4. For details, see *Four Billion Dimes* by Victor Cohn, a monograph published by the Minneapolis Star and Tribune Company in Minneapolis, Minnesota. This 134-page monograph, published circa 1983 and distributed at the 25th anniversary of the March of Dimes Birth Defects Foundation (the organization changed its name in 1979), contains a series of undated articles that Cohn wrote in April 1955 while a science reporter for the *Minneapolis Morning Tribune*.

5. See chap. 2 of *Breakthrough: The Saga of Jonas Salk* by Richard Carter (Trident Press, 1965).

6. See polio researcher John Paul's classic, *A History of Poliomyelitis* (Yale University Press, 1971), p. 261.

7. Lee Salk, a psychologist and author who has focused on family interactions, wrote about his father in *My Father, My Son: Intimate Relationships* (G.P. Putnam's Sons, 1982), p. 10.

8. *Epidemic and Peace, 1918*, by Alfred W. Crosby Jr. (Greenwood Press, 1976), pp. 205–207.

9. From a December 21, 1991, interview with Salk.

10. In *Patenting the Sun: Polio and the Salk Vaccine* (Morrow, 1990), Jane S. Smith quotes 1953 congressional testimony from Dr. Victor Haas, the founding director of the National Microbiological Institute of the National Institutes of Health: "We have felt for many years that the National Foundation for Infantile Paralysis supports research on such a scale that it would not be wise for us to direct our resources away from other important fields which are not so well covered to this one which is." Haas noted that in 1953, the NIH spent $72,000 on polio research, while the National Foundation spent $2 million.

11. See *Breakthrough*, p. 57.

12. Ibid., pp. 95–96.

13. Salk has written about the link between toxoid vaccines and killed virus vaccines in several articles. This quote is from "The Logic of the Magic of Vaccination," *New England and Regional Allergy Proceedings*, 9(6):689–692 (1988), p. 689.

14. See note 4 for Cohn's articles.

15. *Breakthrough*, p. 140.

16. Both *Breakthrough* (p. 156) and *Patenting the Sun* (pp. 185 and 186) cite this item from "On Broadway," the syndicated column by gossip columnist Earl Wilson.

17. *Breakthrough*, p. 162.

18. The committee member was preeminent polio researcher John Paul, who wrote his reactions in *A History of Poliomyelitis* (Yale University Press, 1971). Paul also wrote that "it was immediately made apparent to the members that they were

to have a passive, not active, future participation in the plans. I believe that this meeting, more than any other single factor, was responsible for the belief and the accusation that the foundation had been secretive about its plans and had withheld information from its own scientific advisers. . . . At this point it might have been better for the Immunization Committee to resign, and it might also have saved the foundation some future embarrassment. But such an action would have been ascribed to pique, and this was not what the cause of poliomyelitis deserved." (pp. 423–425)

19. Smith made the mamboing elephants comparison in *Patenting the Sun,* p. 230. Smith was one of the children who participated in the field trial, which is the focus of her book.

20. In *Patenting the Sun* (p. 304), Tom Coleman, who handled press for the University of Pittsburgh during the field trial, is quoted explaining Salk's affection for Murrow. "Ed Murrow was the only person for whom Jonas had stars in his eyes," said Coleman.

21. As Smith explains in *Patenting the Sun* (pp. 220 and 338), Salk never mentioned that National Foundation attorneys had already decided that the killed-virus technology was not patentable because it was not a new invention.

22. A three-part, 65-page paper published in the *American Journal of Hygiene* eight years after the fact ("The Cutter Incident: Poliomyelitis Following Formaldehyde-Inactivated Poliovirus Vaccination in the United States During the Spring of 1955," 78:16–81, 1963) offers the most comprehensive analysis of polio caused by bad lots of the Cutter vaccine. Written by Neal Nathanson and Alexander Langmuir, the study causally links fewer cases to Cutter than commonly is reported.

23. On May 19, 1955, First Secretary of Legation Donald P. Downs in Budapest sent the U.S. Department of State a memo (unclassified "foreign service dispatch," National Archives, Washington, D.C., 911.84/5-1955) about what he said was the first mention in "Budapest or provincial press" of the Salk vaccine. Downs provided a translation of the article, which appeared in the May 4 *Veszpremmegyei Nepujsag* (County Veszprem People's News). Given that the Salk vaccine had just proven effective, the article had a hilarious lead, first noting that "French professor [Pierre] Lepine" had discovered "a serum against polio in Paris, but stated that the use of it would be very dangerous in the present state of the test." The second paragraph then describes the Salk vaccine. "Almost simultaneously an American physician, Dr. Salk, disclosed that he had discovered the bacteria [*sic*] of polio and prepared a vaccine for inoculations. A great advertising campaign started immediately in America. . . . We do not blame Dr. Salk, nor doubt his discovery, but the wary were right in saying that a long series of tests has to take place before this new methods [*sic*] can be used."

24. For detailed discussions of the filtration issue, see Jonas Salk, "Poliomyelitis Vaccine in the Fall of 1955," *American Journal of Public Health,* 46(1):1–14 (1956); and the Interim Report of United States Public Health Service

Technical Committee, *JAMA*, December 10, 1955. The Public Health Service's "Report on the Poliomyelitis Vaccine Produced by the Cutter Laboratories" was issued August 25, 1955, and is on file at the National Library of Medicine's History of Medicine Division ("Reports relating to the Salk polio vaccine").

25. The cornerstone of Sabin's argument was what immunologists call "herd immunity." When part of a community is vaccinated, the number of people carrying the virus should decline. Fewer carriers equals fewer places for the virus to thrive, which equals less virus and less polio. Sabin believed, as he explained in a 1959 National Foundation meeting, that people receiving the Salk vaccine would still be carriers, whereas those who received the live vaccine would not. The essence of the argument was that only the live-virus vaccine could replicate in the gut, stimulating antibodies there and creating "gut immunity." On the other hand, *wild* virus could multiply inside the guts of people who received the killed vaccine; these vaccinees did not become paralyzed because other antibodies prevented the virus from making it to the central nervous system. Sabin also stressed that the live poliovirus vaccine passed through the vaccinee's feces, and once released in the environment, "gives some immunization for free" when the unvaccinated met up with it. Another supposed benefit of the live vaccine was that it could be swallowed rather than injected. Darrell Salk, Jonas's son, later would challenge these assumptions in his 45-page, three-part paper, "Eradication of Poliomyelitis in the United States," *Review of Infectious Diseases*, 2(2):228–273 (March–April 1980); and in his "Herd Effect and Virus Eradication with Use of Killed Poliovirus Vaccine," *Developments in Biological Standardization*, 47:247–255 (1981).

26. Salk recalled this run-in with Sabin during a January 2, 1991, interview.

27. Jonas Salk, "Mechanisms of Immunity in Virus Infections," *Recent Progress in Microbiology,* Eighth International Congress for Microbiology (University of Toronto Press, 1963), p. 398.

28. In a 1993 paper published in *Proceedings of the National Academy of Sciences,* J. W. Drake estimated that poliovirus mutates at the rate of about one mutation per genome per replication cycle. See "Rates of Spontaneous Mutation Among RNA Viruses," 90(9):4171–4175 (1993). John Coffin, in a paper that appeared in the January 27, 1995, issue of *Science* ("HIV Population Dynamics in Vivo: Implications for Genetic Variation, Pathogenesis, and Therapy," 267:483–489), emphasized that HIV does not mutate at a faster rate than other retroviruses (about 10^{-5} to 10^{-4} per base per cycle). Since HIV has 9,000 bases, it has about 0.09 to 0.9 mutations per genome per cycle—a little lower rate than polio. Coffin stressed that HIV's fast replication rate, however, combined with the 10-year average lapse between infection and disease (the "transmission cycle"), allowed many mutations to accumulate. "This extent of replication per transmission cycle is probably without equal among viral (or possibly even bacterial) infections," wrote Coffin.

29. During the hearing, entitled "Oversight of Biomedical and Behavioral

Research in the United States," Salk proposed a mini–Manhattan Project be organized that would be dedicated to "solving the influenza problem" with vaccines. He also called for reintroducing a killed polio vaccine to eradicate that disease. He suggested that a "council of elder advisers on health researchers and practices" be established to evaluate public health proposals like these. "The members of such an advisory council presumably would be experienced and old enough to know how to function through the influence of wisdom rather than through the exercise of power," said Salk.

CHAPTER 3 EMPIRICISTS VERSUS REDUCTIONISTS

1. Before being tapped for the MacArthur Fellowship in 1985, Joan Abrahamson earned a B.A. at Yale (she was in the first class of women ever enrolled), a master's in education at Stanford, and a Ph.D. from the Learning Environment program at Harvard. A painter and a fifth-generation San Franciscan, she also was the guiding force behind the establishment of San Francisco's Fort Mason Center, a collection of old Army warehouses converted into art studios, galleries, theaters, and restaurants. She went on to earn a law degree from the University of California at Berkeley's Boalt Hall. After law school, Abrahamson did human rights work for the United Nations in Geneva and then for UNESCO in Paris. Returning stateside, she won a slot as a White House fellow, where she worked for Vice President Walter Mondale and then for Vice President George Bush. When that ended, Bush hired her on as his assistant chief of staff. Abrahamson met Salk at a party in 1983. In 1985, the same year she won a $180,000 MacArthur Fellowship, Abrahamson founded the Jefferson Institute, a Los Angeles–based, public policy think tank of sorts that aims to solve societal problems. She was all of 34 years old.

2. "Salk played one more of his diplomatic roles by keeping me hidden in the men's lavatory until my adrenaline level returned to normal and the press went away, and then arranging a planned discussion with reporters in a manner much less likely to result in an emotional response from me," wrote Gallo in his autobiography, *Virus Hunting: AIDS, Cancer, & the Human Retrovirus—A Story of Scientific Discovery* (Basic Books, 1991), p. 213.

3. *Louis Pasteur: Free Lance of Science* by René J. Dubos (Little Brown and Co., 1950) covers Pasteur's history well, including a passage on pages 356 and 357 about the first killed vaccine. I am basing my comment about the mechanism of rabies virus on a conversation I had with Harvard's Bernard Fields, a leading authority on rabies and other viruses, shortly before he died in 1995.

4. Each chapter in *Vaccines*, the 1988 book (W. B. Saunders Company) edited by Stanley Plotkin and Edward Mortimer Jr., covers a different vaccine. The chapters were mostly written by the vaccinologists who developed the products, including one chapter coauthored by Jonas Salk. The Plotkin quote comes from a March 25, 1992, interview.

5. "Current Status and Strategies for Vaccines Against Diseases Induced by Human T-Cell Lymphotropic Retroviruses (HTLV-I, -II, -III)," *Cancer Research* (Supplement), 45:4694s–4699s (September 1985). Authored by Peter J. Fischinger, W. Gerard Robey, Hilary Koprowski, Robert C. Gallo, and Dani Bolognesi, the paper explains several ways a weakened HIV vaccine theoretically could induce cancer, including the insertion of its genes next to cancer genes.

6. Scientists can develop vaccines to thwart infections that people do not naturally defeat: the scientific literature records only a handful of documented cases of people who became infected with rabies virus and survived, yet a vaccine to prevent that disease has existed for more than a century. Still, finding such cases of natural protection surely increases the confidence of vaccine developers.

7. Within an hour of Genentech's initial public offering in 1980, the stock shot from $35 a share to $89, setting a Wall Street record.

8. Marilyn Chase, "Long Shots: The Race to Develop Vaccine Against AIDS Mobilizes Researchers, A Nobel Prize to the Winner?" *Wall Street Journal* (September 4, 1984), p. 1. Interestingly, Chase quotes William Haseltine suggesting that an AIDS vaccine consist of "the whole virus minus the genes inside of it," an intriguing idea that no one would pursue for several years.

9. In 1977, William Rutter's lab at the University of California, San Francisco, was part of a team that had first cloned the rat insulin gene. It was a critical passage in the stormy regatta to genetically engineer human insulin. Stephen S. Hall recounts the details in *Invisible Frontiers: The Race to Synthesize a Human Gene* (p. 142): "When Rutter and [UCSF's Howard] Goodman reported to the assembled press on May 23 that they had succeeded in cloning the rat insulin gene, it sent the right kind of ripples through the rest of the country. Accounts of the work appeared on the front pages of the *New York Times*, *Washington Post*, and *Los Angeles Times* the next day. The *San Francisco Chronicle* said the work `offers proof that the much-debated field of combining genes from unrelated organisms can in fact provide crucial insights into the ways by which chemistry governs heredity.' "

10. Rutter himself had to face down a Senate subcommittee that year after *Science* magazine revealed how UCSF researchers had conducted an insulin cloning experiment without first winning the approval of the NIH's Recombinant DNA Advisory Committee. As Hall writes in *Invisible Frontiers* (pp. 169–170), "Rutter conceded that 'the argument for not informing NIH was . . . related to the inflamed social and political climate that existed with respect to recombinant DNA technology at that time.' . . . Rutter would later refer to the [Senate] hearings as 'inquisitional' and a 'witch hunt.' "

11. As further evidence that the cloned hepatitis B vaccine had a powerful influence on AIDS vaccine developers in 1984, a paper recounting AIDS vaccines presentations given at a symposium held at the NIH on December 7, 1984, pro-

motes the hepatitis B model at some length. See Fischinger et al., "Current Status and Strategies for Vaccines Against Diseases Induced by Human T-Cell Lymphotropic Retroviruses (HTLV-I, -II, -III)."

12. On July 23, 1986, Merck's Recombivax HB became the first genetically engineered human vaccine to receive FDA approval. For a good overview of the development of the hepatitis B vaccine, see "Hepatitis B Vaccine" by Saul Krugman, in *Vaccines*, pp. 458–473.

13. Shiu-Lok Hu, S. G. Kosowski, and J. M. Dalrymple, "Expression of AIDS Virus Envelope Gene in Recombinant Vaccinia Viruses," *Nature*, 320:537–540 (April 10–16, 1986).

14. Vaccinia is not actually cowpox virus, but likely a mixture of different viruses from the orthopox family, which includes cowpox, smallpox, monkeypox, and camelpox. Joseph Esposito and J. C. Knight, "Orthopoxvirus DNA: A Comparison of Restriction Profiles and Maps," *Virology*, 143(1):230–251 (May 1985).

15. Dennis Panicali and Enzo Paoletti first developed the recombinant vaccinia technology in the early 1980s while they worked at the New York State Board of Health in Albany. Their first publication was "Construction of Poxviruses as Cloning Vectors: Insertion of the Thymidine Kinase Gene from Herpes Simplex Virus into the DNA of Infectious Vaccinia Virus," *Proceedings of the National Academy of Sciences*, 79:4927–4931 (1982). Right on their heels came a group led by Bernard Moss of the NIH, which published "Vaccinia Virus: A Selectable Eukaryotic Cloning and Expression Vector" in *PNAS*, 79:7415–7419 (1982).

16. Robert Teitelman, *Gene Dreams*, pp. 42 and 43.

17. According to a 1990 study by the Center for the Study of Drug Development at Tufts University, bringing a new medicine to market takes 12 years and costs $231 million. On the other hand, M. Groves in the July 1988 issue of *Pharmaceutical Technology* ("AIDS vaccine: Do We Know Where We Are Going?") predicts that R&D costs for an AIDS vaccine would only be $10 million.

18. Marilyn Chase, "Moving to the Fore in AIDS Research," *Wall Street Journal* (August 19, 1987).

19. Scott D. Putney et al., "HTLV-III/LAV-Neutralizing Antibodies to an *E. coli*–Produced Fragment of the Virus Envelope," *Science*, 234:1392–1395 (1986).

20. Donald P. Francis and John C. Petricciani, "The Prospects for and Pathways Toward a Vaccine for AIDS," *New England Journal of Medicine* (December 19, 1985), pp. 1586–1590.

21. Salk published this idea the next year: "Prospects for the control of AIDS by immunizing seropositive individuals," *Nature*, 327(6122):473–476 (June 11, 1987).

22. From a March 7, 1993, *Los Angeles Times* profile of Jonas Salk, "Hero with Something to Prove," by Sheryl Stolberg.

23. Henderson told me this during an August 6, 1992, interview I had with him about polio vaccines. He was then working in the White House's Office of Science and Technology Policy as the associate director for life sciences.

CHAPTER 4 MOVING INTO HUMANS

1. Zagury received funding through a $1.5 million, U.S. Army contract to Biotech Research Laboratories (see contract DAMD-17-86-C-6284).

2. "Investigation of Noncompliance with DHHS Regulations for the Protection of Human Research Subjects Involving the National Institutes of Health Intramural Research Program," Office for Protection from Research Risks, Division of Human Subject Protections, Final Report (March 26, 1993) p. 6.

3. *Chicago Tribune* reporter John Crewdson interviewed a coworker of Zagury who claimed that the children were vaccinated "several weeks before" his self-vaccination. See "AIDS Vaccine Fails to Save Researcher," *Chicago Tribune*, September 5, 1993.

4. Jean-Yves Nau, "Le danger des essais «sauvages»," *Le Monde*, December 19, 1986 [page unknown]. Translation by Michael Balter (many thanks).

5. Robert Gallo, *Virus Hunting: AIDS, Cancer, and the Human Retrovirus—A Story of Scientific Discovery* (Basic Books, 1991), pp. 311–318.

6. Fauci called these "National Cooperative Vaccine Discovery Groups"; NIAID later renamed them Development Groups.

7. B. M. Nkowane et al., "Vaccine-Associated Paralytic Poliomyelitis. United States: 1973 Through 1984," *JAMA*, 257(10):1335–1340 (March 13, 1987).

8. Both articles appeared on March 19, 1987: Marilyn Chase, "AIDS Scientist's Self-Inoculation Sparks Debate," *Wall Street Journal*, p. C1; and Philip J. Hilts, "French Doctor Testing AIDS Vaccine on Self, *Washington Post*, p. A1.

9. Robert Cooke's *Newsday* article ran June 4, 1987, p. 4.

10. The story actually broke in the *New Haven Register* on August 16 and was picked up by several other publications on the 18th, making the press conference a confirmatory event rather than an announcement.

11. Scientists chose homosexual men as the study subjects because of the belief that they had a different "immunologic profile" than the general population. Gay men also were the most likely people to use the vaccine if it existed, NIAID's August 18, 1987, "backgrounder" explained.

12. The *Wall Street Journal* and *Los Angeles Times* stories ran on March 19, 1987. The *New York Times* article appeared on March 20, 1987.

CHAPTER 5 ANIMAL ILLOGIC

1. S. L. Hu, P. N. Fultz, H. M. McClure, et al., "Effect of Immunization with a Vaccinia-HIV env Recombinant on HIV Infection of Chimpanzees," *Nature*, 328(6132):721–723 (1987).

2. Ronald Desrosiers, "The Simian Immunodeficiency Viruses," *Annual Review of Immunology*, 8:557–558 (1990).

3. From "Prospects for Vaccines Against HIV Infection" (National Academy Press, 1988), p. 17. The steering committee for the Institute of Medicine conference was chaired by Nobel laureate David Baltimore, then head of the Whitehead Institute of Biomedical Research at the Massachusetts Institute of Technology. The other committee members were Dani Bolognesi, Duke University; David Karzon, Vanderbilt University; Larry Lasky, Genentech; Philippa Marrack, National Jewish Center; Wade Parks, University of Miami.

4. Hearings before the Subcommittee on Health and the Environment of the Committee on Energy and Commerce, House of Representatives, Serial No. 100-140, Part 3, GPO, Washington, D.C.

5. "Loophole May Allow Trade in African Chimps," *New Scientist* (October 20, 1983).

6. Gallo's quote appeared in an article, "The Chimpanzee Thriller," published in the German newspaper *Express*. I have an undated translation of it. I believe the article appeared in mid-March 1986. Wyngaarden's comments come from a June 30, 1986, letter he wrote when asked to explain Gallo's interactions with Immuno.

7. The *Profil* article appeared on November 9, 1987. The *American Medical News* story was in the December 4, 1987, issue.

8. Gallo and Immuno signed a formal Cooperative Research and Development Agreement on June 1, 1988.

9. The quote from the Immuno-backed CCC comes from a press release the group issued on July 14, 1987, at a CITES meeting in Ottawa.

10. CITES Deputy Secretary-General, Jaques Berney, wrote Gallo about Immuno's 20 chimps on January 24, 1989.

11. This is from a September 6, 1989, letter from NIH Acting Director William Raub to Geza Teleki, a collaborator with Goodall and head of the Committee for Conservation and Care of Chimpanzees.

12. Anthony Lewis, "Abusing the Law," in his May 10, 1991, column hammered Immuno for filing its suit against Jan Moor-Jankowski, the editor of the *Journal of Medical Primatology* and a prominent chimpanzee researcher himself.

13. See Judge Kaye's opinion in *Immuno AG* v. *J. Moor-Jankowski*, No. 253, N.Y. Court of Appeals, decided December 14, 1989.

14. The passive immunity chimpanzee experiment was headed by Alfred Prince at the New York Blood Center. Prince formally reported the results on June 14, 1988, at the international AIDS conference in Stockholm, Sweden, abstract number 3062.

15. Annamari Ranki presented her talk, "T-cell Epitopes on HIV Envelope Are Recognized by Some Noninfected Partners but Not Infected Men" (abstract 2109), on

June 13, 1988, in a cell-mediated immunity session at the Stockholm AIDS conference.

16. Larry Thompson, "Why an AIDS Vaccine May Never Work," *Washington Post*, June 21, 1988, Health section, pp. 11–15.

17. NERPRC's Muthiah Daniel made the presentation at the Montreal meeting.

18. As the *Wall Street Journal*'s Marilyn Chase darkly reported at some length in her June 16, 1988, dispatch from the Stockholm AIDS conference ("Salk AIDS-Vaccine Research Draws Criticism"), Jorg Eichberg of the Southwest Foundation for Biomedical Research, the main tester of AIDS vaccines in chimpanzees for many years, publicly questioned whether the vaccine used in Salk's chimp experiments contained live virus—and whether the same vaccine was used in humans. Eichberg confronted Salk and the NIH's Joseph Gibbs, who carried out the chimp tests, on this point following their presentation. How could the chimp show antibodies to HIV eight months after receiving one injection of the vaccine? Eichberg asked. "That is impossible for a one-shot deal," said Eichberg.

"We have no evidence of infectivity, of infectious agent, of any untoward reactions, any toxicity or anything in this material, and I can assure you we looked very carefully," replied Gibbs, noting that he had made other killed vaccines that triggered similarly powerful immune responses.

Chase, usually one of the most astute of AIDS reporters, did not report that Salk effectively put the question to rest by pointing out that any live virus would require an intact version of HIV's envelope protein in order to copy itself and survive. The making of the vaccine had stripped off HIV's envelope protein, Salk noted, and indeed the chimp had no antibodies to it.

The issue was resurrected again by Philip Nobile, who wrote the cover story for the October 23, 1990, *Village Voice*, "Jonas Salk's Quest for an AIDS Vaccine." Nobile went so far as to write that a source at UC Davis "asserts that their disenchantment with the project was what they viewed as Salk's intellectual dishonesty and breach of scientific ethics in failing to disclose the bad batch in Stockholm." While Salk indeed appears to have too quickly dismissed UC Davis's concerns about live virus, Nobile, like Chase, makes no mention of Salk's persuasive counterargument to Eichberg's claim.

19. Several newspaper and magazine articles recount the making and passage of Assembly Bill 1952, including "Emotional Pleas of Gay Group Sparked the AIDS Drug Bill" (*Los Angeles Times*, September 30, 1987, p. 1), "The Making of a Bill" (*Frontiers*, Nov. 18–Dec. 2, 1987, p. 20), and "California Acts to Speed AIDS Drug Testing" (*New York Times*, September 30, 1987).

20. Carlo gave me a copy of this log.

21. Here's a sampling of headlines from the major dailies on June 9, 1989: "Salk Says Vaccine Test on Chimps Shows Halt of AIDS Development" (*New York Times*, p. A8), "Two AIDS Vaccine Reports Stir Interest: Approaches Taken by Salk,

Repligen-Merck Group Pass Preliminary Tests" (*Wall Street Journal*, p. B2), "Salk Reports Test Success of AIDS Virus Treatment" (*Los Angeles Times*, p. 3). I also gave the study too much credence in a cover story I wrote for the October 13, 1989, *Washington City Paper*.

22. I received a copy of the reviews from the Immune Response Corporation, not from *Science*.

23. "Observations after human immunodeficiency virus immunization and challenge of human immunodeficiency virus seropositive and seronegative chimpanzees" was published in the April 1991 *Proceedings of the National Academy of Sciences* (88:3348–3352). It was submitted to *PNAS* by Gibbs's boss, Nobel laureate D. Carleton Gajdusek (who, incidentally, in 1997 pleaded guilty to child abuse for molesting a 15-year-old boy). Gajdusek, like other members of the National Academy of Sciences, is allowed to submit papers to the journal after he has sent it out for what amounts to an informal version of peer review.

24. "AIDS Vaccine Proves Effective on Monkeys: Achievement Hailed as Step Toward Protecting Humans From Virus" appeared in the December 8, 1989, *Washington Post*.

25. Robert Bazell's article, "Medicine Show," appeared in the January 22, 1990, *New Republic*, pp. 16–19. The thrust of the article, ironically, was an argument to better organize research.

26. Allan Goldstein's vaccine, dubbed HGP-30, was backed by Viral Technologies, a 50/50 joint venture between a company he founded, Alpha 1 Biomedicals Inc., and Cel-Sci Corp.

27. Phillip Berman and Timothy Gregory.

28. Statistical significance refers to a mathematical calculation that assesses whether chance might explain a given outcome. Scientists typically consider data "significant" when they can say with 95% certainty that something is not due to chance; data are "highly significant" when the so-called confidence interval is 99%.

29. Marc Girard from the Pasteur Institute.

30. John Spouge of the National Library of Medicine wrote a computer program that evaluated experiments like the Genentech challenge and found them statistically insignificant. He described his problems with many such trials in "Statistical Analysis of Sparse Infection Data and Its Implications for Retroviral Treatment Trials in Primates," a paper published in the *Proceedings of the National Academy of Sciences*, 89(16):7581–7585 (August 15, 1992).

31. Phillip Berman spoke for Genentech, while Gale Smith represented MicroGeneSys.

CHAPTER 6 MARKET FORCES

1. Repligen, second quarter report, September 30, 1992.
2. Marc Girard, "Immune Responses of Chimpanzees to Live Recombinant,

Inactivated, Subunit and Synthetic Prototype HIV Vaccines," *4e Colloque Des Cent Gardes* (Foundation Marcel Merieux/Pasteur Vaccins, October 26–28, 1989), pp. 265–269.

3. Marc Girard et al., "Immunization of Chimpanzees Confers Protection Against Challenge with Human Immunodeficiency Virus," *Proceedings of the National Academy of Sciences*, 88:542–546 (January 15, 1991).

4. "A Promising New Assault on AIDS," *Fortune* (February 26, 1990), p. 105; "Firms Make Headway Toward AIDS Vaccine," *Investor's Daily* (May 17, 1990), p. 1; "AIDS: A Break in the Gloom," *Business Week* (June 25, 1990), p. 22.

5. First Boston published this Repligen report by Amy Bernhard Berler on November 20, 1989.

6. Fradd's May 2, 1990, report on Repligen and his August 2, 1990, report entitled "AIDS Vaccines 1990" both contain the same analysis of the marketplace.

7. Institute of Medicine, "The Children's Vaccine Initiative: Achieving the Vision" (National Academy Press, 1993).

8. Cooper obviously was referring to Joseph DiMasi's widely cited figure of $231 million, which he and coauthors later published in "Cost of Innovation in the Pharmaceutical Industry," *Journal of Health Economics*, 10:107–142 (1991). This figure was later upped to $359 million by the Office of Technology Assessment of the U.S. Congress in its February 1993 report *Pharmaceutical R&D: Costs, Risks and Rewards*. Note that neither the DiMasi study nor the one done by the OTA specifically address the cost of vaccine R&D—both focus on drug development.

9. The Frost & Sullivan figures appear in the July 1993 *Market Intelligence* report. Supporting the supposition that one drug could bring in more money than the entire vaccine industry, IMS America, a market research firm, provided me with the 1993 sales figures for drugs and vaccines in the United States. The best-selling drug, Zantac, brought in $2.1 billion. (Interestingly, Zantac was made by Glaxo, which had no AIDS vaccine program but did develop a best-selling AIDS drug, 3TC.) Vaccine sales in the United States totaled $366 million that year. Indeed U.S. Zantac sales were greater than the Frost & Sullivan estimate for the entire 1993 worldwide vaccine market, $2.03 billion.

10. "Keystone AIDS Vaccine Liability Project," Final Report, Keystone Center, May 1990.

11. As of August 6, 1997, the National Vaccine Injury Compensation Program also covers alleged injuries from vaccines for hepatitis B, *Haemophilus influenzae* type b, and varicella.

12. Gerhard Hunsmann of the German Primate Center in Goettingen, Germany, moderated the plenary session on vaccine development held on June 21, 1991, at the Seventh International Conference on AIDS held in Florence, Italy. In his introduction of Vanderbilt University's Barney Graham, who presented data about the Oncogen/MicroGeneSys combination vaccine, Hunsmann said, "This is certain-

ly the basis for progress and is probably the most promising vaccine I can see to help us learn what to do in the human system."

13. The company, Virogenetics, was working with Pasteur Mérieux Connaught.

14. I first quoted Hu anonymously making these statements in an April 10, 1992, article I wrote for *Science*, "Is Liability Slowing AIDS Vaccines?" (256:168–170). In 1998, by which time Hu had left Oncogen and was employed by the University of Washington, he agreed to my identifying him by name.

15. According to a report written for UNICEF in 1993 by Mercer Management Consulting, the "major players" in the vaccine industry were SmithKline Beecham, Merck, Pasteur Mérieux Serums and Vaccines, Lederle, Hoechst, Sclavo, and Medeva. In 1990, Ciba-Geigy and Chiron, which already had a joint AIDS vaccine program, purchased Sclavo. Wellcome and Glaxo by 1992 had both sold their vaccine businesses to Medeva.

16. See the reference in note 14.

17. In addition to giving AIDS vaccine manufacturers a special liability status, the 1986 California state bill, AB 4250—which was enacted into law that year—created a guaranteed state purchase of vaccine, provided money for AIDS vaccine research, and established a compensation fund for people injured by a California-made, marketed AIDS vaccine.

18. *Brown* v. *Abbott Laboratories* (44 Cal. 3d 1049) highlighted an obscure, but visionary and influential, legal argument made by the American Law Institute. The so-called comment k to the Restatement (Second) of Torts, in a flurry of double negatives, offered the rabies vaccine as an example of an unavoidably unsafe product that was not unreasonably dangerous. "The seller of such products, again with the qualification that they are properly prepared and marketed, and proper warning is given, where the situation calls for it, is not to be held to strict liability for unfortunate consequences attending their use, merely because he has undertaken to supply the public with an apparently useful and desirable product, attended with a known but apparently reasonable risk." AB 2892 in 1988 recognized that this language offered AIDS vaccine manufacturers more protection than the 1986 bill, and it was repealed.

19. Sam Donta, the University of Connecticut's chief of infectious diseases who would have been the principal investigator in that trial, brought up the school's liability fears at the September 10, 1990, Institute of Medicine meeting.

20. MicroGeneSys and American Home Products announced on September 18, 1990 that they had agreed to jointly develop the biotech firm's vaccine. (The *New York Times* ran an Associated Press story about the deal the next day.) No details of the deal were revealed.

21. J. Michael Epstein, an attorney for MicroGeneSys, and company lobbyist Charles Duffy were prime movers behind the legislation (Conn. Gen. Stat. ss. 19a-

591-591b). Lobbyists for California AIDS vaccine developer Chiron had convinced California assemblywoman Jackie Speier to introduce copycat legislation there, which also passed—even though the *Brown* ruling made it redundant.

22. Dan Quayle made these comments in Chicago, Illinois, on October 25, 1991, in a speech to the litigation section of the American Bar Association.

23. "Competitiveness Council Under Scrutiny," *Washington Post* (November 26, 1991), p. A19.

24. Senate bill S.640 was introduced by Robert Kasten Jr. (R-WI) on March 13, 1991.

25. Stark was reacting to my article in *Science*, his staff told me. I say this not to boast, but to point out that there was no constituency lobbying him to change the law.

CHAPTER 7 UNWANTED: DEAD OR ALIVE

1. The AIDS Vaccine Clinical Trials Network convened the May 13, 1991, meeting, which took place at Lister Hill Auditorium on the NIH campus.

2. At the International Conference on AIDS held June 16–21, 1991, Alan Schultz of NIAID and Jonathan Warren of the EMMES Corporation presented the listing of every monkey (and chimp) AIDS vaccine study then done, in "Surveillance of AIDS Vaccine Studies in Non-Human Primates," poster MA-1337.

3. Phyllida Brown, "Monkey Tests Force Rethink on AIDS Vaccine," *New Scientist* (September 21, 1991) p. 14.

4. This editorial ran on September 26, 1991 (253:297).

5. I wrote an extensive article for *Science*, published in the December 9, 1994, issue (266:1642–1649), that went through Duesberg's claims one by one—none of which stood up to close scrutiny.

6. The next month, *Science* ran an article that criticized Maddox's editorial (254:376, October 18, 1991). "Duesberg Vindicated? Not Yet," by Joseph Palca, even quoted Duesberg as saying the study had "nothing to do" with his position. (Marginally adding to the confusion of the whole affair, Palca's article contained a minor error, referring to the Stott vaccine as an "attenuated" virus, rather than an inactivated one.)

7. NIAID's International Conference on Advances in AIDS Vaccine Development took place in Marco Island, Florida, October 15–19, 1991.

8. Bolognesi's group published these results in "The Ability of Certain SIV Vaccines to Provoke Reactions Against Normal Cells," *Science*, 255:292–293 (January 17, 1992).

9. At least three labs have reported results from whole, killed vaccines made with human-cell-derived SIV that they challenged with monkey-grown SIV. In each case, the vaccines failed to protect the monkeys. See Roger Le Grand et al., "Specific and Non-specific Immunity and Protection of Macaques Against SIV Infection," *Vaccines*, 10:873–879 (1992); Martin Cranage et al., "Studies on the Specificity of

the Vaccine Effect Elicited by Inactivated Simian Immunodeficiency Virus," *AIDS Research and Human Retroviruses*, 9:13–22 (1993); and Stahl-Henning et al., "Protection of Monkeys by a Split Vaccine Against SIVmac Depends upon Biologic Properties of the Challenge Virus," *AIDS*, 7:787 (1993).

10. See *Science*, 258:1935–1941 (December 18, 1992).

11. The ultrasensitive test, the polymerase chain reaction, can fish tiny amounts of viral genetic material from a blood sample.

12. Because of the possibility of an apparently nonpathogenic HIV reverting into a killer, Desrosiers had deep misgivings about naturally found, weakened strains of the AIDS virus, which typically had only small deletions in their genes. Just such an SIV strain earlier had been found at the University of California at Davis by Marta Marthas and colleagues. See Marta Marthas et al., "Immunization with a Live, Attenuated Simian Immunodeficiency Virus (SIV) Prevents Early Disease but Not Infection in Rhesus Macaques Challenged with Pathogenic SIV," *Journal of Virology*, 64(8):3694–3700 (August 1990). See also Paul Luciw et al., "Genetic and Biological Comparisons of Pathogenic and Nonpathogenic Molecular Clones of Simian Immunodeficiency Virus (SIVmac)," *AIDS Research and Human Retroviruses*, 8(3):395–402 (1992).

13. Larry Arthur actually worked for Program Resources Inc., a contractor to the National Cancer Institute that ran the AIDS vaccine program in Frederick, Maryland.

14. Arthur presented these data on November 2, 1993, in Alexandria, Virginia, at the Conference on Advances in AIDS Vaccine Development, sponsored by the National Institute of Allergy and Infectious Diseases. His talk was entitled "Immunization of Macaques with Cellular Antigens Associated with Simian Immunodeficiency Viruses." He formally published the finding in the May 1995 issue of *Journal of Virology* 69(5):3117–3124, "Macaques Immunized with HLA-DR Are Protected from Challenge with Simian Immunodeficiency Virus."

15. "Alloimmunization as an AIDS Vaccine?" *Science*, 262:161–162, was signed by Gene Shearer and Mario Clerici of the National Cancer Institute and by Angus Dalgleish of London's St. George's Hospital Medical School.

16. The research on the Kenyan prostitutes was headed by Francis Plummer, and, as the October 8, 1993, letter to *Science* notes, he presented the data the preceding summer in Berlin at the Ninth International Conference on AIDS, 1993, poster WS-A07-3.

17. France's Jean-Claude Chermann promoted a similar strategy. Chermann favored a peptide from the cellular protein called ß2-microglobulin, an epitope that he had shown elicited an antibody that could neutralize both laboratory-adapted strains and primary isolates of HIV; Chermann described this epitope at the 12th World AIDS Conference in Geneva, June 28–July 3, 1998, abstract 31110.

18. Ron Desrosiers, who conducted prime-boost experiments in monkeys with

Dennis Panicali of Therion Biologics, first told me about these results in November 1993. Rather than using the "wimpy" SIVmne strain of the virus to challenge the monkeys, they used the much "hotter" SIVmac. They formally published these data the next year. See Muthiah Daniel et al., "High-Titer Immune Responses Elicited by Recombinant Vaccinia Virus Priming and Particle Boosting Are Ineffective in Preventing Virulent SIV Infection," *AIDS Research and Human Retroviruses*, 10(7):839–851 (July 1994).

19. Vanessa Hirsch et al., "Prolonged Clinical Latency and Survival of Macaques Given a Whole Inactivated Simian Immunodeficiency Virus Vaccine," *Journal of Infectious Diseases*, 170:51–59 (July 1994).

20. This work was initiated by Avigdor Shafferman of the Israel Institute of Biological Research. Two papers describe the benefits of this "failed" vaccine: Avigdor Shafferman et al., "Vaccination of Macaques with SIV Conserved Envelope Peptides Suppressed Infection and Prevented Disease Progression and Transmission," *AIDS Research and Human Retroviruses*, 8(8):1483–1487 (August 1992); and Avigdor Shafferman et al., "Prevention of Transmission of Simian Immunodeficiency Virus from Vaccinated Macaques That Developed Transient Virus Infection Following Challenge," *Vaccine*, 11(8):848–852 (1993).

21. Simra Israel et al., "Incomplete Protections, but Suppression of Virus Burden, Elicited by Subunit Simian Immunodeficiency Virus Vaccines," *Journal of Virology*, 68(3):1843–1853 (March 1994).

22. Murphey-Corb reported this in "Efficacy of SIV/DeltaB670 Glycoprotein-Enriched and Glycoprotein-Depleted Subunit Vaccines in Protecting Against Infection and Disease in Rhesus Monkeys," *AIDS*, 5:655–662 (1991). Stott and Cranage's data come from a December 9, 1993, fax that Cranage sent me in response to my questions. Gardner's group reported this in Shabbir Ahmad et al., "Reduced Virus Load in Rhesus Macaques Immunized with Recombinant gp160 and Challenged with Simian Immunodeficiency Virus," *AIDS Research and Human Retroviruses*, 10(2):195–204 (1994). And Desrosiers, who was skeptical about the meaning of this, nonetheless published similar results in Daniel et al., "High-Titer Immune Responses" (see note 18).

23. See Michael Balter, "Europe: AIDS Research on a Budget," *Science*, 280:1856–1858 (June 19, 1998). The top four European funders of AIDS research were the United Kingdom, France, Germany, and Italy.

24. Bush referred to Fauci as one of his heroes during the October 13, 1988, presidential debate with Democratic nominee Michael Dukakis. *ABC News*'s Ann Compton asked both candidates, "Who are the heroes who are there in American life today? Who are the ones that you would point out to young Americans as figures who should inspire this country?" After listing a Los Angeles math teacher, an astronaut, and sports figures in general, Bush said, "I think of Dr. Fauci. You've probably never heard of him. Oh, you did. Ann heard of him. He's a very fine

research—top doctor at National Institute of Health [*sic*]—working hard doing something about research on this disease of AIDS."

CHAPTER 8 ALL MICROGENESYS'S MEN

1. Donald Burke told me this story repeatedly.

2. The MicroGeneSys prospectus was dated November 8, 1988. Company president Franklin Volvovitz notified the Securities and Exchange Commission in a March 2, 1989, letter that "due to market conditions" and other factors, the company was withdrawing its registration statement.

3. I found this quote in Paul Taylor's March 3, 1985, page 1 story in the *Washington Post* about Russell Long's surprising announcement that he planned to retire.

4. Al Kamen, "Laxalt, Long to Join Same Law Firm," *Washington Post* (December 17, 1986), p. A05.

5. "AIDS Grant Proposal Added to Defense Bill," the page 1 *BioWorld Today* article by Michelle Slade, quoted Larry Kurtz, a spokesperson for Chiron, and Robert Abbott, CEO of Viagene.

6. The quotes in this paragraph come from interviews I did while preparing the first two stories I wrote for *Science* about MicroGeneSys and the $20 million appropriation: "Did Political Clout Win Vaccine Trial for MicroGeneSys?" (October 9, 1992), p. 211; and "Lobbying for an AIDS Trial" (October 28, 1992), pp. 536–539.

7. Dingell's letter to Murtha is dated October 1, 1992.

8. Bennett Johnston's speech and the summary he requested be introduced to the record are in the *Congressional Record* (October 5, 1992), S16949–50.

9. The report described 33 people who received VaxSyn and developed antibodies to HIV's gp160 that "were most commonly of weakly reactive intensity." In other words, the antibodies weren't very good at neutralizing HIV. The antibody responses also quickly faded. See Dolin et al., "The Safety and Immunogenicity of a Human Immunodeficiency Virus Type 1 (HIV-1) Recombinant gp160 Candidate Vaccine in Humans," *Annals of Internal Medicine,* 114(2):119–127 (January 15, 1991).

10. In April 1991, the U.S. Army and the Royal Thai Army began a joint study to evaluate the rates of HIV infection among Thai army recruits. The study was aimed at laying the foundation for preventive vaccine studies, and VaxSyn, according to documents provided to me by the U.S. Army through the Freedom of Information Act, was then the vaccine of choice. These plans first surfaced publicly in a September 27, 1991, article in the *Washington Times* (p. A7), "Thai Tests for AIDS Vaccine?" The article quoted Major General Arun Chavanasai, chief of the Thai army's AIDS Center, saying the U.S. Army wanted to test a preventive gp160 vaccine there. "They approached us through their liaison officers here six months

ago, and we have been discussing it since then." Neither the United States nor the Thai army intended to begin vaccine trials without going through a series of scientific and ethical reviews—a detail the article omitted. Still, the article quoted a military spokesman, General Narudol Dejpradiyudh, strongly denouncing the plans. "Thai soldiers are not the mice on which to test any U.S. vaccine," said Narudol. "Let them test it on their own soldiers if they like."

On November 14, 1991, according to other records I received from the U.S. Army through the Freedom of Information Act, Walter Reed's Donald Burke, Robert Redfield, and John McNeil briefed John Shannon, the deputy secretary of defense, about HIV vaccine testing in Thailand. At that point, they had hoped to begin human trials with VaxSyn in Thailand by spring 1992 and move into efficacy trials by 1995.

11. In Fauci's response to Inouye's letter, he explained that NIAID actually had been advised by "two distinguished groups of experts" not to stage efficacy trials with a vaccine unless there was protection shown in animals. Inouye's follow-up letter asked Fauci to name these experts, which Fauci did. One group was a blue-ribbon panel that met on April 17, 1990, to discuss AIDS vaccine development. In addition to NIAID's Fauci, Daniel Hoth, Wayne Koff, Malcolm Martin, John McGowan, and John La Montagne, this group included Nobel laureates David Baltimore and Howard Temin, Duke's Dani Bolognesi, Harvard's Ron Desrosiers, Rochester University's Raphael Dolin, the NCI's Robert Gallo, UC Davis's Murray Gardner, Merck's Maurice Hilleman, Wistar Institute head Hilary Koprowski, former NIAID director Richard Krause, and Tuft University's Sheldon Wolff. The second group was the NIH's AIDS Program Advisory Committee. Its members included Nobel laureate Baruj Benacerraf, Bolognesi, Howard Hughes Medical Institute director Purnell Choppin, Fauci, the University of Arizona's Evan Hersh, Harvard's Martin Hirsch, Yale's Robert Levine, Thomas Malone of the Association of American Medical Colleges, Boston University's Wendy Mariner, NIH acting director William Raub, Emory University's Gary Smith, the University of Miami's Jose Szapocznik, and University of California at San Francisco's Diane Wara and Constance Wofsy.

12. One of the main reasons Fauci and many other researchers believe there never will be a shortage of high-risk populations for efficacy trials in the United States is because among the highest-risk groups are the continually replenished populations of young gay men and young injecting drug users.

13. Quentin Burdick's May 6, 1991, letter was addressed to Gerald V. Quinnan Jr., acting director of the FDA's Center for Biologics Evaluation and Research. As an example of the degree of similarity between this letter and the one sent to Fauci from Senator Daniel Inouye, consider question number 3: "Given that the epidemic is spreading to other groups (i.e., from homosexuals to intravenous drug users to spouses of bisexuals and drug users to infants born of infected mothers to college students), what would prevent subsequent tests from being run on

groups which may be at higher risk in the future?" The Inouye letter contains an identically worded question.

14. Fauci replied to Chow's letter on June 7, 1991, explaining that "it is premature to select a vaccine for large efficacy trials from the products currently available. . . . "

15. See the *Congressional Record* (September 11, 1991), p. S12724.

16. Cardiac Arrhythmia Suppression Trial (CAST) Investigators, "Preliminary Report: Effect of Ecainide and Flecainide on Mortality in a Randomized Trial of Arrhythmia Suppression after Myocardial Infarction," *New England Journal of Medicine*, 321(6):406 (August 10, 1989).

17. Richard Johnston Jr., a pediatrics professor at the University of Pennsylvania School of Medicine, chaired the committee.

18. On November 7, 1991, basketball player Magic Johnson announced that he had AIDS, driving up the stock of many biotechs working on AIDS treatments and preventives. These companies fell especially hard after the meeting the next week of the FDA advisory committees.

19. "How could a government agency that regulates drugs and food start stocks tumbling to one of the market's worst days?" asked *USA Today* writer Chris Wloszczyna in "FDA Unlikely Force in Market Plunge," which ran on page 1 of the "Money" section on November 18, 1991. "By bringing reality back to part of the market that was flying on fantasy."

20. The AIDS Clinical Trials Group, a network of academic institutions that test experimental treatments (as opposed to preventive vaccines), then had sponsored two studies of VaxSyn as a therapeutic, with 52 patients in one study (ACTG 137) and seven in the other (ACTG 148). The ACTG declined an invitation to join in the military's large-scale trial of VaxSyn. (Robert Redfield described this on page 2 of the November 17, 1992, statement he gave as part of the Army's investigation into his conduct, which I detail later in this chapter.) The NIH intramural program also ran a therapeutic VaxSyn study, which involved 10 patients.

21. I confirmed each of these meetings with the respective staffers.

22. Long Law Firm's C. Kris Kirkpatrick, another registered lobbyist for MicroGeneSys, told me this policy after I had phoned Russell Long seven times and sent him a fax with detailed questions.

23. France's Auzias-Turrene first promoted vaccine therapy for syphilis in the mid-1800s. Germany's Robert Koch believed it could work for tuberculosis. And England's Sir Almroth Wright—nicknamed Sir Almost Right—was so devoted to the cause in the early 1900s that his friend George Bernard Shaw wrote a play about the idea, *The Doctor's Dilemma*. Burke, who had carefully studied this history, wrote about it at length in "Vaccine Therapy for HIV: A Historical Review of the Treatment of Infectious Diseases by Active Specific Immunization with Microbe-derived Antigens," *Vaccine*, 11(9):883–891 (1993).

24. *Christians in the Age of AIDS: How We Can Be Good Samaritans Responding to the AIDS Crisis* was written by Shepherd and Rita Moreland Smith (Victor Books, 1990). In the Foreword, Redfield made his own Christian perspective clear: "It is time to reject the temptation of denial of the AIDS/HIV crisis; to reject false prophets who preach the quick-fix strategies of condoms and free needles; to reject those who preach prejudice; and to reject those who try to replace God as judge," wrote Redfield. "The time has come for the Christian community—members and leaders alike—to confront the epidemic with the commitment that comes from Christ's example."

25. This claim is specious, as Colonel Burke told me he was aware of the proposed appropriation before it was passed.

26. Another registered MicroGeneSys lobbyist, Robert Thompson, noted on his June 18, 1991, report that his firm, Thompson & Company, received $10,000 per month from MicroGeneSys to lobby for legislation "relating to AIDS." Thompson, who worked as an assistant to Vice President George Bush and later President Ronald Reagan, did not return my phone calls to discuss his lobbying activities. I never found anyone with whom Thompson spoke on behalf of MicroGeneSys, but he was a well-known lobbyist on Capitol Hill, and became enmeshed in another lobbying controversy involving a savings-and-loan deal. See "Insolvent Bank Bought Cheaply With Help of Former Bush Aide," by Jeff Gerth, *New York Times* (July 22, 1990) p. 1, and "Web of Connections Links S&L Mogul Fail [*sic*] with GOP Campaign Heavyweights," by Glenn R. Simpson, *Roll Call* (May 25, 1992) p. 12.

27. This quote is from the statement Donald Burke gave on November 20, 1992, as part of the Army's Redfield investigation. I received the document through the Freedom of Information Act.

28. The *Hartford Courant*'s three-part series, by Lyn Bixby and Frank Spencer-Molloy, ran on page 1, February 7–9, 1993.

29. The three-page memo was signed by Major Craig Hendrix and Colonel R. Neal Boswell of the Wilford Hall United States Air Force Medical Center, Lackland Air Force Base, Texas.

30. Phyllida Brown, "Memo Suggests US Army HIV Vaccine Oversold," *New Scientist* (October 24, 1992), p. 8.

31. I interviewed Smith repeatedly during this period, but my quotes here are taken from a November 17, 1992, memo he wrote to Burke and gave to me. Burke resigned from ASAP on October 11 "because of Mr. Smith's insistence on seeing the Vahey memo of his call to her," he stated in an April 6, 1993, interview that was part of the Redfield investigation.

32. "Special Rules for AIDS Drugs," *Washington Post* (November 30, 1992), p. A18.

33. After I filed a Freedom of Information Act request, the Department of

Defense provided me with a January 26, 1993, memorandum addressed to Commander, USAMRDC, that recounted the meeting of the advisory group.

34. The February 11, 1993, issue of *Nature* (361:503) ran a scientific correspondence from John Moore, George Lewis, and James Robinson entitled "Which gp160 Vaccine?" The letter restates Moore's earlier charge that VaxSyn is "the worst choice" from the current envelope-based vaccines to stimulate antibodies capable of neutralizing HIV. Volvovitz and MicroGeneSys scientist Gale Smith replied in the April 8th *Nature* (362:504), contending that their gp160 by "design" was denatured, as envelope proteins in their native form "would be unlikely to be of clinical benefit to an HIV-infected individual." Moore and cowriters had the last word. "The conversion of MicroGeneSys to the virtues of denatured immunogens is one of Damascene proportions," they wrote. "Is it cynical to suppose that, far from 'designing' its gp160 to be denatured, MicroGeneSys has rewritten history and grasped the concept of vaccine immunotherapy as a way to exploit their denatured product?"

35. Volvovitz told Colin Macilwain at *Nature* that MicroGeneSys and Wyeth pulled out of the trial because they "disagreed with the primary and secondary end points." See "MicroGeneSys Drops Out of NIH Trial for AIDS Vaccine," 362:277 (March 25, 1993).

36. Both Kessler's and Healy's letters were addressed to William Natcher, chair of the House's Committee on Appropriations.

37. The Army in fact spelled this out to Senator Bennett Johnston in an April 2nd letter signed by Lieutenant Colonel D. William Atwood. As Atwood explained, he was writing in response to a letter from Johnston, who had asked whether other products were to be included in the phase III trial and whether they were ready for such tests. Johnston also wanted to know how researchers planned to pay for a multiproduct trial and what, if anything, had "changed the professional opinions" about VaxSyn of the Army researchers who testified at the February 1992 congressional hearings. "The Army plans to study only gp160," wrote Atwood, noting that the Army researchers' opinions about "the potential value of vaccine therapy" had not changed.

38. Frank Spencer-Molloy, "Official Attacks Exclusive Testing of AIDS Vaccine," *Hartford Courant* (April 6, 1993), p. A2.

39. Specifically, Carol Rasco, Clinton's Domestic Policy Advisor.

40. Sally Squires, "Controversial Plan to Test Single AIDS Drug Said to Be Canceled," *Washington Post* (April 9, 1993), p. A4.

41. Sally Squires, "Bowing to Pressure, Defense Dept. Agrees to Drop AIDS Vaccine Test," *Washington Post* (April 15, 1993), p. A4.

42. See my story in the August 13, 1993, issue of *Science*, "Army Clears Redfield—but Fails to Resolve Controversy," 261(5123):824–825.

43. Harriet Rabb, the HHS general counsel, wrote the DoD's general counsel

Jamie Gorelick on August 20, 1993, to reiterate that HHS was still willing to run a multiproduct trial if MicroGeneSys donated vaccine. Rabb notes in this letter that a lawyer from DoD told her office that MicroGeneSys was supplying vaccine only for the first year of the three-year trial. Rabb also noted that in previous negotiations with HHS, "MicroGeneSys was demanding $10 million as the purchase price of its vaccine."

44. *Congressional Record* (September 30, 1993), p. H7295.

45. Moore was the second chair of the panel selected by the American Institute of Biological Sciences. The first one, the University of Maryland's Gerald Cole, resigned because he did not think the other members of the panel were qualified. Upon hearing the news of his resignation, the AIBS executive coordinating the panel had a minor stroke, and the job was passed to a former Cambridge University schoolmate of Moore's, who selected him. AIBS selected a different panel, too, which included Cole.

46. I, too, received a heavily redacted version of the Redfield investigation in response to my Freedom of Information Act request. I appealed this decision, but to no avail.

47. The CNN story by Jeff Levine aired on June 7. The print media stories ran the next day.

48. An Army memorandum dated December 2, 1994, and signed by Colonel John Scovill explains that Wyeth announced on April 11, 1994, it was cutting its funding of the phase II trial. This was, interestingly, before the peer-review panel even held its first meeting, raising the question of why the Army waited, as the memo details, until July 28 before deciding to scale back the number of grants supported by the $20 million.

49. Public Citizen wrote Waxman on October 14 and Defense Secretary William Perry on November 17.

50. I gleaned this information from a May 5, 1995, letter sent by the Army to Public Citizen's Peter Lurie in response to his Freedom of Information Act request.

CHAPTER 9 A MANHATTAN PROJECT FOR AIDS

1. "Democrats' Efforts to Lure Gay Voters Are Persistent but Subtle," by Todd S. Purdum, *New York Times* (April 7, 1992), p. A24.

2. One of the seven bulleted points on the campaign sheet distributed in Amsterdam, "Bill Clinton on AIDS," said "Appoint an AIDS policy director to head a 'Manhattan Project' that coordinates federal AIDS policies. . . . " By November, the "Clinton/Gore on AIDS" had been modified to read "Appoint an AIDS policy director to coordinate federal AIDS policies. . . . " The other points were identical.

3. I first read this distinction between use and understanding in *Pasteur's Quadrant: Basic Science and Technological Innovation* by Donald E. Stokes

(Brookings Institution Press, 1997, pp. 27–30), who traces the dichotomy back to the ancient Greeks.

4. Vannevar Bush, *Science: The Endless Frontier* (National Science Foundation, 1945; reprinted July 1960) p. 83.

5. John Walsh, "NIH: Demand Increase for Applications of Research," *Science* (July 8, 1966), pp. 149–152; See also Julius H. Comroe Jr. and Robert D. Dripps, "Scientific Basis for the Support of Biomedical Science," *Science*, 192:105–107 (April 9, 1976).

6. Robert Bazell, "Cancer Research Proposals: New Money, Old Conflicts," *Science*, 171:877 (March 5, 1971).

7. Nicholas Wade, "Cancer Politics: NIH Backers Mount Late Defense in House," *Science* (October 8, 1971) pp. 127–131.

8. All of the excerpted quotes from the March 1974 "Zinder Committee Report" come from a copy of the 25-page document that Norton Zinder gave to me.

9. Comroe and Dripps, "Scientific Basis for the Support of Biomedical Science."

10. The quotes from Stokes's *Pasteur's Quadrant* in this and the following two paragraphs are from pp. 58–89.

11. Roughly 80% of the NCI's budget went to universities, medical schools, and other independent research institutions. Of that, about 65% supported investigator-initiated grants known as R01s. Cancer centers, which run trials of experimental treatments, received another 16%, while about 17% went to contracts that supported targeted research projects.

12. NCI supplied me with an executive summary of the report, *An Assessment of the Factors Affecting Critical Cancer Research Findings*, which the NIH published in May 1990 (No. 90-567). The contractor, CHI Research/Computer Horizons Inc. (contract NO1-CO-339933), gave me a copy of the entire final report, which was dated June 1, 1987.

13. R01 grants, specifically, funded 20.1% of the work that led to major advances. Contracts accounted for 16.9%. Collaborations known as "program projects"—which, by definition, were "directed toward a range of problems having a central research focus in contrast to the usually narrower thrust of the [R01]"—made up another 15.5% of the total. Research done on the NIH campus, which does not go through the peer-review mechanism of R01 grants, accounted for 19.8% of the major advances. The rest of the advances were ascribed to "core" grants that pay for shared resources and facilities (5.7%), cooperative clinical research grants that support human tests of treatments and preventives (5.1%), and "other" grants that pay for such things as training and fellowships (16%).

A breakdown of these advances by type—basic, clinical, and epidemiological—further showed that the equivalence of contract and R01 grants was especially true in the basic research arena. In advances deemed clinical, R01s only contributed 10%, while program projects explained 30% of them.

14. Richard J. Wurtman and Robert L. Bettiker, "The Slowing of Treatment Discovery, 1965–1995," *Nature Medicine*, 1(11):1122–1125 (November 1995).

15. I quoted Hilleman saying this in "A 'Manhattan Project' for AIDS?", which ran in *Science*, 259:1112–1114 (February 19, 1993).

16. Wayne Koff gave me the text from the NIH's AIDS Program Advisory Committee meeting.

17. I have a copy of the December 19, 1990, "confidential" letter that Koff sent to Jonas Salk.

18. I received this list from Wayne Koff. According to Koff, the other scientists who received the letter were Australia's Gordon Ada, Belgium's Arsene Burny, France's Jean-Claude Chermann, the CDC's Walter Dowdle, the Rockefeller Foundation's Scott Halstead, Vanderbilt's David Karzon, Ron Kennedy of the Southwest Foundation for Biomedical Research, the NCI's Werner Kirsten, Finland's Kai Krohn, and Sweden's Erling Norrby. The one abstention was Nobel laureate Gerald Edelman.

19. Salk already had started a similarly named foundation with epidemiologist Philippe Stoeckel, who ran the Forum for the Advancement of Immunization Research. FAIR supported studies of new-generation killed poliovirus vaccines; Salk continued to hope that the world would come to its senses and see that the killed vaccine worked just as well as Sabin's live polio vaccine—if not better—and had fewer risks.

20. Maurice R. Hilleman, "Impediments, Imponderables and Alternatives in the Attempt to Develop an Effective Vaccine Against AIDS," *Vaccine*, 10(14):1053–1058 (1992). The paper was based in part on a talk Hilleman gave in Honolulu, Hawaii, at the Fourth Joint Scientific Meeting of the AIDS Panel, US-Japan Cooperative Medical Science Program, January 9–12, 1992.

21. Hilleman specifically was referring to the U.S. Armed Forces Board of Epidemiological Research and the Vaccine Committee of the National Foundation for Infantile Paralysis.

22. The meeting, which was held on May 19, 1992, and focused on AIDS vaccines, was of the Combined Division Advisory Committee, which included the Advisory Subcommittee and the AIDS Liaison Subcommittee of NIAID's Division of AIDS.

23. TAG issued "AIDS Research at the NIH: A Critical Review" on July 20, 1992. Part I of the review was a summary, while Part II, "The NIH, A User's Guide," detailed each NIH institute's AIDS research funding.

24. Kramer's op-ed, "Name an AIDS High Command," ran in the November 15, 1992, *New York Times* sec. 4, p. 19. On January 15, 1993, Kramer wrote Bob Hattoy on Clinton's transition team about the "equivalent of a Manhattan Project."

In a February 4, 1993, letter to Donna Shalala, Kramer complained that "neither you nor anyone else has responded to any of my letters or FAXes of the past weeks."

25. For a fuller, if shamelessly aggrandized, accounting of Martin Delaney's early years as an AIDS activist, see *Acceptable Risks* by Jonathan Kwitny (Poseidon Press, 1992).

26. John S. James, editor and publisher of the comprehensive newsletter *AIDS Treatment News*, ran the full transcript of this press conference ("Research, Political Leaders Plan Future Directions") in the August 6, 1993, issue.

The reason for barring media from this meeting was to allow for the free exchange of ideas. I think the premise is faulty—scientists routinely exchange controversial ideas when I am present. And I find barring the media particularly wrongheaded when U.S. government scientists are part of a process that aims to influence how public monies are spent and policy is set. The public must trust the scientists, activists, and policy makers leading the search for AIDS treatments and vaccines. Closed-door meetings, convenient as they are for the participants, erode that trust.

27. In a March 3, 1994, fax to me, TAG's Gregg Gonsalves called the Accelerated AIDS Research Initiative "Marty's vague proposal" and criticized him for not running it by the working-group members he suggested had helped draft it.

28. Moore's "viewpoint" ran in the March 8, 1994, issue, on page 8, but he had completed the editorial in January and circulated it widely.

29. Nadler first introduced this bill, H.R.3310, on October 19, 1993. He tried again on May 10, 1994 with H.R.4370, which met the same fate.

30. These quotes from Bernard Fields come from "AIDS: Time to Turn to Basic Science," *Nature*, 369(6476):95–96 (May 12, 1994).

CHAPTER 10 THE DAIRYMAIDS OF AIDS

1. In *The History of the Peloponnesian War*, Thucydides offers a vivid description of an epidemic of plague that struck Athens (see Richard Crawley's translation, bk. 2, chap. 7). Medical historians long have debated what the disease actually was, with the possibilities ranging from typhus to measles to Marburg. Whatever its cause, Thucydides wrote in his 431 B.C. account of the epidemic that "the same man was never attacked twice—never at least fatally." This not only led these immune people to care for the sick and dying; they also "in the elation of the moment, half entertained the vain hope that they were for the future safe from any disease whatsoever."

2. For an excellent overview of variolation, see Nicolau Barquet and Pere Domingo, "Smallpox: The Triumph over the Most Terrible of the Ministers of Death," *Annals of Internal Medicine*, 127:635-642 (October 15, 1997).

3. See *Scientific Papers: Physiology, Medicine, Surgery, Geology, with Introductions, Notes and Illustrations*, The Harvard Classics, vol. 38 (P. F. Collier & Son, c. 1910). The text of all three of Jenner's landmark papers is reprinted

here. Jenner's papers also are posted on the Internet, in searchable form, at www.fordham.edu/halsall/mod/1798jenner-vacc.html.

4. Many experts believe that the Soviets, interested in smallpox as a bioweapon, dispersed viral samples to other sites. See Laurie Garrett, "Inside Russia's Germ Warfare Labs," *Newsday* (August 10, 1997), p. 5; David Brown, "Smallpox's Threat as Weapon Is Weighed," *Washington Post* (March 15, 1999), p. A1; and Ken Alibek and Stephen Handelman, *Biohazard: The Chilling True Story of the Largest Covert Biological Weapons Program in the World—Told from the Inside by the Man Who Ran It* (Random House, 1999).

5. In 1989, Annamari Ranki, Kai Krohn, and coworkers did publish a study in *AIDS* (3:63–69) that described T-cell responses to HIV in five uninfected sexual partners of HIV-infected men. But this study primarily was analyzing how HIV infection itself rendered T cells "anergic," a state in which they cannot function. The authors used the data from the uninfected partners to bolster their argument that anergy is the result of an "active infection" rather than the presumably transient one in the EUs. Nowhere does the paper suggest that the EU data might point the way to an immunization strategy.

6. Mario Clerici, Janis Giorgi, et al., "Cell-Mediated Immune Response to Human Immunodeficiency Virus (HIV) Type 1 in Seronegative Homosexual Men with Recent Sexual Exposure to HIV-1," *Journal of Infectious Diseases*, 165(6):1012–1019 (June 1992).

7. Beginning in September 1995, the NIH awarded Giorgi $219,528 a year to study "cellular immunology of resistance to HIV infection." The grant was slated to last for four years.

8. John Robbins and Rachel Schneerson, prominent researchers at the NIH's National Institute of Child Health and Human Development who developed the tremendously successful *Haemophilus influenzae* type B vaccine, contend that all licensed vaccines work by establishing a critical level of antibodies that can inactivate the invader. For a detailed version of their argument, see John Robbins, Rachel Schneerson, and Shousun Chen Szu, "Hypothesis: Serum IgG Antibody Is Sufficient to Confer Protection Against Infectious Diseases by Inactivating the Inoculum," *Journal of Infectious Diseases*, 171(6):1387–1398 (June 1995).

9. Accomplished scientists often complain about being robbed of a Nobel, but I first heard about Shearer's near miss from Bill Paul, the NIAID immunologist who edited the standard textbook for the field. Shearer was one of the first researchers to discover the elaborate mechanism that explains how cytotoxic T lymphocytes—killer cells—know how to identify an infected cell and eliminate it. For a detailed description of the finding (and Shearer's contribution to it), see "The Discovery of MHC Restriction" in *Immunology Today*, 18(1):14–17 (January 1997). The article is written by Rolf Zinkernagel and Peter Doherty, the two researchers who won the Nobel for the work.

10. Shearer and Clerici described their IL-2 assay in detail in Clerici et al., "Interleukin-2 Production Used to Detect Antigenic Peptide Recognition by T-helper Lymphocytes from Asymptomatic HIV-Seropositive Individuals," *Nature*, 339(6223):383–385 (June 1, 1989).

11. For a complete description of this trial, see Clerici et al., "Immunization with Subunit Human Immunodeficiency Virus Vaccine Generates Stronger T Helper Cell Immunity Than Natural Infection," *European Journal of Immunology*, 21(6):1345–1349 (June 1991). Note that the proliferation assay actually measures the response of peripheral blood mononuclear cells, a mix that includes, but is not limited to, CD4s.

12. Clerici, Shearer, and collaborators Jay Berzofsky and Carol Tacket, described their findings in "Exposure to Human Immunodeficiency Virus Type 1-Specific T Helper Cell Responses Before Detection of Infection by Polymerase Chain Reaction and Serum Antibodies," *Journal of Infectious Diseases*, 164:178–182 (July 1991).

13. Major daily newspaper typically shun anonymous quotes from scientists who are merely offering an educated opinion about a new study—which is just what happened in "Tests Find `Silent' AIDS Infections," by Robert Steinbrook, *Los Angeles Times* (June 1, 1989), p. 1.

14. C. R. Parish, "Immune Response to Chemically Modified Flagellin. II. Evidence for a Fundamental Relationship Between Humoral and Cell-Mediated Immunity," *Journal of Experimental Medicine*, 134(1):21–47 (July 1, 1971). The next year, Parish published a more comprehensive overview, "The Relationship Between Humoral and Cell-Mediated Immunity," *Transplantation Review*, 13:35–66 (1972).

15. Janis Giorgi, Mario Clerici, Jay Berzofsky, and Gene Shearer, "HIV-Specific Cellular Immunity in High-Risk Seronegative Homosexual Men," Seventh International Conference on AIDS, Florence, Italy, June 16–21, 1991, WA-1209. At the time the authors submitted their abstract, they did not have the PCR data.

16. I was at the Florence meeting, but I do not have this in my notes. Both Shearer and Clerici recounted the story to me.

17. "HIV-1 in Seronegative Homosexual Men," *New England Journal of Medicine*, 325(17):1250–1251 (October 24, 1991).

18. The symposium, "A Tribute to Melvin Cohn," was held in honor of Bretscher's former mentor at the Salk. Bretscher's talk was entitled "Decision Criteria and Coherence in the Regulation of the Immune Response: Their Theoretical and Practical Importance."

19. Bretscher et al., "Establishment of Stable, Cell-Mediated Immunity that Makes 'Susceptible' Mice Resistant to Leishmania major," *Science*, 257(5069): 539–542 (July 24, 1992).

20. The Seventh International Conference on AIDS was held July 19–24, 1992, in Amsterdam, The Netherlands.

21. Vaslin Bruno et al., "Purified Inactivated SIV Vaccine: Comparison of Adjuvants," Seventh International Conference on AIDS, Amsterdam, Netherlands, July 19–24, 1992, poster PoA-2239.

22. Jonas Salk, "Immunological Paradoxes: Theoretical Considerations in the Rejection or Retention of Grafts, Tumors, and Normal Tissue," *Annals of the New York Academy of Sciences*, 164(2):365–380 (October 14, 1969).

23. Cell-mediated immunity also appeared to play a critical role at the other end of the infection spectrum: people who had just become infected. HIV levels skyrocket shortly after a person becomes infected, and then, within weeks, precipitously drop and are maintained at a relatively low level for many years. Several researchers, Levy included, looked at people within the first few weeks of becoming infected by HIV and found that cell-mediated immunity—not neutralizing antibodies—contained the initial infection. Potential cell-mediated mechanisms responsible for the drops included Levy's CAF and killer cells. I described some of these studies in an article I wrote for *Science,* "AIDS Research Shifts to Immunity," which came out on the eve of the Amsterdam meeting (257:152–154, July 10, 1992). Elizabeth Connick of the University of Colorado and colleagues presented a poster in Amsterdam, "HIV-1 Envelope and Gag Specific Cellular Immunity in Primary HIV Infection" (poster PoA-2172), that described one such study in detail.

24. Mario Clerici et al., "Changes in Interleukin-2 and Interleukin-4 Production in Asymptomatic, Human Immunodeficiency Virus-Seropositive Individuals," *Journal of Clinical Investigation*, 91(3):759–765 (March 1993). See also Mario Clerici and Gene M. Shearer, "A TH1—>TH2 Switch Is a Critical Step in the Etiology of HIV Infection" *Immunology Today*, 14(3):107 (March 1993).

25. Shearer and Clerici agreed with one of the criticisms: That people could have high levels of a cytokine that were not produced by a Th cell; IL-4, for example, could be produced by a macrophage rather than a Th2 cell. But to them this criticism missed the point. Shearer was less interested in which cells produced the cytokines than the observation that a Th1-like cytokine profile existed in healthy people, while those who progressed to disease had a more Th2-like profile. They thus stopped using the Th1/Th2 designations, referring instead to type 1 and type 2 cytokine profiles. They explained their reasoning in "The Th1-Th2 Hypothesis of HIV Infection: New Insights," *Immunology Today*, 15(12):575–581 (December 1994).

26. Salk made these comments during his November 3, 1993 talk, "Immunological Paradoxes," at NIAID's Conference on Advances in AIDS Vaccine Development in Alexandria, Virginia.

27. Swedish virologist Sven Gard.

28. *Breakthrough: The Saga of Jonas Salk*, by Richard Carter (Trident Press, 1965) p. 204.

29. I first reported the work of New York University's Fred Valentine and Mindell Seidlin in "AIDS Research Shifts to Immunity," *Science*, 257:152–154, (July

10, 1992). The researchers published their findings, "Lymphocytes from Some Long-term Seronegative Heterosexual Partners of HIV-infected Individuals Proliferate in Response to HIV Antigens," in *AIDS Research and Human Retroviruses*, 8(8):1355–1359, (August 1992).

30. M. Schlesinger et al., "A Distinctive Form of Soluble CD8 Is Secreted by Stimulated CD8+ Cells in HIV-1-Infected and High-Risk Individuals," *Clinical Immunology and Immunopathology*, 73(2):252–260 (November 1994).

31. Scott Tenenbaum and coworkers first reported this at the Eighth International Conference on AIDS held in Berlin, Germany, June 6–11, 1993. See S. A. Tenenbaum et al., "Virological and Serological Studies in HIV-Exposed/HIV-Negative Hemophiliacs," poster PO-C17-3004. Specifically, the researchers compared blood from eight, presumably exposed, hemophiliacs and eight normal controls. CD8 cells from the hemophiliacs proved better at suppressing HIV replication, implying that these people had developed cell-mediated immunity to HIV, and the difference between the groups was statistically significant.

32. Sarah Rowland-Jones et al., "HIV-Specific Cytotoxic T-Cell Activity in an HIV-Exposed but Uninfected Infant," *Lancet* 341(8849):860–861 (April 3, 1993).

33. Sarah Rowland-Jones et al., "HIV-Specific Cytotoxic T-Cells in HIV-Exposed but Uninfected Gambian Women," *Nature Medicine*, 1(1):59–64 (January 1995).

34. P. Langlade-Demoyen et al., "Human Immunodeficiency Virus (HIV) nef-Specific Cytotoxic T Lymphocytes in Noninfected Heterosexual Contact of HIV-Infected Patients," *Journal of Clinical Investigation*, 93(3):1293–1297 (March 1994).

35. The conference, "Immunologic and Host Genetic Resistance to HIV Infection and Disease," was held at the Holiday Inn in Bethesda, Maryland, February 25–27, 1993. Paul Rowe wrote about the meeting in the *Lancet*, 341(8845):624 (March 6, 1993).

36. "A Strategy for Prophylactic Vaccination Against HIV," *Science*, 260:1270–1272 (May 28, 1993). Peter Salk, Jonas's eldest son, was the fifth author of this paper. A researcher himself, Peter Salk began working closely with his father in 1991.

37. Mario Clerici et al., "HIV-Specific T-Helper Activity in Seronegative Health Care Workers Exposed to Contaminated Blood," *Journal of the American Medical Association*, 271(1):42–46 (January 5, 1994).

38. Lawrence K. Altman, "H.I.V. Immunity Discussed at Berlin Conference," *New York Times* (June 9, 1993) p. A7; David Brown, "Immunology: Lessons in Resistance to AIDS Virus," *Washington Post* (December 13, 1993), p. A3; and Sheryl Stolberg, "Some Immune to AIDS, Study Hints," *Los Angeles Times* (December 10, 1993), p. A3.

39. The Immune Response Corporation held this hastily organized press conference at a hotel away from the meeting headquarters, shuttling over journalists in a bus. Many journalists took great umbrage at the whole spectacle, as did

researchers who witnessed it. Making matters worse, leading scientists sharply criticized the IRC data presented at the meeting. See my article, "Somber News from the AIDS Front," *Science*, 260:1712–1713 (June 18, 1993); and Sheryl Stolberg's "Salk Report on AIDS Vaccine Meets Skepticism at Convention," *Los Angeles Times*, p. A7, June 10, 1993.

40. Mario Clerici et al., "T-Cell Proliferation to Subinfectious SIV Correlates with Lack of Infection After Challenge of Macaques," *AIDS*, 8:1391–1395 (October 1994).

41. Maria Salvato et al., "Cellular Immune Responses in Rhesus Macaques with Low Dose SIV Infection," Symposium on Nonhuman Primate Models for AIDS, September 19–22, 1993, abstract 11. For a fuller description of the work, see Parul Trivedi et al., "Intrarectal Transmission of Simian Immunodeficiency Virus in Rhesus Macaques: Selective Amplification and Host Responses to Transient or Persistent Viremia," *Journal of Virology*, 70(1):6876–6883, (October 1996).

42. U. Dittmer et al., "Repeated Exposure of Rhesus Macaques to Low Doses of Simian Immunodeficiency Virus (SIV) Did Not Protect Them Against the Consequences of a High-Dose SIV Challenge," *Journal of General Virology*, 76(pt. 6):1307–1315 (June 1995). See also U. Dittmer et al., "Cell-mediated Immune Response of Macaques Immunized with Low Doses of Simian Immunodeficiency Virus (SIV)," *Journal of Biotechnology*, 44(1–3):105–110 (January 26, 1996).

43. Gunnel Biberfeld presented these data at the 11th International Conference on AIDS held in Vancouver, Canada, July 7–12, 1996. See "Protection Against SIVsm in HIV-2 Exposed, Seronegative Macaques but Not in Macaques Exposed i.r. to Subinfectious Doses of SIVsm," abstract WeA-283. For a detailed description of the HIV-2 study, see Per Putkonen et al., "Protection of Human Immunodeficiency Virus Type 2-Exposed Seronegative Macaques from Mucosal Simian Immunodeficiency Virus Transmission," *Journal of Virology*, 71(7): 4981–4984 (July 1997).

44. Researchers do sometimes check to see whether volunteers have CTLs against HIV at the start of a trial. But this test is not done as a screen to determine whether someone has preexisting immunity to HIV. Rather, it simply provides a baseline so that the scientists can determine whether the vaccine boosts CTL production. A CTL screen also is too specific; it likely would fail to detect cell-mediated immunity to HIV in many people who would respond to the more sensitive IL-2 or proliferation assays.

CHAPTER 11 PERPETUAL UNCERTAINTY

1. "Advances in AIDS Vaccine Development," the Sixth Annual Meeting of the National Cooperative Vaccine Development Groups for AIDS, ran from October 30 to November 4, 1993.

2. NIAID's Lew Barker, associate director of the clinical research program at the Division of AIDS, announced these plans at the April 17, 1992, AIDS Program Advisory Committee meeting. "What we hope to be ready for is the initiation of efficacy trials late in 1993 in this country and some time in 1994 in international sites," said Barker. NIAID indeed gave this advisory committee a proposed research initiative at its May 19–20, 1992, meeting that said the institute "has a goal of initiating HIV vaccine efficacy trials in the United States by December 1993 and internationally by May 1994." (From p. A-118 of the proposal distributed to the advisers.)

3. Eric Daar et al., "High Concentrations of Recombinant Soluble CD4 Are Required to Neutralize Primary Human Immunodeficiency Virus Type 1 Isolates," *Proceedings of the National Academy of Sciences*, 87:6574–6578 (September 1990).

4. These data were presented by Duke's Tom Matthews, Chiron's Kathelyn Steimer, and Walter Reed's John Mascola. For a detailed description of their presentations, see the article I wrote about the meeting for *Science*, "Jitters Jeopardize AIDS Vaccine Trials," 262(5136):980–981 (November 12, 1993).

5. November 12, 1993, the same day that my article appeared in Science about the meeting, the *New York Times*, the *Wall Street Journal*, the *Los Angeles Times*, *USA Today*, the Associated Press, and Reuters all carried the story. Tellingly, they all were building off my story because none of them, as usual, had sent a reporter to the NIAID AIDS vaccine meeting. Indeed, the meeting, then the most important annual gathering for the field, rarely attracted more than a couple of reporters. This long struck me as a reflection of how little these media organizations thought their audiences cared about AIDS vaccines. These same outfits, in contrast, gave heavy coverage to meetings on AIDS treatments and epidemiology, such as the annual National Conference on Human Retroviruses and Other Related Infections, which routinely was so full that it had to turn press away. And the international AIDS conferences consistently attracted more than 1,000 journalists.

6. Evidence existed that antibodies by themselves could protect against HIV. In "passive immunization" studies, researchers infused chimps with a mixture of antibodies taken from HIV-infected humans and then showed that the animals remained virus-free after being challenged with infectious doses. See Alfred Prince et al., "Prevention of HIV Infection by Passive Immunization with HIV Immunoglobulin," *AIDS Research and Human Retroviruses*, 7(12):971–973 (December 1991).

7. Proponents of the gp120 approach emphasized that the simian and human versions had pronounced differences. Many scientists believed that the Biocine and Genentech vaccines worked because they triggered production of antibodies that could attach to the V3 loop, the Achilles heel of HIV's gp120. Since scientists studying SIV had not found a corollary to the V3 loop, some maintained that simian trials of gp120 vaccines held little meaning.

8. See my story "Is NIH Failing an AIDS 'Challenge'?" *Science*, 251:518–520 (February 1, 1991).

9. Joe Palca, "Errant HIV Strain Renders Test Virus Stock Useless," *Science*, 256:1387–1388 (June 5, 1992).

10. Gregory J. LaRosa et al., "Conserved Sequence and Structural Elements in the HIV-1 Principal Neutralizing Determinant," *Science*, 249:932–935 (August 24, 1990). The researchers later discovered that the paper contained many errors, which they corrected and clarified in a technical comment (*Science*, 251:811, February 15, 1991). They wrote that the corrections "in no way affect any conclusions drawn from the data presented."

11. According to Pasteur Mérieux Connaught literature, the New York State Department of Health owns the other 15% of Virogenetics. The collaborators on this vaccine, called the AGIS group, included the Pasteur Institute's Marc Girard, Duke's Dani Bolognesi, NCI's Robert Gallo, Jefferson Medical College's Hilary Koprowski, and Pasteur Mérieux's Philippe Kourilsky. The group is listed in a paper about a different vaccine strategy: Dominique Salmon-Céron et al., "Safety and Immunogenicity of a Recombinant HIV Type 1 Glycoprotein 160 Boosted by a V3 Synthetic Peptide in HIV-Negative Volunteers," *AIDS Research and Human Retroviruses*, 11(12):1479–1486 (December 1995).

12. I'm quoting from a written copy that Mary Lou Clements gave me of her presentation.

13. Rather than growing the SF2 in an immortalized cell line, researchers kept it alive by feeding it peripheral blood mononuclear cells, the name given to a mixture that includes CD4s. The reason why growing a freshly harvested isolate of HIV in an immortalized cell line would alter the virus did not become clear until two years later.

14. For a thorough discussion of the R&D for *Haemophilus influenzae* type b and how it relates to the AIDS vaccine search, see William Heyward, Kathleen MacQueen, and Karen Goldenthal, "HIV Vaccine Development and Evaluation: Realistic Expectations," *AIDS Research and Human Retroviruses*, 14:(suppl. 3):205–210 (October 1998).

15. Although there is wide agreement that the vote was nearly unanimous, many versions of the actual count have made the rounds. The WHO's Jose Esparza, who took notes, says it was not actually a vote, but 21 people clearly said go forward, 8 were maybes, and 2 clearly said no. Esparza abstained, as did the FDA representative.

16. The meeting, entitled "HIV Preventive Vaccines: Social, Ethical, and Political Considerations for Domestic Efficacy Trials," was sponsored by the AIDS Action Foundation, with funding provided by the Henry J. Kaiser Family Foundation and the NIH's Office of AIDS Research. The AIDS Action Foundation (which backed

a lobbying group, the AIDS Action Council) distributed at the meeting a "working draft" of a monograph with the same title. The monograph summarized discussions held by a multidisciplinary working group prior to the May meeting.

17. Research anthropologist Kathleen MacQueen presented the CDC study, while Liza Solomon described the Johns Hopkins data.

18. James Jones, *Bad Blood: The Tuskegee Syphilis Experiment—A Tragedy of Race and Medicine* (The Free Press, 1981).

19. This activist was Eric Ciasullo, commissioner of the San Francisco Delinquency Prevention Commission and acting HIV prevention coordinator at the Lavender Youth Recreation and Information Center.

20. I asked Fields specifically what he thought about staging efficacy trials of the gp120 vaccines. "I wouldn't fund a very large, expensive trial," Fields said. "I think the likely outcome is, it's very unlikely to be positive."

21. See my story, "Will Media Reports KO Upcoming Real-Life Trials?" *Science*, 264:1660 (June 17, 1994).

22. In addition to the Biocine and Genentech gp120 vaccines, breakthrough infections had occurred with vaccines made by MicroGeneSys, Bristol-Myers, and Viral Technologies.

23. "Who Put the Lid on gp120?" by Jesse Green, *New York Times Magazine* (March 16, 1995), p. 50.

24. I do not know when and where Frederickson said or wrote this; the earliest reference I can find for the quote is Curtis L. Meinert, *Clinical Trials: Design, Coordination and Analysis* (Oxford University Press, 1986), p. 18.

25. The letter, dated June 13, 1994, was written by Prasert Thongcharoen and signed by members of a group he chaired called the THAIVEG Consortium. Prasert warned that if the ARAC and Fauci decided not to proceed, companies might decide not to make Thai-based vaccines, which "could have devastating effects on our population." While Prasert said the Thais did not want to influence Fauci's decision making, they asked that he clearly distinguish the needs in the United States from those of developing countries that had been harder hit by HIV.

26. The ACT UP New York speaker was Luis Santiago. The other speakers were Garance Frank-Ruta of the Treatment Action Group, Bill Snow of ACT UP Golden Gate, and Boston's David Scondras of the Massachusetts AIDS Fund.

27. Although I do not have any evidence—nor do I believe—that Bolognesi spoke against the efficacy trials because of his relationship with Pasteur Mérieux Connaught, many researchers in the field noted the inconsistency. "I had the same problem: Why one and not the other?" Bolognesi told me in 1998. "Somehow I got clearance." Jack Killen, then director of NIAID's Division of AIDS, remembered it differently. Killen, who stressed that "being a 'competitor' does not equate with prohibited conflict of interest," said Bolognesi chose to attend one meeting but not the other.

Both times, Killen said he or someone else from NIAID told Bolognesi that "we would like him to participate and that he could from a legal point of view, but that it was his call to do what he felt was right, and that we would understand either way."

28. See my article "Naked DNA Points Way to Vaccines," *Science*, 259(5102):1691–1692 (March 19, 1993).

29. The VaxGen vaccine mixed the original gp120 made from the MN strain of HIV with a gp120 derived from a primary isolate of HIV. The new HIV strain was selected specifically because antibodies against its gp120 could, in test-tube experiments, neutralize HIV that had managed to infect one of the breakthrough cases.

CHAPTER 12 NEW WORLD ORDER

1. "Panel Votes to Delay Real-World Vaccine Trials," *Science*, 264:1839 (June 24, 1994).

2. Donald Kennedy in 2000 became editor-in-chief of *Science*. He has had no influence on anything I have reported, and indeed, when he took the job, we still had never communicated with each other about AIDS vaccines or anything else.

3. Jean-Paul headed France's Agence Nationale de Recherches sur le SIDA.

4. For an overview of the concerns about enhancing antibodies and HIV vaccines, see Deborah Barnes, "Another Glitch for AIDS Vaccines?" *Science*, 241:533–534 (July 29, 1988).

5. John Moore and Roy Anderson, "The WHO and Why of HIV Vaccine Trials," *Nature*, 372:313–314 (November 24, 1994).

6. Anderson later published a paper about this model, "Low-Efficacy HIV Vaccines: Potential for Community-Based Intervention Programmes," *Lancet*, 348:1010–1013 (October 12, 1996). In this paper, Anderson and coauthor G. P. Garnett describe how if a vaccine had 20% efficacy and was used by everyone in a community, it could reduce the prevalence of HIV-infected people in a population from 27% to 16%. The paper does not specify the time required for this drop, but makes the point that "the general principle seems to be robust to parameter variation—namely, low-efficacy vaccines at high coverage can act to greatly reduce the endemic prevalence of HIV-1."

7. This malaria vaccine, SPf66, failed to demonstrate efficacy in subsequent human trials and was abandoned. See Shoklo SPf66 Malaria Vaccine Trial Group, Francis Nosten et al., "Randomised Double-Blind Placebo-Controlled Trial of SPf66 Malaria Vaccine in Children in Northwestern Thailand," *Lancet*, 348(9029): 701–707 (September 14, 1996).

8. Dorman attracted well-heeled potential investors with help from Acrogen's chairman of the board, former U.S. Postmaster General Anthony Frank.

9. "Whatever Happened to Sidney Weniger?" by George Waldon, *Arkansas Business* (December 10, 1990), p. 16. This article also discusses how Weniger

declared bankruptcy in 1987 after the failure of a savings and loan that financed many of his Arkansas projects.

10. Robert Tietelman, *Gene Dreams* (Basic Books, 1989), p. 38.

11. "Striking It Rich in Biotech," by Stuart Gannes, *Fortune* (November 9, 1987), p. 131.

12. Malcolm Martin lab's at NIAID first presented data about a disease-causing SHIV that contained the HIV envelope at the Second National Conference on Human Retroviruses and Related Infections, January 29–February 2, 1995, abstract 397.

13. Bruce Weniger et al., "The Epidemiology of HIV Infection and AIDS in Thailand," *AIDS*, 5(suppl. 2):S71–85 (1991).

14. Kenrad Nelson et al., "Risk Factors for HIV Infection Among Young Adult Men in Northern Thailand," *Journal of the American Medical Association*, 270(8):955–960 (August 25, 1993).

15. The American epidemiologist, Chris Beyrer of Johns Hopkins University, stressed that the gay men in Thailand actually faced as much risk as heterosexual men. Still, Anek's perception of risk jarred me.

16. UNAIDS, which stands for the United Nations Programme on HIV/AIDS, combined the WHO's AIDS efforts with those at UNICEF, UNESCO, the UN Development Programme, the UN Population Fund, and the World Bank. This joint program did not formally begin operating until January 1996.

17. The Third International Conference on AIDS in Asia and the Pacific held in conjunction with the Fifth National AIDS Seminar in Thailand was held at the Pang Suan Keaw Hotel in Chiang Mai, Thailand, September 17–21, 1995.

18. The quotes in this and in the next three paragraphs come from "HIV Vaccines: Accelerating the Development of Preventive HIV Vaccines for the World," summary report and recommendations of an international meeting, sponsored by the Rockefeller Foundation, Bellagio, Italy, March 7–11, 1994.

19. IAVI published reports about both of these meetings. The first meeting took place in Paris on October 27–28, 1994, and had 16 scientists participating. Its 24-page report is entitled *HIV Vaccines—Accelerating the Development of Preventive HIV Vaccines for the World*, published February 1995 by Le Val de Grace, Paris. The second meeting gathered financial experts, lawyers, and public health experts in New York on August 17, 1995. The resulting 16-page report has the same title as the first report but included the subtitle *Financial and Structural Issues* (New York, December 1995). IAVI formally became incorporated in the United States on January 26, 1996.

20. The Correlates of HIV Immune Protection group, funded by a contract from NIAID, included John Moore and his boss, David Ho; George Shaw, Beatrice Hahn, and Sten Vermund from the University of Alabama at Birmingham; Steven

Wolinsky, Northwestern University; David Montefiori, Duke University; Bruce Walker, Massachusetts General Hospital; and Bette Korber and Avidan Neumann from the Los Alamos National Laboratory.

21. Alexandra Levine, the University of Southern California oncologist who pioneered the testing of Salk's therapeutic AIDS vaccine, chaired PACHA's research committee.

22. Included in this group were Don Francis, Harvard's Max Essex—chair of the AIDS Institute there and Francis's graduate school adviser—Acrogen's Burt Dorman, IAVI's Seth Berkley, Chiron's Anne-Marie Duliège, the WHO's José Esparza, the CDC's William Heyward, and Wayne Koff of United Biomedical Inc. To my surprise, my name, too, was on Weniger's list.

23. This November 17, 1995, letter came from Yichen Lu, Bryan Roberts, and Lendon Payne of the Virus Research Institute, a small company in Cambridge, Massachusetts that was testing AIDS vaccines in monkeys.

24. The Eighth Annual Conference on Advances in AIDS Vaccine Development was held in the NIH's Natcher Building on February 11–15, 1996.

25. The journal *Science* is divided into journalism and original research reports, and, for the most part, the two staffs have a church-and-state relationship. As editor-in-chief, Floyd Bloom, a neuroscientist at the Scripps Research Institute, oversaw the entire publication, but he did not edit my copy nor influence its content.

26. Bill Paul phoned research committee chair Alexandra Levine about his concerns. She, in turn, organized the "emergency" conference call of PACHA's research committee, on May 14, 1996.

27. Weniger's May 16, 1996 e-mail included anonymous "community" comments that he had solicited about Paul's concerns. The NIH was being "sensitive" and "somewhat defensive," said one, while another suggested that "your group should not be overly influenced by that one person." Weniger did suggest that they add a few sentences about the NIH contributing to a coordinated federal effort.

28. The Eleventh International Conference on AIDS was held on July 7–12, 1996, in Vancouver, Canada.

29. This collaboration was formed on April 19, 1993.

30. "The End of AIDS?" written by Andrew Sullivan, appeared on p. 52 of the December 2, 1996, issue of the *New York Times Magazine*, which included a disclaimer in the subhead: "Not Yet—But New Drugs Offer Hope." *The Wall Street Journal* ran David Sanford's essay, "Last Year, This Editor Wrote His Own Obituary," on November 8, 1996, p. 1. *Time* magazine named David Ho its Man of the Year in the December 30, 1996, issue.

31. AVAC, founded in December 1995, consisted of Sam Avrett, Chris Collins, Garance Franke-Ruta, David Gold, and William Snow.

32. *New York Times* (January 10, 1997), p. A32.

33. On June 22, 1997, following three days of talks, the Denver Summit of

The Eight issued a "communiqué" that included an item about AIDS vaccines. "We will work to provide the resources necessary to accelerate AIDS vaccine research, and together will enhance international scientific cooperation and collaboration," it read in part.

34. The press release about the President and AIDS vaccines was issued by the White House's Office of National AIDS Policy on the day of his Morgan State speech.

35. Paul first raised the idea of a center to Clinton himself on December 1, 1996, World AIDS Day, in a White House meeting that included Tony Fauci and Harold Varmus.

CHAPTER 13 RUNNING IN PLACE

1. H. Kulaga et al., " Infection of Rabbits with Human Immunodeficiency Virus 1: A Small Animal Model for Acquired Immunodeficiency Syndrome," *Journal of Experimental Medicine*, 169(1):321–326 (January 1, 1989).

2. B. W. Snyder et al., "Developmental and Tissue-Specific Expression of Human CD4 in Transgenic Rabbits," *Molecular Reproduction and Development*, 40(4):419–428 (April 1995).

3. The $3.5 million per year figure and the quote come from a draft document of the OAR's Etiology and Pathogenesis Area Review Panel. The panel's final report made the same points but was a bit more diplomatic, referring to the rabbit work as "low priority" and "disappointing."

4. Laurie Garrett, "Mi$$pent," *Newsday* (March 13, 1996), p. A5.

5. Yu Feng et al., "HIV-1 Entry Cofactor: Functional cDNA Cloning of a Seven-Transmembrane, G Protein-Coupled Receptor," *Science*, 272(5263):872–877 (May 10, 1996). I wrote an accompanying article, "Likely HIV Cofactor Found," p. 809.

6. William Paxton et al., "Relative Resistance to HIV-1 Infection of CD4 Lymphocytes from Persons Who Remain Uninfected Despite Multiple High-Risk Sexual Exposures," *Nature Medicine*, 2(4):412–417 (April 1996).

7. Rong Liu et al., "Homozygous Defect in HIV-1 Coreceptor Accounts for Resistance of Some Multiply-Exposed Individuals to HIV-1 Infection," *Cell*, 86:367–377 (August 9, 1996).

8. Roberto Speck et al., "Rabbit Cells Expressing Human CD4 and Human CCR5 Are Highly Permissive for Human Immunodeficiency Virus Type 1 Infection," *Journal of Virology*, 72(7):5728–5734 (July 1998).

9. Burton spoke to the NIH's AIDS Vaccine Research Committee on September 30, 1997.

10. Dennis Burton and John Moore, "Why Do We Not Have an HIV Vaccine and How Can We Make One?" *Nature Medicine*, 4(5 suppl.):495–498 (May 1998).

11. For an overview of these three monoclonal antibodies, see A. Trkola et al., "Cross-Clade Neutralization of Primary Isolates of Human Immunodeficiency Virus

type 1 by Human Monoclonal Antibodies and Tetrameric CD4-IgG," *Journal of Virology*, 69(11):6609–6617 (November 1995).

12. Although researchers attempting to make a mouse model for AIDS had yet to figure out how to make HIV replicate inside of mouse cells, they had succeeded in making transgenic mice that had human CD4 and CCR5 receptors. This transgenic mouse had no immune responses to these human receptors, and thus allowed Nunberg to inject them with his vaccines and harvest antibodies that were directed at the viral protein.

13. This quote comes from the grant announcement published in the *NIH Guide*, 26(7):PAR-97-04 (March 7, 1997).

14. Rachel A. LaCasse et al., "Fusion-Competent Vaccines: Broad Neutralization of Primary Isolates of HIV," *Science*, 283:357–362 (January 15, 1999). David Montefiori and John Moore wrote the perspective, "HIV Vaccines: Magic of the Occult?" (p. 336).

15. Behringwekre AG, a German company, in 1993 published results of a failed chimpanzee challenge experiment with a poorly characterized whole, killed HIV vaccine: M. Niedrig, "Immune Response of Chimpanzees After Immunization with the Inactivated Whole Immunodeficiency Virus (HIV-1), Three Different Adjuvants and Challenge," *Vaccine*, 11(1):67–74 (1993). A group led by Marc Girard at the Pasteur Institute protected chimps with a whole, killed HIV vaccine, but they boosted the animals with envelope proteins, making it impossible to determine what accounted for the animals' resisting the challenge virus. See Marc Girard et al., "Immunization of Chimpanzees Confers Protection Against Challenge with Human Immunodeficiency Virus," *Proceedings of the National Academy of Sciences*, 88(2):542–546 (January 15, 1991).

16. John Oxford, a virologist at the Royal London Medical College and scientific director of the company Retroscreen, did make a whole, killed HIV that retained its envelope, but he too had great difficulty advancing it and eventually assigned patents to an Italian vaccine maker, ISI, in exchange for research funding. In January 1999, Oxford told me that Retroscreen hoped to begin testing the vaccine as a therapeutic in about one year's time. He had no immediate plans to test it as a preventive.

17. For a thorough, but polemical, overview of how the California Endowment/California Healthcare Foundation came to be, see "Saving Their Assets: How to Stop Plunder at Blue Cross and Other Nonprofits," by Judith Bell, *American Prospect*, 26:60–66 (May–June 1996).

18. J. L. Rossio et al., "Inactivation of Human Immunodeficiency Virus Type 1 Infectivity with Preservation of Conformational and Functional Integrity of Virion Surface Proteins," *Journal of Virology*, 72(10):7992–8001 (October 1998).

19. Eric Rosenberg et al., "Vigorous HIV-1-Specific CD4+ T Cell Responses Associated with Control of Viremia," *Science*, 278:1447–1450 (November 21, 1997).

20. In more technical terms, Walker's laboratory had identified epitopes on

HIV's core proteins that stimulated CD4 help. Peptimmune had a technology for stringing together these key epitopic protein pieces, called peptides. By testing different peptides in mice, they hoped to find the combination of epitopes that could most potently stimulate HIV-specific CD4 help.

21. These figures appear on p. 20 of *HIV Vaccines: Accelerating the Development of Preventive HIV Vaccines for the World*, a summary report of the landmark meeting held in Bellagio, Italy. The Rockefeller Foundation, the biggest backer of what became the International AIDS Vaccine Initiative, published the report in June 1994.

22. Thomas Lehner et al., "Protective Mucosal Immunity Elicited by Targeted Iliac Lymph Node Immunization with a Subunit SIV Envelope and Core Vaccine in Macaques," *Nature Medicine*, 2:767–775 (July 1996).

23. Lehner called this approach "targeted iliac lymph node" vaccination.

24. Bolstering the role of chemokines in protection, France's Daniel Zagury, in collaboration with Robert Gallo, published a paper in 1998 that described 128 hemophiliacs who repeatedly had received HIV-contaminated lots of clotting factor prior to the development of a blood test to screen for the virus. Strikingly, most everyone resisted HIV infection for a period of time, which correlated with levels of the same chemokines that Gallo's lab had made famous. And the 14 people who completely resisted infection by and large overproduced these same chemokines. Daniel Zagury et al., "C-C chemokines, Pivotal in Protection against HIV Type 1 Infection," *PNAS*, 95(7):3857–3861 (March 31, 1998).

A separate study done by Frank Plummer's group on exposed, uninfected prostitutes in Nairobi, Kenya, did not find a relationship between chemokine levels and resistance. K. R. Fowke, "HIV Type 1 Resistance in Kenyan Sex Workers Is Not Associated with Altered Cellular Susceptibility to HIV Type 1 Infection or Enhanced Beta-Chemokine Production," *AIDS Research and Human Retroviruses*, 14(17):1521–1530 (November 20, 1998). Negative data do not, of course, disprove positive data, but they do suggest that different mechanisms of protection might be responsible for different populations, especially considering that the main routes of infections differ for these prostitutes and hemophiliacs.

25. Sandra Mazzoli et al., "HIV-Specific Mucosal and Cellular Immunity in HIV-Seronegative Partners of HIV-Seropositive Individuals," *Nature Medicine*, 3(11):1250–1257 (November 1997).

26. In 1998 the annual AIDS conference changed its name from "International Conference on AIDS" to "World AIDS Conference."

27. Chris Beyrer et al., "Epidemiologic and Biologic Characterization of a Cohort of Human Immunodeficiency Virus Type 1 Highly Exposed, Persistently Seronegative Female Sex Workers in Northern Thailand," *Journal of Infectious Diseases*, 179:59–67 (January 1999). The diverse group of researchers came from Johns Hopkins University, the Walter Reed Army Institute of Research, the Henry M.

Jackson Foundation, NIAID, Chiang Mai University, Thailand's Department of Communicable Disease Control, Siriraj Hospital, the U.S. Armed Forces Research Institute for Medical Sciences, and University College London Medical School.

28. Lehner also had found mucosal antibody-producing cells in lymph nodes.

29. MacDonald's work focused on markers known as "major histocompatibility complexes." She presented her findings in an oral presentation on June 30, 1998: "Class I MHC Polymorphism and Mother-to-Child HIV-1 Transmission in Kenya," abstract 31131.

30. Basically, the vaccines worked because the researchers had grown SIV in human cells. The surface of the monkey virus thus incorporated protein pieces of the human cells. Antibodies to these human cells, Stott showed, could defeat a challenge with SIV grown in human cells, but not SIV grown in monkey cells.

31. Ligia Pinto et al., "Alloantigen-Stimulated Anti-HIV Activity," *Blood*, 92(9):3346–3354 (November 1, 1998).

32. Shearer's collaborator, Sebastiano Gattoni-Celli, studied cancer immunology at the Medical University of South Carolina.

33. Canada's HIV Vaccine Development Meeting was held February 4–5, 1997, at Chateau Cartier, Aymer, Québec. This quote appears in the official proceedings.

34. The Denver Summit of Eight, after three days of meetings, issued their final communiqué on June 22, 1997.

CHAPTER 14 BETTER WAYS

1. The quotes here and in the next two paragraphs come from *9 Years and Counting: Will We Have an HIV Vaccine by 2007? An Agenda for Action for an HIV Vaccine* (AIDS Vaccine Advocacy Coalition, May 1998). Except where otherwise noted, the quotes are taken from pp. 5–17 of this 55-page report.

2. Ibid., p. 34. AVAC estimated that Walter Reed spent about half of its $25 million budget on AIDS vaccine work.

3. The *Wall Street Journal* article about the AVAC report, written by Michael Waldholz, ran on May 15, 1998, p. B6.

4. David Gold, AVAC's founder, edited IAVI's newsletter.

5. The Integrated Preclinical/Clinical Program was, as IAVI noted, "a redesign" of the former National Cooperative Vaccine Development Groups. The big difference was that the new program explicitly required human trials, which NIAID stipulated "must be feasible, realistic, and achievable within the period of the award." For more details, see the NIH's Program Announcement 97-056, issued April 25, 1997, and its April 3, 1998 addendum.

6. The quotes here and in the next paragraph come from *Scientific Blueprint for AIDS Vaccine Development* (International AIDS Vaccine Initiative, June 1998), p. 12.

7. Other IAVI donors included the Rockefeller Foundation, the Starr Foundation, Until There's a Cure Foundation, UNAIDS, the Elton John AIDS Foundation, Glaxo Wellcome, Foundation Marcel Merieux, the U.K.'s National AIDS Trust, and the Alfred P. Sloan Foundation.

This last funder, independent of IAVI's grant, gave me a grant to complete this book. In no way did the Sloan Foundation attempt to influence my opinions or my coverage of IAVI.

8. It helped, too, that IAVI had recently hired Victor Zonana, a press person for Health and Human Services Secretary Donna Shalala—and a former AIDS reporter at the *Los Angeles Times*.

9. The *Newsweek* online story "Shot Boosters" ran on July 1, 1998, and was all the more flattering because it was written by one of the more critical science reporters around, Oliver Morton, who for several years trenchantly covered AIDS for the *Economist*.

10. Nathanson and Francisco Gonzalez-Scarano's dour AIDS vaccine essay, published in *Advances in Veterinary Science and Comparative Medicine* (33:397, 1989), appeared prior to anyone protecting an animal from SIV or HIV. "It is highly questionable whether a conventional vaccine, even one employing recombinant DNA methodology, will be capable of preventing AIDS," they wrote, suggesting that researchers should, as Salk suggested in 1987, focus on vaccine and drug strategies that could prevent an infected person from developing disease.

11. See Chapter 5. Hilleman was not the only prominent scientist to advocate this strategy. The year before Nathanson became the OAR director, NIAID virologist Malcolm Martin had drafted a detailed proposal to stage a massive monkey experiment. In an April 7, 1997, e-mail to then OAR director Bill Paul, Martin asked, "Do you have a spare $10,000,000?" Martin then outlined an experiment that would use 550 macaques to compare 18 different SIV vaccine strategies. Martin went so far as to specify how many people it would take to do the job (5 veterinarians, 10 animal caretakers, and 20 laboratory technicians), which strains of SIV to use as challenge viruses, and the timing of the challenge itself. He emphasized that in addition to collecting blood samples and lymph nodes from the animals, the experiment should run long enough to determine the cause of death in each animal. As far as I can determine, Martin's proposal never received any serious discussion, and I only learned of it from Martin himself.

12. The reformulated VaxGen vaccine contained an additional gp120 from a primary isolate, which stimulated antibodies capable of neutralizing a strain of virus that had infected one of the breakthrough cases in the earlier studies.

13. On October 29, 1999, VaxGen announced in a press release that the Centers for Disease Control and Prevention had decided to contribute $2 million a year for four years to the company's efficacy trials. "The CDC funds will support additional epidemiological, social and behavioral research, designed to provide a

more powerful analysis of vaccine effectiveness and evaluation of factors contributing to participation and risk behaviors as the trial progresses," the release stated.

14. McMichael's collaborators were J. J. Bwayo (and, later, Omu Anzala) of the University of Nairobi and the German company Impfstoffwerk Dessau-Tornau, GmbH.

15. "Salvation in a Snippet of DNA?" by Gary Taubes, *Science*, 278:1711 (December 5, 1997).

16. According to Apollon press releases, tests of the vaccine in infected people began on June 16, 1995, and the company received FDA approval to stage trials in uninfected volunteers on March 25, 1996.

17. Jean Boyer et al., "Protection of Chimpanzees from High-Dose Heterologous HIV-1 Challenge by DNA Vaccination," *Nature Medicine* 3(5):526–532 (May 1997).

18. "HIV DNA Vaccines Move Slowly into Human Trials," by David Gold and Sam Avrett, *IAVI Report*, 3(3):1 (July–September 1998).

19. Norman Letvin et al., "Potent, Protective Anti-HIV Immune Responses Generated by Bimodal HIV Envelope DNA Plus Protein Vaccination," *Proceedings of the National Academy of Sciences*, 94(17):9378–9383 (August 19, 1997).

20. Vanessa Hirsch et al., "Patterns of Viral Replication Correlate with Outcome in Simian Immunodeficiency Virus (SIV)-Infected Macaques: Effect of Prior Immunization with a Trivalent SIV Vaccine in Modified Vaccinia Virus Ankara," *Journal of Virology*, 70(6):3741–3752 (June 1996). Although the vaccine did not prevent infection, it did appear able to lower viral loads and delay disease. Gunnel Biberfeld of the Swedish Institute for Infectious Disease Control presented similar data at NIAID's 1997 Conference on Advances in AIDS Vaccine Development: "Outcome of Rectal SIVsm Challenge of Macaques Vaccinated with Modified Vaccinia Ankara Expressing SIVsm env and gag-pol," on p. 72 of the program book from the meeting.

21. Zagury and colleagues wrote a letter to the editor about cutaneous necrosis that occurred in three patients given a therapeutic HIV vaccine that contained vaccinia. O. Picard et al., "Complication of Intramuscular/Subcutaneous Immune Therapy in Severely Immune-Compromised Individuals," *Journal of Acquired Immune Deficiency Syndrome*, 4(6):641–643, (May 1991).

Although this live virus vaccine intentionally was given to HIV-infected people, it is easy to imagine how an infected person might inadvertently receive such a vaccine: with mass vaccination campaigns, it's unlikely that people would first be tested for their HIV status, which inevitably would lead to HIV-infected people unwittingly taking a vaccine. Just such a case of vaccinia disease occurred in the U.S. Army, which still vaccinated soldiers against smallpox in the 1980s. See Robert Redfield et al., "Disseminated Vaccinia in a Military Recruit with Human Immunodeficiency

Virus (HIV) Disease," *New England Journal of Medicine*, 316(11):673–676 (March 12, 1987).

Another problem with vaccinia is that the virus can pass from an immunized person to an unimmunized one. If an AIDS vaccine containing vaccinia were used in a population at high risk for HIV infection, presumably this "weakened" virus could harm immunocompromised people in the community.

22. A. Seth et al., "Recombinant Modified Vaccinia Virus Ankara-Simian Immunodeficiency Virus gag pol Elicits Cytotoxic T Lymphocytes in Rhesus Monkeys Detected by a Major Histocompatibility Complex Class I/Peptide Tetramer," *PNAS*, 95(17):10112–10116 (August 18, 1998).

23. Gallo's lab meeting, sponsored by his Institute for Human Virology, ran on August 23–28, 1998 in Baltimore, Maryland. For a detailed description of Johnston's presentation, see the unbylined article "Key Vaccine Reports at Human Virology Meeting," *IAVI Report*, 3(4):8 (October–December 1998).

24. After leaving NIAID in 1992, Koff headed development of an AIDS vaccine project at New York's United Biomedical Inc.

25. *Newsday's* Laurie Garrett suggested the idea to me of calling this organization the March of Dollars.

26. Actually, I suspect that breeders in the United States now have enough monkeys to conduct this experiment. Although AIDS researchers cannot purchase all of the monkeys they need—which I wrote about in the February 11, 2000, *Science* (287:959–960)—that's a function of cost, not supply. The NIH simply does not allocate enough money for them to purchase monkeys at the going rate. A well funded philanthropic organization could, theoretically, afford to spend $5,000 an animal, which is roughly twice the amount that an NIH-funded researcher can afford. If the experiment used 1,000 monkeys, which is a high estimate, it then would cost $5 million for the purchase of the animals. That's a fortune to an NIH researcher—but should not hurt the budget of the organization that I am proposing.

CHAPTER 15 DISPARATE MEASURES

1. I first wrote about the ethical debates surrounding these trials nearly two years before Public Citizen claimed to have "revealed" the dilemma, based on documents, the group wrote in its press release, that it had "obtained from the government." For details, see my article, "Bringing AZT to poor countries," *Science*, 269:624–626 (August 4, 1995). NIH director Harold Varmus indeed made this point when he testified before the congressional subcommittee hearing on May 8, 1997, stating, "The issues that were raised by Public Citizen and brought to your attention are not new ones."

2. The other signatories were clinician Wilbert Jordan of the Oasis Clinic and AIDS Program in Los Angeles, Boston University bioethicists George Annas (a

lawyer) and Michael Grodin (a clinician), and Yale University clinician George Silver.

3. *World Medical Association Declaration of Helsinki: Recommendations Guiding Physicians in Biomedical Research Involving Human Subjects* (18th World Medical Assembly, Helsinki, 1964; revised 1975, 1983, 1989).

4. The hearing was held by the U.S. House of Representatives, Committee on Government Reform and Oversight, Subcommittee on Human Resources.

5. The first quote comes from a letter written by pediatrician Robert Broadhead of the University of Malawi, while the other quotes were in a letter written by J. D. Chiphangwi and L. Mtimavalye. Both letters were addressed to collaborators at Johns Hopkins University, who forwarded them to Varmus.

6. This six-page letter was signed by 15 Hopkins researchers, including Neal Halsey, Donald A. Henderson, and Kenrad Nelson.

7. Interestingly, "The Ethical Design of an AIDS Vaccine Trial in Africa" by Nicholas A. Christakis (in the June–July 1988 *Hastings Center Report*, 18:31–37), did explore the Zagury tests, but it offered neither new information about what actually occurred in Zaire nor any overt criticism. Wendy Mariner's "Why Clinical Trials of AIDS Vaccines Are Premature," published in the January 1989 issue of *Public Health and the Law* (79:86–91), makes no mention whatsoever of Zagury's trials. As for the small meetings, the WHO's Global Programme on AIDS issued a "Statement from the Consultation on Criteria for International Testing of Candidate HIV Vaccines" following a conference held February 27–March 2, 1989. The Institute of Medicine's International Forum for AIDS Research on February 12–13, 1991, held a meeting at the National Academy of Sciences entitled "Preparation for Vaccine Trials and Their Management," which discussed ethical issues in some depth. I was barred from most of the meeting, but did write an article about it, "AIDS Vaccine Trials: Bumpy Road Ahead," *Science*, 251:1312–1313 (March 15, 1991).

8. Crewdson's story quotes from "an NIH report obtained by the *Chicago Tribune*." I could not find the exact language that Crewdson quoted in the story in the official documents that OPRR supplied me with after I filed a Freedom of Information Act request. But this passage, from a January 24, 1991, memo written by F. William Dommel, closely matched what Crewdson reported.

9. Carol Levine of New York's Citizens Commission on AIDS.

10. Seymour Klebanoff, a University of Washington clinician who chaired the panel of consultants.

11. This critique, dated June 21, 1991, was cosigned by Zagury's main collaborator, Odile Picard, but he wrote it in the first person.

12. France's Bruno Durieux, from the office of the minister of health, wrote the OPRR in August 1991 that its request to come to France to investigate Zagury "is not within the framework of a judicial procedure and there is no Franco-American or multilateral agreement that provides a sufficient legal basis to justify carrying out,

by foreign agents, an administrative inquiry on French soil." I obtained this through the Freedom of Information Act, which cataloged the letter as Attachment B of the OPRR's final report on the Zagury matter, issued March 26, 1993.

13. See the article I wrote for *Science*, "The Rise and Fall of Projet SIDA," 278:1565–1568 (November 28, 1997).

14. See pp. 101 and 102. The book was published by Indiana University Press.

15. The April 20, 1996, editions of two local newspapers covered the scare: Sean Gardiner, "Primate Center Takes No Chances with Suspicious Box," *Middlesex News* (page number unknown: Ron Desrosiers gave me a copy of this, but it did not have a page number, and it's not listed in the database Nexus); and Mary Frain, "Package Scare Was False Alarm," *Telegram & Gazette*, p. A5.

16. NIAID's Conference on Advances in AIDS Vaccine Development, Seventh Annual Meeting of the National Cooperative Vaccine Development Groups for AIDS, was held November 6–10, 1994, in Reston, Virginia.

17. Reinforcing the notion that delta-3 had not become more pathogenic, when Ruprecht gave blood from that monkey to another newborn and its mother, the newborn suffered severe CD4 drops, yet the mother remained healthy.

18. Ruprecht later published her findings in *Science*. Timothy Baba et al., "Pathogenicity of Live, Attenuated SIV After Mucosal Infection of Neonatal Macaques," 267(5205):1820–1825 (March 24, 1995).

19. Frank Kirchhoff et al., "Brief Report: Absence of Intact nef Sequences in a Long-Term Survivor with Nonprogressive HIV-1 Infection," *New England Journal of Medicine*, 332(4):228 (January 26, 1995).

20. Nicholas Deacon et al., "Genomic Structure of an Attenuated Quasi Species of HIV-1 from a Blood Transfusion Donor and Recipients," *Science*, 270:988–991 (November 10, 1995). I wrote an accompanying story that ran on p. 917 of the same issue.

21. The Australian Broadcasting Corporation's science show, *Quantum*, first ran its documentary, "Billion Dollar Blood," on August 7, 1997.

22. "Attenuated HIV Vaccine: Caveats," *Science*, 271:1790–1792 (March 29, 1996). The Australian researchers offered a strong counterargument that indeed the patient had not died from AIDS and stressed that they stood by their contention that this attenuated HIV might be the basis for a vaccine.

23. Michael Wyand et al., "Resistance of Neonatal Monkeys to Live Attenuated Vaccine Strains of Simian Immunodeficiency Virus," *Nature Medicine*, 3:32–36 (January 1997).

24. Charles Farthing, "A Call to Physicians," *Journal of the International Association of Physicians in AIDS Care*, 3(8):10 (August 1997).

25. This session took place on July 2, 1998.

26. T. C. Greenough, John Sullivan, and Ron Desrosiers, "Declining CD4 T-

Cell Counts in a Person Infected with nef-Deleted HIV-1," *New England Journal of Medicine*, 340(3):236–237 (January 21, 1999).

27. Berman technically still was with Genentech at the time.

28. Ho e-mailed me on May 20, 1997, from China, where he visited after Thailand, and confirmed the gist of these remarks. The *Bangkok Post* article, written by Aphaluck Bhatiasevi, appeared on p. 3 of the *Bangkok Post* (May 17, 1997).

29. José Esparza, William Heyward, and Saladin Osmanov, "HIV Vaccine Development: From Basic Research to Human Trials," *AIDS*, 10(suppl. A):S123–132 (1996).

30. Weniger shared this letter with me, despite Moore's having warned him that if he did not treat this communication as private, "I will make a vigorous official complaint to your superiors." I asked Moore for his permission to use it; he said that as long as I made it clear how I received it, he had no objection.

31. Weniger followed up the next year by lodging a formal complaint against Moore to Nobel laureate Torsten Wiesel, head of Rockefeller University, an affiliate of the Aaron Diamond AIDS Research Center. "I believe Dr. Moore's letter has overstepped even the most extreme boundaries of freedom of speech into the proscribed areas of threat and libel," wrote Weniger in his March 18, 1998, letter. Wiesel replied on April 7 that this was a matter for Aaron Diamond, not Rockefeller, and he forwarded the complaint to David Ho "for whatever action and response he deems appropriate." And that was the end of it.

32. Mann spoke to the International Issues subcommittee of PACHA at the Madison Hotel, Washington, D.C.

33. The letter appeared on p. 803. Mann limply replied in the May 29 issue (p. 1327), basically repeating the same arguments he made at the PACHA meeting.

34. Bloom spelled out his thinking in a *Science* editorial, "The Highest Attainable Standard: Ethical Issues in AIDS Vaccines," 279:186–188 (January 9, 1998).

35. Angell's editorial, "The Ethics of Clinical Research in the Third World," ran on pages 847–849, while Lurie and Wolfe's "sounding board," entitled, "Unethical Trials of Interventions to Reduce Perinatal Transmission of the Human Immunodeficiency Virus in Developing Countries," ran on pages 853–856.

36. One of the major trials that Angell and Public Citizen criticized would later reveal the weakness of their argument to stage "equivalency" tests of new treatment strategies against the complicated, expensive, proven one called 076. This UNAIDS "PETRA" trial—which involved 1,792 HIV-positive pregnant women in South Africa, Tanzania, and Uganda—tested various treatment schedules with AZT and 3TC in HIV-infected pregnant women and their babies. One arm of the study, in which the mother was treated when she went into labor but the baby was given no treatment, saw a transmission rate of 17.7%—better than the 25.5% rate seen in the untreated arm of the 076 study. Had 076 been used as a comparison, it would

have appeared that this new strategy worked. But because PETRA had a placebo group, it became clear that the transmission rate in the untreated group was 17.2%, which meant this treatment strategy had no impact whatsoever. For more details, see my article "Cheap Treatment Cuts HIV Transmission," *Science*, 283:916–917 (February 12, 1999).

37. The WHO and CIOMS issued this document, "International Ethical Guidelines for Biomedical Research Involving Human Subjects," in 1993.

38. Lawrence Altman of the *New York Times* and I were the only two journalists at the meeting. (Adrian Ivinson, editor of *Nature Medicine* wrote about it too, but he was a participant.) Altman's story, "Ethics Panel Urges Easing of Curbs on AIDS Vaccine Tests," appeared on June 28, 1998, p. 4. My story, "No Consensus on Rules for AIDS Vaccine Trials," appeared in *Science*, 281:22–23, on July 3, 1998.

39. Peter Lurie and eight other meeting attendees on September 14 wrote UNAIDS head Peter Piot a letter of complaint about this release, asking them to remove it. Piot wrote back and agreed to do so, but as of March 3, 1999, the last time I checked, it was still on the website.

40. Written by Adrian Ivinson, the journal's editor and a participant at the meeting, the article (4:874) so starkly differed from my coverage that I wrote a letter to the editor. My letter, and Ivinson's response, appear in the October 1998 *Nature Medicine*, "Clarifying AIDS Vaccine Trial Guidelines," 4(10):1091.

41. According to the NIH, the program, which by 1999 received more than $10 million a year, spends more on bioethics research than any other federally supported effort.

POSTSCRIPT: BREAKING THE SILENCE

1. "Intensifying Action Against HIV/AIDS in Africa: Responding to a Development Crisis" (World Bank, June 1999), pp. 3–6.

2. Press conference of the Development Committee, a forum of the World Bank and the International Monetary Fund, Washington, D.C., April 17, 2000.

3. As a May 11, 2000, press release from UNAIDS explained, Boehringer Ingelheim, Bristol-Myers Squibb, Glaxo Wellcome, Merck & Co., and Hoffmann–La Roche had agreed to work with the United Nations "to explore ways to accelerate and improve the provision of HIV/AIDS-related care and treatment in developing countries."

4. With help from the Turner Foundation, UNICEF launched pilot studies of such interventions in eight countries in sub-Saharan Africa. France, aided by Glaxo Wellcome (which markets three anti-HIV drugs), launched a similar program in Côte d'Ivoire called the Fund for International Therapeutic Solidarity. More support for drugs to thwart mother-to-infant transmission came from the Pediatric AIDS Foundation in March, which announced that it would award $800,000 in grants to seven programs in five African countries.

5. James Wolfensohn, "Free from Poverty, Free from AIDS," January 10, 2000.

6. "Report on the Global HIV/AIDS Epidemic" (Joint United Nations Programme on HIV/AIDS, June 2000). The vaccine quote comes from page 68. The quote about the magnitude of the epidemic appeared in "AIDS and Population," a fact sheet that accompanied the release of the report.

7. "First Vaccine Designed for Africa Is Cleared for Human Testing," reads a July 11, 2000, press release from IAVI. The idea that it was "designed for Africa" rests on the fact that this DNA vaccine contains HIV from subtype A, the most common strain in Kenya, where the first African test of the vaccine would, IAVI hoped, begin three to six months later.

Glossary

adjuvant An additive (such as alum) to a vaccine that increases the vaccine's ability to stimulate the immune system.

allogenic Genetically dissimilar.

antibodies Y-shaped proteins made by the immune system that latch onto invaders, such as viruses, to prevent them from infecting cells.

antigen A molecule that stimulates an immune response.

applied research Scientific investigations that have a practical goal, such as developing a vaccine.

assay A laboratory test that measures the proportional amount of a substance in a sample.

attenuated virus A weakened form of an otherwise dangerous virus. Also called live virus. Attenuated viruses, which occur naturally but also are made in labs, serve as vaccines by causing a harmless infection that trains the immune system. Killed-virus vaccines, in contrast, do not cause infections.

basic research Scientific investigations that aim to uncover knowledge that may have no immediate practical application.

B cells A class of white blood cells that secretes antibodies; also called B lymphocytes.

boost To restimulate a primed immune system; giving a vaccine to a person who already has received at least one dose of a similar vaccine.

CCR5 A chemokine receptor that HIV uses to infect cells.

CD4 cells White blood cells that have CD4 receptors on their surfaces, and that HIV selectively targets and destroys; also known as T4 lymphocytes.

CD8 cells White blood cells that have CD8 receptors on their surfaces and that can be trained to kill already-infected cells (*see* cytotoxic T lymphocytes); also known as T8 lymphocytes.

cell line Typically a cancerous cell that can grow indefinitely.

cell-mediated immunity An arm of the immune system that clears cells infected with HIV and other unwanted invaders.

challenge To test the worth of a vaccine by injecting a vaccinated animal with the disease-causing pathogen and assessing whether the animal can resist infection or disease.

chemokines Immune system messengers that turn on an inflammatory response.

clinical trial A human experiment of an experimental preventive or treatment.

cohort A group of people whom researchers follow over time, typically in a clinical trial but often just to establish, say, the rate of new HIV infections in a population.

control group Untreated people or animals used as a comparison with a group that receives the drug or vaccine.

core proteins The proteins inside a pathogen; in HIV, refers to p17 and p24.

correlates of protection Typically immune responses that explain why a person or animal has resisted infection or disease from a pathogen.

culture An artificial medium, such as isolated white blood cells, used to grow HIV in a test tube or petri dish.

cytokines Chemical messengers used by immune system cells to communicate with each other; for example, interleukin-2.

cytotoxic T lymphocytes White blood cells that clear the body of other cells that have become infected by an invader; also called killer cells.

efficacy trial A phase III trial typically involving thousands of people that is designed to determine whether a preparation is safe and effective.

endpoint The parameter chosen to assess the impact of a drug or vaccine, such as disease or death.

envelope proteins The proteins on the surface of a pathogen; with HIV: gp41, gp120, and gp160.

enzyme A substance (typically a protein) that starts or speeds a chemical reaction.

epitope A specific part of a protein that stimulates an immune response.

exposed, uninfecteds (EUs) People or animals that have resisted infection despite being confronted with HIV.

genetic engineering The process of artificially combining genes, often from one organism to another.

glycoproteins Proteins that have sugars coating them.

gp41 A glycoprotein that weighs 41 kilodaltons and spans HIV's membrane.

gp120 A glycoprotein that weighs 120 kilodaltons and resides on HIV's surface.

gp160 A glycoprotein that weighs 160 kilodaltons and is made up of gp120 and gp41.

heterologous challenge Injecting a vaccinated animal with a viral strain different from the one used to make the vaccine.

homologous challenge Injecting a vaccinated animal with a viral strain similar to the one used to make the vaccine.

humoral immunity The arm of the immune system that relies on antibodies.

immortalized cells Typically cancer cells, which are capable of endless divisions.

immunization The process of training an immune system to recognize an invader.

immunocompromised Having a weakened immune system.

immunodeficient Having a severely weakened immune system.

immunologic memory The phenomenon by which immune cells recognize pathogens they have met before.

inactivated virus *See* killed virus.

interleukins A family of cytokines.

investigator-initiated research Experiments proposed by investigators.

killed vaccine A vaccine that makes use of the killed-virus technique.

killed virus A virus that scientists render harmless by "inactivating" or "killing" its genetic material with heat, light, or chemicals to create a safe version of the pathogen for use in a vaccine; also called inactivated virus.

killer cells *See* cytotoxic T lymphocytes.

lab-adapted strains Strains of HIV that typically have been grown outside of the body in immortalized cells.

lentiviruses A subfamily of retroviruses, which includes HIV, that cause slow, progressive diseases.

live vaccine *See* attenuated virus.

long-term nonprogressors Humans or animals that remain immunologically intact despite being infected with an AIDS virus.

lymphocytes White blood cells.

major histocompatibility complex A family of molecules on the surface of every cell that helps the immune system distinguish self from nonself.

membrane The outer layer of a cell or pathogen.

mucosal immunity The arm of the immune system that protects mucosal tissue, which lines the vagina, rectum, and other surfaces.

mutation A change in a genetic sequence.

naked DNA A vaccine strategy that uses a piece of a bacteria called a plasmid to hold foreign genetic material; also called a gene vaccine or genetic immunization.

neutralizing antibodies Antibodies capable of preventing a virus from infecting a cell.

nonself versus self A specific immune response that is triggered when a foreign substance (nonself), such as another person's cells or tissues, enters a body (self). Graft rejection works by this mechanism.

oncogenes Genes that can induce cells to endlessly divide, causing cancer.

p17 A protein weighing 17 kilodaltons found inside HIV.

p24 A protein weighing 24 kilodaltons found inside HIV.

pathogen A disease-causing virus, bacterium, or fungus.

pathogenic Capable of causing disease.

peptide A piece of a protein.

phases I, II, III Various stages of clinical trials that begin with small numbers of people to study safety (phase I), expand to evaluate safety and immune responses in many more people (phase II), and end with a large number of people to evaluate safety, immune responses, and efficacy (phase III).

placebo A "dummy" (inert) injection or pill.

plasmid A circular piece of bacterial DNA.

polymerase chain reaction assay A test used to find minute amounts of genetic material.

primary isolates HIV strains obtained directly from people and not put into laboratory cultures.

prime The first round in a series of vaccinations.

prime-boost The process of using several vaccinations.

proliferation assay A test that measures the ability of cells to copy themselves when confronted with a foreign substance; typically used to assess cell-mediated immunity.

protease An enzyme used by HIV and other similar viruses during the process of copying themselves. Specifically, after HIV infects the nucleus of a cell and begins producing viral proteins, protease cuts these proteins into their proper size, a key step in the process of manufacturing a new virus.

protease inhibitors Drugs that block the action of the enzyme protease, which viruses such as HIV need in order to copy themselves.

protein A fundamental structural unit of all living things that is composed of amino acids.

receptors "Docking" molecules found on the surfaces of cells.

recombinant DNA DNA altered through genetic engineering.

replication The process whereby a pathogen or cell copies itself.

retroviruses The group of viruses, among them HIV and SIV, that contain RNA and copy themselves by first using the enzyme reverse transcriptase and converting into a DNA form.

reverse transcriptase An enzyme used by all retroviruses, including HIV, to convert their genetic material from RNA into DNA, a key step in replication. Specifically, only the DNA form of the virus can integrate with the DNA of a host cell. Once integrated, the virus can then direct the cell to manufacture viral proteins, forming new HIVs.

reverse transcriptase inhibitors Drugs that block the action of the enzyme reverse transcriptase. Without a functional reverse transcriptase enzyme, retroviruses such as HIV cannot convert from RNA into a DNA form, which prevents them from copying themselves.

Salk vaccine The killed poliovirus vaccine named after Jonas Salk. Also called inactivated poliovirus vaccine.

seronegative Having no detectable antibodies against HIV.

seropositive Having detectable antibodies against HIV.

serotypes A classification of viral strains based on the ability of antibodies to recognize them.

statistically significant A mathematical test used to determine that an observation is not due to chance.

sterilizing immunity The immunologic state used to describe a vaccinated animal that completely defeats a viral challenge, such that scientists cannot find any virus in the animal. The ideal immune response.

strain A specific isolate of SIV or HIV that is genetically distinct; $HIV_{MN,}$ for example, is a viral strain. *See also* subtype.

study sections Groups of scientists who gather to evaluate grant proposals submitted by their peers.

subinfectious exposure An amount of virus that is too small to establish an infection in an animal.

subtype A genetic classification system used to designate different families of a pathogen; the strain HIV_{MN}, for example, is from subtype B, in the HIV group M, one of several groups that are part of HIV type 1.

subunit vaccine A preparation that contains proteins (or pieces of them) from a pathogen.

surface proteins *See* envelope proteins.

surrogate marker An indicator that correlates with a biological response; the decline in CD4 cells, for example, is a surrogate marker for HIV infection and AIDS.

targeted research Experiments with a practical goal.

T cells A class of white blood cells that forms the cell-mediated arm of the immune system; also called T lymphocytes.

therapeutic vaccination versus preventive vaccination Giving a vaccine to an infected person versus an uninfected person. Therapeutic vaccines aim to treat the disease by fortifying the assaulted immune system. Preventive vaccines try to protect a person from becoming infected or, failing that, from developing disease.

Th1/Th2 theory The seesaw relationship between stimulation of cell-mediated immunity (Th1) and antibodies (Th2), such that when one is high, the other is low.

titer The concentration of a substance in a solution, such as the amount of antibody in a blood sample.

T4 lymphocytes *See* CD4 cells.

T8 lymphocytes *See* CD8 cells.

toxoids Toxins secreted by bacteria that scientists disable and use as the basis for vaccines.

transgenic animals Biologically engineered animals that have new DNA spliced into their own genes.

transient infection A state in which the AIDS virus temporarily establishes an infection and is aborted.

type The broadest genetic classification used to distinguish strains of HIV. By 2000, scientists had identified only two types, designated HIV-1 (by far the more common) and HIV-2.

V3 loop The tip of gp120.

vaccinee A vaccinated person or animal.

vaccinia The scientific name for the smallpox vaccine. Although likely derived from cowpox virus, vaccinia is distinct and, in immune-damaged people—such as those infected by HIV—can cause its own disease, which is called by the same name.

variability The difference between various strains and subtypes.

variola The scientific name for both the smallpox disease and the virus that causes it.

vector A bacteria or virus used to hold genes from another organism.

viral load The total amount of virus in the blood of a person or animal.

virion A single complete copy of a virus.

virus A tiny pathogen made of proteins and genetic material that needs living cells to copy itself.

white blood cells The fundamental cells of the immune system; also called lymphocytes.

wild type virus Isolates of the virus taken directly from a person and never kept in a laboratory culture (which can change the wild type virus's properties).

Acronyms

ACT UP	AIDS Coalition to Unleash Power
AIDS	acquired immunodeficiency syndrome
AmFAR	American Foundation for AIDS Research
ARAC	AIDS Research Advisory Committee
ASAP	Americans for a Sound AIDS/HIV Policy
AVAC	AIDS Vaccine Advocacy Coalition
AZT	azidothymidine
BCG	Bacillus Calmette-Guérin (tuberculosis) vaccine
CDC	Centers for Disease Control and Prevention (before October 1992: Centers for Disease Control)
CIOMS	Council for International Organizations of Medical Sciences
CITES	Convention on International Trade and Endangered Species
CTLs	cytotoxic T lymphocytes
DNA	deoxyribonucleic acid
DoD	U.S. Department of Defense
EUs	exposed, uninfected people
FAIR	Forum for the Advancement of Immunization Research
FDA	U.S. Food and Drug Administration
FDB	Food and Drug Board of California
HIV	human immunodeficiency virus
HHS	U.S. Department of Health and Human Services
HTLV-I, HTLV-II, HTLV-III	human T-cell leukemia (sometimes "lymphotropic") virus strains I, II, and III

IAPAC	International Association of Physicians in AIDS Care
IAVI	International AIDS Vaccine Initiative
INDs	investigational new drugs
INRB	National Institute of Biomedical Research
IOM	Institute of Medicine
IRC	Immune Response Corporation
JAMA	*Journal of the American Medical Association*
NAS	National Academy of Sciences
NCI	National Cancer Institute
NEJM	*New England Journal of Medicine*
NERPRC	New England Regional Primate Research Center
NIAID	National Institute of Allergy and Infectious Diseases
NIH	National Institutes of Health
OAR	Office of AIDS Research
OMB	Office of Management and Budget
OPRR	Office for Protection from Research Risks
OTA	Office of Technology Assessment
PACHA	Presidential Advisory Council on HIV/AIDS
PCR	polymerase chain reaction
PNAS	*Proceedings of the National Academy of Sciences*
R01s	investigator-initiated NIH grants
RNA	Ribonucleic acid
SIV	simian immunodeficiency virus
TAG	Treatment Action Group
Th	T helper cells: Th1=cell-mediated, Th2=antibody-mediated
UBI	United Biomedical Inc.
UC Davis	University of California, Davis
UCLA	University of California, Los Angeles
UCSF	University of California, San Francisco
UNAIDS	United Nations Programme on HIV/AIDS
USC	University of Southern California
WHO	World Health Organization
World FAAIR	World Foundation for the Advancement and Application of Immunization Research

Index

Aaron Diamond AIDS Research Center, 296, 297
Abrahamson, Joan, 43, 44
Accelerated AIDS Research Initiative, 195–97, 199
acquired immune deficiency syndrome, *see* AIDS
Acrogen, 260, 262–63, 292, 303
Ada, Gordon, 87, 121, 185, 241
adjuvants, 20, 26, 58, 88, 252, 329
Africa, HIV infection rates in, xv, 361–65, 366
AIDS (acquired immune deficiency syndrome):
  African epidemics of, xv, 230, 270, 361–65, 366
  cofactor theory of, 125
  community activism on, 98–99, 171, 175–76, 190–94, 241–42, 252, 260, 288, 369
  development and spread of, 4–5, 64, 65
  drug treatments for, xv, 99, 110, 114, 149, 150, 192, 194, 280, 287–88, 333, 334–35, 354–56, 363, 366, 367
  in gay male population, 4–7
  high-risk populations for, 130
  international conferences on, 44, 49, 74, 86–88, 89, 91, 98–100, 117, 121, 194, 213, 215, 223, 272–75, 286, 366–69
  potential market for vaccine against, 106–7
  public health protection protocols on, 15, 61, 99
  religious views on, 153
  sociopolitical marginalization of Western victims of, 39–40
  theories on causes of, 9–10, 125
  transmission of, 15
  U.S. rates of, xvi, 250
AIDS Coalition to Unleash Power (ACT UP), 98–99, 100, 171, 175, 190, 252
  legislation proposed by, 193–94, 197
AIDS Research Advisory Committee (ARAC), 231, 235, 248–57, 258, 259, 260, 261, 266, 267, 289, 348
AIDS Vaccine Advocacy Coalition (AVAC), 260, 288, 317–18, 327, 328
AIDS Vaccine Evaluation Group, 239, 246, 313
AIDS vaccine research:
  in Africa, 65–68, 69, 74, 229, 322–23, 336–39, 364–65, 366
  animal models in, 69, 72, 74, 75, 78–101, 128–31, 136–37, 145, 146, 222–23, 232–33, 237, 238–39, 253, 283, 284, 295–96, 297–98, 305, 311–12, 320–21, 328–31, 343, 345–46
  in Asia, 230, 252, 258, 260, 262, 266–67, 269–70, 293
  attenuated live-virus approach in, 11–12, 46, 47–48, 115, 128–31, 340–47
  breakthrough infection cases in, 246–48, 253, 274–75, 276, 347–48
  on cell-mediated immunity, 208–23, 224–25, 234, 283, 305–9, 312–15, 323–26
  cellular proteins in, 131–33
  congressional appropriations on, 138–74
  cooperation on, 100
  coordination of, 70–71, 127, 137–38, 183–99, 282, 298–99, 317, 327–32
  curtailment of large-scale efficacy trials in, 227–58
  development schedule of, 8–9, 10–14, 129–30, 270–71, 273–74, 285, 289–92, 317

effectiveness vs. urgency in, 117, 253–54, 261–62, 272, 321
empiricists vs. reductionists in, 45–46, 49, 62–63, 121, 260, 264, 275–79, 286, 293–94, 322, 330, 351
ethics of human testing in, 11, 12–13, 65–68, 69, 75, 78, 79, 86–87, 109, 130, 236, 243–45, 246, 251, 332–57
funding strategies in, 180, 182–84, 282, 295–316, 318–19, 321, 359–60
genetic engineering in, 45, 47, 48–49, 50–60, 62, 68, 75, 94–95, 101, 121–22, 127–28
government safety checks on, 139–40, 141, 246–47
on infection prevention vs. disease limitation, 133–36, 333
killed-HIV approaches to, 11–12, 42, 43–44, 46, 88–94, 95–96, 99–100, 115, 120–27, 215–16, 220, 260, 262–63, 289, 292, 298, 303–5
liability concerns in, 72–73, 96, 103–4, 109–12, 114–17, 118, 263
live-virus vs. killed virus in, 11–12, 46
market analysis on, 102, 103, 104–7, 272–73
neutralization assay controversy in, 227–34, 238, 245
patent concerns in, 111, 325, 326–27, 328
polio vaccine research vs., 38–42, 66, 198, 264
for poor countries, 261–62, 270, 271–74, 319, 323, 327, 333, 336–39, 347–50, 352–56
post-infection testing strategies in, 62–63, 67–68, 69, 88–91, 103–4, 107, 139, 149–50
on pregnant women, 153–54, 314, 334–35, 353
prime-boost approach in, 113, 114, 127–28, 134, 237
private sector in, 272–73, 276, 277, 280, 283–84; *see also* biotechnology industry, research limitations of; pharmaceutical industry
shifts of organizational prominence in, 260–94
trial costs of, 239, 250, 254
of virologists vs. immunologists, 207
volunteer pool used in, 145, 147, 212, 242–43
V3 loop as focus of, 103, 105, 108, 234

AIDS Vaccine Research Committee, 282, 287, 292–93, 297, 299, 309, 310
Allen, Paul, 360
AlphaVax, 325, 326
Altman, Lawrence, 69–70
American Foundation for AIDS Research (AmFAR), 302–3, 309, 310
American Medical Association (AMA), on polio immunization programs, 36, 37, 41
Americans for a Sound AIDS/HIV Policy (ASAP), 152, 159, 160, 161, 170
Ammann, Arthur, 278
anal intercourse, 6–7
Anderson, Roy, 262
And the Band Played On (Shilts), 244, 248
Angell, Marcia, 353
animals:
  AIDS vaccine research on, 69, 72, 74, 75, 78–101, 128–31, 136–37, 145, 146, 222–23, 232–33, 237, 238–39, 253, 283, 284, 295–96, 297–98, 305, 311–12, 320–21, 328–31, 343, 345–46
  endangered status of, 80, 82–84
antibodies:
  immune system release of, 54
  maternal, 21–22
  monoclonal, 300–301
antigens, 299
Apollon, 324
Arthur, Larry, 131–32, 305
Aspin, Les, 168, 169, 171
azidothymidine (AZT), 99, 150, 194, 272, 280, 334

Baltimore, David, 282, 288, 289–90, 292–93, 297, 304, 310, 322, 330–32, 351–52
Barr, David, 175, 176
Bazell, Robert, 93, 179
Belli, Melvin, 35
Berger, Edward, 296
Berkley, Seth, 270–74, 326, 364–65
Berler, Amy Bernhard, 102, 105–6
Berman, Phillip, 50, 238, 258, 265, 348
Biberfeld, Gunnel, 224–25
Bill and Melinda Gates Foundation, 359–60, 362, 367
Biocene, 227–34, 236, 238–39, 240, 246, 249, 252, 266, 270, 274, 280
biotechnology industry, research limitations of, 103, 104, 235, 272–73

Blech, David, 55
Blech, Isaac, 55
Blick, Gary, 160
Bloom, Barry, 142, 333–34, 352–53, 354
Bloom, Floyd, 282
Boehringer Ingelheim, 367
Bohr, Neils, 182
Bolognesi, Dani, 59, 60, 103, 125, 128, 169, 194
  on human vaccine trials, 74, 87, 232, 233, 234, 237, 253
  on NIH reorganization, 184, 189
  on progress of AIDS vaccine research, 92, 93, 130, 184, 254, 279, 281
  on SIV vs. HIV, 121, 122
  on Th1/Th2 theory, 218
Brandt, Edward, Jr., 6, 7, 8, 10–12, 13
Breakthrough (Carter), 25, 37, 220
Bretscher, Peter, 214–15, 216, 221–22
Bristol-Myers, 56, 109, 265
Bristol-Myers Squibb, 78, 109, 113–14, 118, 149, 248, 362
Broder, Sam, 161
Brodie, Maurice, 23, 41
Brown, Jerry, 175
Brown ruling, 116, 117
Burdick, Quentin, 148
Burke, Donald, 161, 184, 262, 272
  as MicroGeneSys funding, 139, 152–53, 166–67, 171, 174
  on Thai AIDS epidemic, 230, 252
  on VaxSyn trial analysis, 159, 160
Burroughs Wellcome, 99, 272
Burton, Dennis, 299–301
Bush, George, 116, 138, 153, 164
Bush, Vannevar, 178, 179, 181, 182, 183
Busingye, Rose, 361

Calder, Bill, 148
California:
  Food and Drug Branch (FDB) OF, 90, 91, 96
  liability reforms in, 115–16
California, University of, at Davis, 88–89, 90, 91, 99–100
California Endowment/California Healthcare Foundation, 304
Cameron, Edwin, 366, 369
canarypox, 313–14, 331
cancer:
  attenuated HIV vaccine and, 129
  research policies on, 177, 179–80, 182, 189, 199
Cantor, Eddie, 23–24
Carlo, Dennis, 90, 121
Carlson, James, 88–89, 90
Carter, Jimmy, 164, 270
Carter, Richard, 25, 31, 37, 220
CD8 antiviral factor (CAF), 217, 221, 312, 315
cellular proteins, 131–33
Centers for Disease Control (CDC), 5, 8, 260
Centers for Disease Control and Prevention, *see* Centers for Disease Control
Centre for Applied Microbiology and Research, 119
challenge stock, 233
Chanock, Robert, 95
chemokines, 280, 296–97, 301–2, 312
Chen, Irvin, 303–5, 309, 310
Cheney, Richard, 156
chimpanzees, endangered status of, 80, 82–84
Chiron, 51–53, 151, 163, 274, 299–300, 324, 326
  in Thailand, 266–67, 293, 348
  U.S. efficacy trials of, 227, 228, 229, 230, 235, 239, 244, 248, 251–52, 255, 266
Chow, Jack, 148
Chrétien, Jean, 316
Ciba-Geigy, 227
Clarey, Donald, 154, 155, 157
Clements-Mann, Mary Lou, 237–38, 247–48, 255, 354, 356
Clerici, Mario, 207, 209–14, 216–18, 220–22, 225–26, 311, 312–13, 362, 364
Clinton, Bill, 262
  AIDS research coordination efforts and, 175–77, 187, 190, 191, 192–93, 194, 196–97, 352
  HIV/AIDS council of, 260, 263–64
  in international AIDS efforts, 316, 352, 363, 364
  ten-year vaccine goal set by, ix, 264, 265, 285, 289–92, 317, 318, 348
cloning, genetic, 51, 52
Cochran, Mark, 57, 58, 76
cofactor theory, 125
Cohn, Gordon, 90–91
Cohn, Victor, 30
Comroe, Julius, Jr., 181, 182
Congress, U.S.:
  in early hearings on AIDS vaccine effort, 70–73

MicroGeneSys vaccine appropriation of, 138–74, 250
research coordination legislated by, 187–90, 194, 196–97
Convention on International Trade and Endangered Species (CITES), 82, 83–84
Cooper, Theodore, 110–11, 161
Coovadia, Hoosen (Jerry), 366
Coppola, Francis Ford, 192, 193
Corey, Larry, 162, 169, 187, 254
Council on Competitiveness, 116–17
Cowan, Max, 360
cowpox, 204, 205–6
Cranage, Martin, 119, 125, 135, 136
Crawford, Cindy, 160
Crewdson, John, 96, 247, 337, 338
Crosby, David, 235
Cutter Laboratories, 35–36, 38, 41, 48, 320

Dalgleish, Angus, 122
Davis, Nancy, 325–26
Decker, Bruce, 271
Defense Department, U.S., Army AIDS research funded by, 66, 139, 141, 143, 155–57, 165–73, 325–26
Delaney, Martin, 191–93, 194, 195–97, 216, 218
dengue virus, 79, 261
Desrosiers, Ronald, 100, 142, 184, 185
animal vaccine testing of, 84–86, 87, 92, 93, 95, 128–31, 135, 339–47
background of, 85
on human vaccine trials, 129–30, 275
live, attenuated vaccine research of, 128–31, 339–47
on MicroGeneSys funding, 142
on Stott's whole, killed vaccines, 122, 124, 127
DeStefano, Paul, 115–16
Detels, Roger, 210, 213–14
DeVita, Vincent, Jr., 7–8, 9, 182
Dina, Dino, 229, 230
Dingell, John, 111, 143, 171, 191, 193
Dobson, Jesse, 162, 169
Dodd, Christopher, 152, 169
Dorman, Burton, 263, 292, 303–4
Douglas, Gordon, 95
Dowdle, Walter, 95
Dripps, Robert, 181, 182
drug users, HIV exposure of, 39
Duesberg, Peter, 125
Dulbecco, Renato, 184, 304
Duliège, Anne-Marie, 244, 252, 266
Dwip Kitayoporn, 354–55
Dylan, Bob, 358

Eddy, Gerald, 135, 136
Edison, Thomas, 182
Eibl, Martha, 84
Eichberg, Jorg, 233
Eisenhower, Dwight, 36
Emini, Emilio, 114, 234
Enders, John, 27, 30, 41, 198
epitopes, 94–95, 299
Ericson, Brad, 57, 58
Esparza, José, 230, 261, 267, 272
Essex, Max, 289, 292
Evans, John, 360
exposed, uninfecteds (EUs), 203–4, 206–9, 215, 216, 220–21, 225, 296–97, 311, 312–13

Farthing, Charles, 344, 345–47
Fast, Patricia, 136, 241, 246, 248, 279
Fauci, Anthony, 121, 260, 308, 339
on animal model vs. human trials, 72, 74, 145
on assay credibility, 228
background of, 138
diplomatic leadership style of, 138, 142, 147, 148
on large-scale efficacy trials, 228, 235, 237, 241, 243, 244, 245–46, 248, 249, 252, 254, 255, 256, 267, 281, 350
on live, attenuated vaccine, 342–43, 344
on MicroGeneSys appropriation, 138, 142–43, 145–48, 152, 161, 163, 164, 165, 172
on prevention of infection vs. disease, 133, 134
on progress of vaccine research, 93, 133, 149, 169, 278, 279, 280, 282, 289–90
on research coordination, 71–72, 137–38, 187, 188, 189, 190, 194
on Th1/Th2 theory, 218, 219, 220
Fields, Bernard, 198, 224, 245
Fischinger, Peter, 81–82
Fletcher, William, 262–63
Foege, William, 185
Food and Drug Administration (FDA):
community activism on, 99
on human vaccine trials, 58, 75, 78, 80, 90, 91

Food and Drug Administration (FDA) (*continued*)
surrogate markers evaluated by, 149–51, 163
vaccine advisory committee of, 101
Ford, Gerald, 110
Fradd, R. Brandon, 102, 104–5, 106, 107
Francis, Donald, 60–63, 91, 92, 263
background of, 60–61
on drug therapy for trial participants, 354
Gallo criticized by, 61
Genentech financial support arranged by, 257, 265–66, 273
IAVI begun by, 271, 272, 273
on large-scale efficacy trials, 243–45, 251, 256, 257, 259–60
Moore's conflict with, 258, 286
on Thai trials, 348, 349
Francis, Thomas, Jr., 25–26, 27, 34
Frank, Barney, 175
Frederickson, Donald, 251
Fullilove, Robert, 243

Gallo, Robert, 244, 280
AIDS virus isolated by, 7, 9–10, 64, 71, 73, 143
background of, 17, 40
chemokine work of, 280, 296, 312
conferences organized by, 192, 196, 326
in dispute on AIDS virus discovery, 10, 18–19, 44, 45, 61
genetic engineering espoused by, 45, 49, 59, 60, 103, 237
killed-virus vaccine model rejected by, 43–44, 48
live-virus vaccine opposed by, 48, 128, 129, 346, 347
media skills of, 8, 10, 11, 44–45
professional influence of, 58, 59, 61, 103, 121, 122, 194, 197
on progress of vaccine research, ix, 7, 8–9, 10–14, 358
on research animals, 82, 83, 84
research coordination backed by, 193, 194
Salk vs., 16–17, 19–20, 42
on Zagury's research, 64, 69–70, 74, 337, 338, 339
Gardner, Murray, 89, 99–100, 120, 121, 125, 135, 184
Gates, Bill, Jr., 265, 318, 359, 367
Gates, Bill, Sr., 360
Gates, Melinda, 359
gay liberation movement, 201
gay men:
HIV exposure of, 39, 200–201
sexual promiscuity among, 4, 5–6, 13, 201–2, 207
sociopolitical attitudes on, 5–7, 153
U.S. AIDS cases first noted in, 4–5
Gay Men's Health Crisis, 175, 190
Gebbie, Kristine, 171, 194–95, 197
Geffen, David, 318
Genentech, 49–53, 55, 72, 73, 97–98, 101, 115–16, 150, 151, 163, 170, 235, 274, 299–300
aborted efficacy trials of, 227, 228–40, 243, 249, 250, 251, 252, 256–57, 258
VaxGen financing for, 257, 260, 265–66, 270, 273
Genetic Systems, 55, 56
Gerco, Dirceu, 355
Gere, Richard, 160
Gibbs, Joseph, 91–92
Gilot, Françoise, 19, 62
Giorgi, Janis, 206–7, 210, 211–13, 214, 215, 216
Girard, Marc, 105
Gladstone Institute of Virology and Immunology, 297
Gold, David, 242
Goldsmith, Mark, 297–98, 309–10, 328
Goldstein, Allan, 94
Gonsalves, Gregg, 188
Goodall, Jane, 80, 82, 83
Gore, Al, 190, 363
Grady, Christine, 339, 355
Graham, Barney, 247, 248
Greater New York Hospital Association, 140–41, 157
Groopman, Jerome, 51

Haase, Ashley, 231, 237, 249–50, 256, 281
Haddow, C. McClain (Mac), 6–8, 11
Haigwood, Nancy, 248, 254–55
Harkin, Tom, 145–46, 148
Harrington, Mark, 143, 162, 164, 169, 188, 281
Haseltine, William, 142, 147
Hatch, Orrin, 151, 152
Hatfield, Mark, 147–48, 152
Hattoy, Bob, 191
Health and Human Services (HHS), U.S. Department of, 6, 8, 334, 351
health care, universal, 112
Healy, Bernardine:
on congressional vaccine-trial funding,

138, 142–43, 144, 152, 161–62, 164–69
on research coordination, 187–88, 189
Heckler, Margaret, 6–11, 13, 14, 18, 241, 290, 358
Henderson, D. A., 63
Henry M. Jackson Foundation for the Advancement of Military Medicine, 135, 159, 173, 313
hepatitis B vaccine, 50, 51, 52, 61, 62, 67, 79, 106, 186, 244
herpes simplex vaccine, 50, 293
Heyward, William, 269, 348–49, 350
Hilleman, Maurice, 77, 95, 199, 254
on animal models for human trials, 79, 80, 81, 82, 101, 137, 320
on large-scale gp120 trials, 241
on research coordination, 183, 184, 185–87, 194
Hirsch, Vanessa, 134–35
HIV (human immunodeficiency virus):
blood tests for, 16, 18, 55, 210, 211
daily infection rate of, xv
diverse strains of, 18, 38–39, 73–74, 230, 232, 233–34
drug treatments for, xv, 99, 110, 114, 149, 150, 192, 194, 280, 287–88, 333, 334–35, 354–56, 363, 366, 367
French-U.S. patent dispute on, 18, 44, 45
as HTLV-III, 10, 14
immune system undermined by, 39, 47, 296–97, 307
molecular structure of, 46
natural resistance to, 202–4, 206–9, 210–11, 296–97, 306–8, 314–15
replication speed of, 18, 38–39, 48, 49
SIV vs., 79–80
Hlabisa Research Clinic, 364–65
Ho, David, 142, 161, 168, 169, 228, 238, 259, 281, 288, 353
on ethical issues of vaccine trials, 347–50, 354
Hodel, Derek, 241, 250, 251
Hoffmann-La Roche, 325
Hoover, Edward, 135
Hoth, Dan, 99, 114–15, 117, 236
Hu, Shiu-Lok, 53–56, 68, 113–14, 127–28, 134
Hudson, Rock, 22
human experiments, ethical concerns on, 11, 12–13, 65–68, 69, 75, 78, 79, 86–87, 109, 130, 236, 243–45, 246, 251, 332–57
Human Genome Project, 199, 356
human immunodeficiency virus, *see* HIV
human T-cell leukemia (lymphotropic) viruses (HTLVs), 9–10, 14
Hunsmann, Gerhard, 224

Imagawa, David, 211, 213–14
immortalized cells, 126, 228
Immune Response Corporation (IRC), 88–91, 96, 103, 115, 121, 149, 150, 174, 221–22, 303
immune system:
antibody response vs. cell mediation of, 208–19
HIV subversion of, 39, 47, 296–97, 307
live-virus effects vs. killed-virus responses of, 54–55
maternal antibodies in, 21–22
Immuno AG, 82–83, 84, 98
immunologic enhancement, 79
influenza vaccines, 25, 26, 88, 240, 323, 324
informed consent forms, 246–47
Inouye, Daniel, 141, 146, 148, 157, 158, 171
Institute of Medicine (IOM), 76–77, 79, 80–81, 108, 206
interferons, 57–58
interleukin-2 (IL-2), 14, 210, 212–13
International AIDS Vaccine Initiative (IAVI), 260, 270–74, 278–79, 284, 285, 292, 318–20, 321, 359, 362
African programs of, 322–27, 364–65, 367
International Association of Physicians in AIDS Care (IAPAC), 344, 345

Jefferson, Thomas, 290
Jenner, Edward, 45–46, 53, 78, 204–6
Johnson, Lyndon B., 179
Johnston, Bennett, 141, 143–44, 152, 169
Johnston, Peggy, 272, 318, 321–22
Johnston, Robert, 325

Kaczynski, Ted (Unabomber), 339–40
Kahn, Louis, 16
Kallings, Lars, 184, 185
Karzon, David, 95
Kennedy, Donald, 259, 304
Kennedy, Edward, 41–42, 70–71, 72, 73, 152, 172, 188–89, 190
Kennedy, John F., 264, 290
Kessler, David, 138, 144, 157, 161–62, 168, 169, 191, 194
Keystone AIDS Vaccine Liability Project, 111–12, 115

Killen, Jack, 246, 254
killer cells (cytotoxic T lymphocytes) (CTLs), 208, 209–10, 307
  cell-mediated vaccine research and, 208–23, 224–25, 234, 283, 305–9, 312–15, 323–26
Kindt, Thomas, 295–96, 297
Kingham, Peter, 109–10, 111
Kinsolving, Lester, 3–4, 15
Kisner, Daniel, 235
Koff, Wayne, 99, 186, 271, 327
  on assay credibility, 228
  on progress of vaccine research, 94, 95, 100, 184, 279
  research coordination sought by, 183–85
Kolata, Gina, 86
Kolmer, John, 13, 41
Koop, C. Everett, 191
Koprowski, Hilary, 29, 40
Koup, Richard, 296–97
Kramer, Larry, 190–91, 194, 195, 197
Kupor, Robert, 102, 105
Kurth, Reinhard, 184

Landau, Nate, 297
Lane, Cliff, 58, 300
Lange, Joep, 242
Lasker, Mary, 179
Lasky, Laurence, 49, 50, 51, 74
Lawrence, William, 358
Learmont, Jennifer, 342, 346
Lederberg, Joshua, 184
Lehner, Thomas, 311–12
lentiviruses, 80, 129
Letvin, Norman, 241, 324, 330–31
Levine, Arnold, 276, 281, 282–83, 292–93
Levy, Jay, 217, 221, 312, 315
Lévy, Jean-Paul, 185, 261, 272
Lieberman, Joseph, 152
Liebling, A. J., 141
Liu, Margaret, 324, 326
Loftus, Rick, 171
Long, Huey, 140
Long, Russell, 140–41, 144, 148, 151, 152, 154–58, 169, 172
long-term nonprogressors, 207–8, 217, 221, 306–8, 341–42
Los Angeles Men's Study, 200–203, 210, 212
Luciw, Paul, 52
Lurie, Peter, 172, 334, 335, 352–53, 355
Lurhuma Zirimwabagabo, 65, 66, 67, 68, 73
Luttwak, Edward, 192
Lyme disease, 351

McCarthy, Bill, 159, 160, 161
MacDonald, Kelly, 311, 314–16
Macklin, Ruth, 355
McMichael, Andrew, 323, 324–25
McNamara, Robert, 191
Maddox, John, 125
Madison Project, 195
Mahony, Roger, 96
malaria, 64, 65
Mandela, Nelson, 368–69
Manhattan Project, 176, 178, 182
Mann, Jonathan, 65–66, 69, 70–71, 75, 111, 350–52, 354, 356
March of Dimes (National Foundation for Infantile Paralysis), 16, 328
  birth defects as new focus of, 36
  establishment of, 5, 23–24
  polio vaccine research backed by, 5, 25–28, 29, 30, 31–33, 36, 41, 76, 101, 113, 127
  research coordinated by, 185, 186–87, 193, 198, 352, 358
Mariner, Wendy, 112
Martenson, Don, 89
Martin, David, 72
Martin, Edward, 169, 172
Martin, Malcolm, 58
Mason, James, 6, 8, 9, 153
Massie, Bob, 305–7, 308
Mather, Cotton, 205
Mathews, Jessica, 157–58
Matthews, Tom, 228
Mbeki, Thabo, 368
Mbidde, Edward, 230, 286, 335
Medical Research Council, 119, 311
Meeting Point, 361–62
Meister, Joseph, 63
Melnick, Joseph, 63
Mendez, Enrique, Jr., 155, 161
Merck & Co., 50, 52, 77, 79, 324, 367
  Repligen partnership with, 60, 102, 103, 107–8
MicroGeneSys, 57, 58, 59, 75–76, 78, 83, 94, 101, 113, 116
  congressional appropriation to, 138–74, 250, 276
  lobbying for, 140–41, 144–48, 151–52, 154–55, 157, 167, 172, 259
modified vaccinia Ankara (MVA), 323, 324–25, 326

Montagnier, Luc, 10, 11, 14, 18, 44, 45, 61
Moore, John, 279
on ethics of Thailand trials, 347, 349–50
on gp120 vaccines, 258, 259–60, 261, 262, 274–75, 299
MicroGeneSys funding opposed by, 142, 160, 168, 170, 171, 172, 173, 259, 276
on vaccine research strategies, 96, 275–78, 286
Morgan, Isabel, 27, 40
Moss, Bernard, 58
Murphey-Corb, Michael, 92–94, 95, 120, 121, 125, 126, 135
Murrow, Edward R., 33, 34
Murtha, John, 143

Nabel, Gary, 332
Nader, Ralph, 172, 334
Nadler, Jerrold, 197
naked DNA, 254, 323–24
Nalin, David, 77, 206
Nathanson, Neal, 320–21, 322, 327, 328, 330–31, 352, 366
National Academy of Sciences (NAS), 76–77, 92, 108
National Childhood Vaccine Injury Compensation Act, 72–73, 112
National Foundation for Infantile Paralysis, *see* March of Dimes
National Institute for Biological Standards and Control, 119
National Institute of Allergy and Infectious Diseases (NIAID):
advisory council of, 146–47, 231
AIDS vaccine research coordinated by, 71–72, 94–97, 231, 279–81
Genetech backed by, 98
HIV Vaccine Working Group of, 231, 236–41, 249–50, 252–53, 255, 256, 258
human vaccine trials begun by, 75–76
industry partnerships with, 280–81
Innovation grants of, 297, 301–10, 315, 318
large-scale efficacy trials arranged by, 227–58, 259, 276, 322
product development teams of, 367–68
vaccine research strategy reassessed at, 260–61, 279, 281, 321–22
waning prominence of, 260
National Institute of Biomedical Research (INRB), 66
National Institutes of Health (NIH):
AIDS budget at, 8, 70, 71, 137, 177, 188, 281, 283, 284, 285, 288, 292, 320, 327
Bethesda campus of, 19, 369
critics of, 317, 321–22, 351–52, 369
directors of, 189, 351, 352
ethics standards of, 334, 335, 337–39
funding strategies of, 180, 182, 183–84, 282, 297–98, 302–10, 318–19, 321
growth of, 179
industry relations with, 229, 235, 237, 282, 283–84, 287, 319
limitations of, 303, 309, 331–32
Office of AIDS Research at, 137, 188–91, 194–99, 260, 275–76, 278, 281–85, 288, 293, 295–96, 317, 320, 352
peer-review system of, 174, 177, 282, 283, 322
protests at, 99, 369
reassessments of AIDS vaccine efforts at, 282–86, 288, 292, 309
research coordination of, 137–38, 288–89, 291, 292–93, 300–301, 317
on sources of experimental animals, 83–84, 283
Vaccine Research Center of, 321, 331–32
VaxSyn appropriation reviewed by, 161–65
WHO collaboration with, 229, 230
Natth Bhamarapravati, 350
Nelmes, Sarah, 204, 205
nevirapine, 367
New England Regional Primate Research Center (NERPRC), 84–86, 87, 339–40
Novello, Antonia, 153
Nowinski, Robert, 265–66
Nunberg, Jack, 301–3, 304, 309, 310
Nunn, Sam, 141, 152, 155–57, 158
Nuremberg Code, 78, 334

Obijeski, Jack, 229, 238
O'Connor, Basil, 5, 16, 22, 26–27, 30, 32, 33, 40, 96, 101, 138, 328, 352
Oncogen, 53, 55–56, 68, 78, 79, 94, 109, 113–14, 115, 265, 325
Osborn, June, 13
O'Shaughnessy, John, 140, 154, 155, 157
O'Shaughnessy, Michael, 140, 154

Panetta, Leon, 196–97
Parish, Chris, 211–12, 214

passive immunity experiments, 86, 300
Pasteur, Louis, 29, 46, 63, 67, 182, 274, 338
Pasteur Institute:
AIDS virus discovery and, 10
Gallo's conflicts with, 10, 14, 18
HIV blood test of, 55
Pasteur Mérieux Connaught, 114, 237, 261, 280, 315–16, 324, 325, 331
Pasteur's Quadrant (Stokes), 181–82
Paul, William, 194, 198, 284, 285–86, 288, 289–90, 292, 320
Pauza, David, 222, 223
Pelosi, Nancy, 195, 196–97
Petricciani, John, 60–63
pharmaceutical industry:
anti-HIV therapies of, 99, 110, 114, 118, 149, 287, 363
liability concerns of, 72–73, 96, 109–12, 114
reluctance toward AIDS vaccine research in, 50, 56, 103, 104, 108–9, 113–14, 118, 270–71, 272–73
in Third World markets, 273, 363
philanthropic organizations, 310, 359–60
Philip, Prince, 82, 83
Phipps, James, 205
Piot, Peter, 269–70, 366
Plotkin, Stanley, 47–48, 77, 95, 115
Plummer, Frank, 216, 220, 313, 314
Pneumocystis carinii pneumonia, 5
poliomyelitis, 20–22
public support of research on, 22–24, 28, 39
U.S. levels of, 22, 24, 27, 30, 32, 33, 36, 37
polio vaccines, 5, 21, 358
AIDS vaccine research vs., 38–42, 66, 198, 264
animal trials of, 101
cooperative research on, 127
human trials of, 23, 28, 29–34, 35–36, 76, 320
killed virus used in, 20, 23, 27–38, 54, 94
live-virus type of, 27, 36–38
manufacturers' liability concerns on, 72–73
1930s failures of, 23, 28, 41
polymerase chain reaction (PCR) test, 210, 211, 213
Powell, Jody, 164, 169, 170
Praphan Phanuphak, 347–48
Prasert Thongcharoen, 267
Pratoom Thajorn, 268–69
Prayura Kunasol, 267
Presidential Advisory Council on HIV/AIDS (PACHA), 260, 263–64, 275–78, 285–86, 289–90, 350–52
primary isolates, 228, 232, 234, 298
Prince, Alfred, 82
Project Inform, 191, 192
Projet SIDA, 65, 66, 269, 350
proliferative response, 224
prostitutes, HIV in, 267–68, 269, 313
protease inhibitors, 280, 287
Public Citizen, 172–73, 334–35, 352–53, 355
Putney, Scott, 58–60, 103, 234

Quayle, Dan, 116–17
Quinnan, Gerald, 80, 81, 95

Rabb, Harriet, 170
rabies vaccines, 46, 63, 67
Raske, Kenneth, 157
Reagan, Nancy, 164
Reagan, Ronald, 4, 5, 6, 7, 36, 61, 140, 153, 154
Redfield, Robert, Jr., 148–49, 152–55, 158–61, 163, 165, 166, 170, 172
Redman, Glen, 200–203, 206, 207, 208
Repligen, 58–60, 94–95, 102–3, 104, 105, 106, 107–8, 273
Rich, Alexander, 59
Rivers, Thomas, 31–32, 40
Robbins, Fred, 161
Robinson, Max, 328
Robinson, Randall, 328
Rockefeller Foundation, 260, 270, 271, 310
Romagnani, Sergio, 218, 219, 220
Roosevelt, Franklin D., 5, 6, 22–23, 178, 187, 244, 264
Ruprecht, Ruth, 340–41, 343–44, 345–46
Ruranga, Rubaramira, 355
Rutter, William, 51–52

Saag, Mike, 235
Sabin, Albert, 27, 31, 32, 36–37, 40, 72
Sachs, Jeffrey, 368
Sadruddin Aga Khan, Prince, 82, 83
Salaun, Jean-Jacques, 66, 68, 69, 73
Salinas, Ed, 206–7, 208, 212, 214
Salk, Daniel, 24
Salk, Dolly, 24
Salk, Donna Lindsay, 25

Salk, Jonas Edward:
  on AIDS vaccine development, 19, 20, 42, 43, 44, 45, 49, 60, 62–63, 67, 74, 77, 85, 88–92, 94, 95, 96, 103, 112–13, 115, 121, 149, 184–85, 207, 208, 214–24, 225, 264, 271, 279, 360
  background of, 24–26
  celebrity of, 32, 34, 88, 215
  on cell-mediated immunity vs. antibody protections, 208, 214–17, 218–19, 221–22
  death of, 224, 316
  empirical strategies of, 62–63, 121, 264
  Gallo's press skills aided by, 44–45
  on natural immune protections, 207
  personality of, 17–18, 19–20, 26, 219–20, 226
  polio vaccine developed by, 16–17, 27–38, 40, 41–42, 48, 54, 66, 94, 112–13, 198, 264, 320
  publicity resisted by, 88, 89–91, 96
  on research coordination, 184–85
  on Sabin vaccine, 37–38
Salk, Lee, 24
Salk Institute for Biological Studies, 16, 19, 214
Scheele, Leonard A., 35
Schoofs, Mark, 362
Schooley, Robert, 168
Schultz, Alan, 128, 134, 253, 330, 331, 344, 345
scientific research, basic vs. applied, 178–83, 198–99, 358–59
Senate, U.S.:
  early AIDS research addressed by, 70–73
  see also Congress, U.S.
Shalala, Donna, 165, 168, 188, 189, 190, 191, 334, 352
Sharp, Phillip, 282
Shearer, Gene, 132, 133, 207, 208, 209–16, 217–18, 220, 221, 225–26, 312–13, 315
Shepherd, H. R., 271
Shilts, Randy, 61, 248
Siegel, Jacqueline G., 102, 106
simian immunodeficiency virus (SIV), 78–81, 85–86, 87, 92–93, 95, 98–101
  attenuated vaccine for, 128–31, 339–47
  cell-mediated immunity studies on, 222–23, 224–25
  whole, killed vaccine tests on, 119–27, 132, 215–16
smallpox, 46, 78, 204–6, 237, 351
Smith, Mark, 304
Smith, W. Shepherd, Jr., 152, 153, 159–61, 163, 170
SmithKline Beecham, 317
Soros, George, 318, 360
Southern California, University of (USC), 90–91
Speakes, Larry, 3–4, 5–6, 14–15
Stark, Pete, 117
sterilizing immunity, 134–35, 136, 250
Stokes, Donald E., 181–82
Stone, David K., 102
Stott, James, 119, 120–21, 122, 123–27, 131, 132, 133–34, 135, 284, 315
Studds, Gerry, 153
Sullivan, John, 345, 347
Sullivan, Louis, 99
surrogate markers, 149–51, 163
Szilard, Leo, 220

targeted research, 127, 177, 180–82, 183, 198–99, 283–84, 293–94, 358–59
Tate, Sheila, 164
T cells, 54, 64
  therapeutic results of vaccine on, 68
Teitelman, Robert, 55
Temin, Howard, 189, 304
Thailand:
  AIDS epidemic in, 252, 258, 263–64, 266, 267–69
  vaccine trials in, 230, 252, 258, 260, 262, 266–67, 269–70, 293, 322, 347–50
Th1/Th2 theory, 215, 216–19, 220, 221–23, 224–25, 234, 312–13
Thucydides, 204
Todaro, George, 56, 108–9, 114
Tramont, Ed, 151, 152, 253–54
Travis, Richard, 165
Treatment Action Groups (TAG), 143, 170–71, 188, 190, 196, 197, 252
Truman, Harry S., 178
Tulane University, 92–93
Turner, Ted, 360

Uganda, AIDS in, 230, 270, 335, 361–62
Unabomber (Ted Kaczynski), 339–40
UNAIDS, 270, 333, 334, 350, 353–56, 362, 363
United Biochemical Inc., 186, 228
United for AIDS Action, 175
Upjohn Company, 110

vaccine development:
animal tissue-culturing in, 27
manufacturers' liability in, 72–73
three trial phases of, 67
see also AIDS vaccine research; polio vaccines
vaccinia, 53–54, 55, 58, 323, 325
Vagelos, P. Roy, 191, 282
Vahey, Maryanne, 159–60, 161
VanCott, Thomas, 313
Vanderbilt University, 116
variolation, 204–5
Varmus, Harold, 189, 195–96, 197, 289, 317, 335, 351, 352
VaxGen, 257, 260, 266, 270, 293, 294, 322, 348, 349
VaxSyn, 139–40, 143, 144–45, 148–49, 151, 158–59, 174
Vázquez, Robert, 241–42, 252
vectors, 53–55, 323–27, 328–29
Venezuelan equine encephalitis virus (VEE), 325–26
Vermund, Sten, 239, 240, 244–45, 256
viral loads, 135, 158, 306
Viral Technologies, 94, 235
Virogenetics, 237
Volvovitz, Franklin, 57–58, 76, 150, 160, 168, 259
on congressional MicroGeneSys appropriation, 144, 151, 154–55, 161–62, 164, 169, 170, 171, 174

Walker, Bruce, 305–9, 310
Walter Reed Army Institute of Research, 135, 139, 148, 260, 263, 313, 317–18
Wara, Diane, 167
Warner, John, 141, 152, 156, 157, 158
Warren, Robert Penn, 140
Washington, George, 205
Watson, James, 191
Waxman, Henry, 152, 171, 172, 173
Weaver, Harry, 26–27, 28, 32, 40, 138
Weiner, David, 324
Weniger, Bruce, 263–67, 275–78, 285–86, 289, 292, 293, 348, 349, 350, 352
Weniger, Sidney, 264
Widdus, Roy, 79, 108
Wilfert, Catherine, 353
William H. Gates Foundation, 319, 359
Wolfe, Sidney, 172, 334, 352–53
Wolfensohn, James, 363
Wolinsky, Steven, 347, 348, 350
World Bank, 273, 319, 362, 363, 367, 368
World Health Organization (WHO), 350
on ethics of vaccine tests, 70–71, 229, 337, 353
worldwide vaccine testing overseen by, 229, 230, 258, 260, 261–62, 269, 270
Wurtman, Richard, 182–83
Wyeth-Ayerst, 116, 168, 170, 172, 173, 324, 326
Wyngaarden, James, 7–8, 9, 83

Yarborough, Ralph, 179–80
Young, Frank, 90

Zagury, Daniel, 64–70, 73–75, 78, 86–87, 94, 208, 229, 232, 325
ethics of Zairian trials conducted by, 336–39, 356
Zaire, AIDS vaccine research in, 65–68, 69, 74, 229, 336–39, 356
Zinder, Norton, 180, 199
Zolla-Pazner, Susan, 255–56